Operator Theory: Advances and Applications
Vol. 177

J. López-Gómez
C. Mora-Corral

Algebraic Multiplicity of Eigenvalues of Linear Operators

Birkhäuser
Basel · Boston · Berlin

Authors:

J. López-Gómez
Department of Applied Mathematics
Universidad Complutense de Madrid
Ciudad Universitaria de Moncloa
28040 Madrid
Spain
e-mail: Lopez_Gomez@mat.ucm.es

C. Mora-Corral
Mathematical Institute
University of Oxford
24-29 St Giles'
Oxford OX1 3LB
UK
e-mail: mora-cor@maths.ox.ac.uk

2000 Mathematics Subject Classification 15A18, 15A22, 47A10, 47A11, 47A53, 47A56, 47A60, 47H11, 47J15

Library of Congress Control Number: 2007924801

Bibliographic information published by Die Deutsche Bibliothek
Die Deutsche Bibliothek lists this publication in the Deutsche Nationalbibliografie; detailed bibliographic data is available in the Internet at <http://dnb.ddb.de>.

ISBN 978-3-7643-8400-5 Birkhäuser Verlag AG, Basel - Boston - Berlin

Part of Springer Science+Business Media
Printed on acid-free paper produced from chlorine-free pulp. TCF ∞
Cover design: Heinz Hiltbrunner, Basel
Printed in Germany

ISBN 978-3-7643-8400-5
e-ISBN 978-3-7643-8401-2

9 8 7 6 5 4 3 2 1
www.birkhauser.ch

—Calla te digo otra vez, Sancho
—dijo Don Quijote—; porque te hago
saber que no sólo me trae por
estas partes el deseo de hallar al
loco, cuanto el que tengo de hacer
en ellas una hazaña, con que he
de ganar perpetuo nombre y fama en
todo lo descubierto de la Tierra;
y será tal, que he de echar
con ella el sello a todo aquello
que puede hacer perfecto y
famoso a un andante caballero.

Don Quijote de la Mancha (Capítulo 25)
Miguel de Cervantes Saavedra

'Peace, I say, Sancho, once again,'
said Don Quixote: 'for know, that it is
not barely the desire of finding the
madman that brings me to these parts,
but the intention I have to perform an
exploit in them, whereby I shall
acquire a perpetual name and renown
over the face of the whole earth:
and it shall be such an one as shall
set the seal to all that can render
a knight-errant complete and famous.'

Translated by Charles Jarvis.
(Oxford University Press.
Hong Kong, 1999.)

Contents

Preface

This book analyzes the existence and uniqueness of a generalized algebraic multiplicity for a general one-parameter family $\mathfrak{L}$ of bounded linear operators with Fredholm index zero at a value of the parameter λ_0 where $\mathfrak{L}(\lambda_0)$ is non-invertible. Precisely, given $\mathbb{K} \in \{\mathbb{R}, \mathbb{C}\}$, two Banach spaces U and V over $\mathbb{K}$, an open subset $\Omega \subset \mathbb{K}$, and a point $\lambda_0 \in \Omega$, our admissible operator families are the maps

$$\mathfrak{L} \in \mathcal{C}^r(\Omega, \mathcal{L}(U, V)) \tag{1}$$

for some $r \in \mathbb{N}$, such that

$$\mathfrak{L}(\lambda_0) \in \mathrm{Fred}_0(U, V);$$

here $\mathcal{L}(U, V)$ stands for the space of linear continuous operators from U to V, and $\mathrm{Fred}_0(U, V)$ is its subset consisting of all Fredholm operators of index zero. From the point of view of its novelty, the main achievements of this book are reached in case $\mathbb{K} = \mathbb{R}$, since in the case $\mathbb{K} = \mathbb{C}$ and $r = 1$, most of its contents are classic, except for the axiomatization theorem of the multiplicity.

The book consists of three parts, and, although the necessary background to understand the book differs from each part, the entire book might become easily accessible to a multidisciplinary audience by establishing all its results in a finite-dimensional setting, which is not a serious restriction. Part I covers the classic finite-dimensional spectral theory, and assumes hardly any previous knowledge by the reader, except some basic facts in complex analysis. In fact, its contents are intended to be taught at an undergraduate level, as we have considerably tidied up and shortened most of the available material in the literature. Part II introduces a generalized concept of algebraic multiplicity, which substantially extends all previous available concepts, to cover the case of families $\mathfrak{L}$ as in (1). Actually, this objective is accomplished by adopting several different perspectives, each of them of great interest in its own right. For a comfortable reading of this part, some basic background in Functional Analysis is appropriate; essentially, the Hahn–Banach theorem and the most important properties of the Fredholm operators. Finally, Part III, which consists of a single chapter, introduces some pivotal concepts in nonlinear analysis (like the concept of nonlinear eigenvalue) and characterizes the nonlinear eigenvalues of $\mathfrak{L}$ through the value of $(-1)^{\mathfrak{m}[\mathfrak{L};\lambda_0]}$, where $\mathfrak{m}[\mathfrak{L}; \lambda_0]$ denotes the generalized algebraic multiplicity of $\mathfrak{L}$ at λ_0. Although it would be desirable to have a good knowledge of the topological degree for a comfortable reading of this part, this is far from necessary, because a self-contained introduction to the topological degree is provided therein. Going beyond would take us outside the general scope of this book.

A great proportion of Part II consists of versions and new findings of recent results attributable to the authors of this book and coworkers, except for the classic case when $\mathbb{K} = \mathbb{C}$ and $r = 1$, where most of the results are attributable to the Russian school. Among the most important results of Part II, there is the next axiomatization theorem of the multiplicity. We denote by $\mathcal{S}^{\infty}_{\lambda_0}(U)$ the set of functions of class $\mathcal{C}^{\infty}$ defined in a neighborhood of λ_0 with values in $\mathrm{Fred}_0(U,U)$.

Theorem 0.0.1. *There exists a unique map*

$$\mathfrak{m}[\cdot;\lambda_0] : \mathcal{S}^{\infty}_{\lambda_0}(U) \to [0,\infty]$$

satisfying the following axioms:

A1. *For all* $\mathfrak{L}, \mathfrak{M} \in \mathcal{S}^{\infty}_{\lambda_0}(U)$,

$$\mathfrak{m}[\mathfrak{L}\mathfrak{M};\lambda_0] = \mathfrak{m}[\mathfrak{L};\lambda_0] + \mathfrak{m}[\mathfrak{M};\lambda_0].$$

A2. *There exists a rank one projection* $P_0 \in \mathcal{L}(U)$ *such that*

$$\mathfrak{m}[\mathfrak{L};\lambda_0] = 1,$$

where

$$\mathfrak{L}(\lambda) = (\lambda - \lambda_0)P_0 + I_U - P_0, \qquad \lambda \in \mathbb{K}.$$

The value $\mathfrak{m}[\mathfrak{L};\lambda_0]$ is called the *algebraic multiplicity* of $\mathfrak{L}$ at λ_0. Axiom A1 is known as the *product formula*, while Axiom A2 is nothing but a *normalization condition*, as will discussed in Chapter 6.

Before describing the properties of the multiplicity, we should recall the familiar concept of *order* of a zero of a function at a point. Let X be a $\mathbb{K}$-Banach space, $\Omega \subset \mathbb{K}$ an open set, $r \geq 0$ an integer or infinity, $\lambda_0 \in \Omega$, and $f \in \mathcal{C}^r(\Omega, X)$. If there exists an integer $0 \leq k \leq r$ such that $f^{j)}(\lambda_0) = 0$ for all $0 \leq j < k$, and $f^{k)}(\lambda_0) \neq 0$, then we define $\mathrm{ord}_{\lambda_0} f = k$. If $r = \infty$ and $f^{j)}(\lambda_0) = 0$ for all integers $j \geq 0$, then we define $\mathrm{ord}_{\lambda_0} f = \infty$. If $f^{j)}(\lambda_0) = 0$ for all $0 \leq j \leq r$, and f is not of class $\mathcal{C}^{r+1}$ in any neighborhood of λ_0, then we leave $\mathrm{ord}_{\lambda_0} f$ undefined.

Now we describe some properties of the multiplicity. For every $\mathfrak{L} \in \mathcal{S}^{\infty}_{\lambda_0}(U)$, the function $\mathfrak{m}[\cdot;\lambda_0]$ inherits, from Axioms A1 and A2, the following properties:

i) $\mathfrak{m}[\mathfrak{L};\lambda_0] \in \mathbb{N} \cup \{\infty\}$.

ii) $\mathfrak{m}[\mathfrak{L};\lambda_0] = 0$ if and only if $\mathfrak{L}(\lambda_0)$ is an isomorphism.

iii) $\mathfrak{m}[\mathfrak{L};\lambda_0] < \infty$ if and only if there exists an integer $k \geq 0$ such that

$$(\lambda - \lambda_0)^k \mathfrak{L}(\lambda)^{-1}$$

exists and is bounded for λ in a perforated neighborhood of λ_0.

iv) If $\dim U < \infty$, then $\mathfrak{m}[\mathfrak{L};\lambda_0] = \operatorname{ord}_{\lambda_0} \det \mathfrak{L}$.

v) If $\mathbb{K} = \mathbb{C}$ and $U = V = \mathbb{C}^N$, then

$$\mathfrak{m}[\mathfrak{L};\lambda_0] = \operatorname{tr} \frac{1}{2\pi i} \int_\gamma \mathfrak{L}'(\lambda)\mathfrak{L}(\lambda)^{-1}\, d\lambda$$

for every Cauchy contour γ with inner domain D containing λ_0 such that $\mathfrak{L}^{-1}$ is holomorphic in $D \setminus \{\lambda_0\}$ and continuous in $\bar{D} \setminus \{\lambda_0\}$.

vi) For every $\mathbb{K}$-Banach space V and $\mathfrak{M} \in \mathcal{S}^\infty_{\lambda_0}(V)$,

$$\mathfrak{m}[\mathfrak{L} \oplus \mathfrak{M};\lambda_0] = \mathfrak{m}[\mathfrak{L};\lambda_0] + \mathfrak{m}[\mathfrak{M};\lambda_0].$$

vii) If $\mathfrak{L}$ is analytic, Ω is connected, and $\mathfrak{L}(\Omega)$ contains an isomorphism, then

$$\mathfrak{m}[\mathfrak{L};\lambda] < \infty \qquad \text{for all} \quad \lambda \in \Omega.$$

viii) If $A \in \mathcal{L}(U)$ and $\mathfrak{L}$ is the family defined by

$$\mathfrak{L}(\lambda) := \lambda I - A, \qquad \lambda \in \mathbb{K},$$

then

$$\mathfrak{m}[\mathfrak{L};\lambda_0] = \sup_{k\in\mathbb{N}} \dim N[(\lambda_0 I - A)^k].$$

ix) $\mathfrak{m}[\mathfrak{L};\lambda_0] < \infty$ if and only if there exist an open neighborhood $\tilde{\Omega} \subset \Omega$ of λ_0, a topological decomposition $U = U_0 \oplus U_1$ with $n := \dim U_0 < \infty$ and two maps $\mathfrak{E}, \mathfrak{F} \in \mathcal{S}^\infty_{\lambda_0}(U)$ such that $\mathfrak{E}(\lambda_0), \mathfrak{F}(\lambda_0)$ are isomorphisms and

$$\mathfrak{L}(\lambda) = \mathfrak{E}(\lambda)\, [\mathfrak{S}(\lambda) \oplus I_{U_1}]\, \mathfrak{F}(\lambda), \qquad \lambda \in \tilde{\Omega},$$

where I_{U_1} stands for the identity of U_1, and

$$\mathfrak{S}(\lambda) = \operatorname{diag}\left\{(\lambda - \lambda_0)^{\kappa_1}, \dots, (\lambda - \lambda_0)^{\kappa_n}\right\}, \qquad \lambda \in \mathbb{K},$$

for some integers $\kappa_1 \geq \cdots \geq \kappa_n \geq 1$. Moreover, in such a case,

$$\mathfrak{m}[\mathfrak{L};\lambda_0] = \kappa_1 + \cdots + \kappa_n.$$

Owing to Theorem 0.0.1, each of the listed properties, and actually any further property of the multiplicity, must be a consequence of Axioms A1 and A2.

The concept of multiplicity for maps $\mathfrak{L}$ of the form (1), rather than for operators (as is the case in the context of the classic spectral theory) goes back to the Russian school; it began to be developed in the 1940s for holomorphic families, that is, for the special case when $\mathbb{K} = \mathbb{C}$ and $r = 1$ (see Keldyš [66, 67] for probably the first definition, and Markus & Sigal [95], Eni [28], Gohberg [42], Gohberg & Sigal [54], Gohberg, Kaashoek & Lay [44], as well as the references therein, for later generalizations and expositions). Most of the available underlying theories are based on generalized Jordan chains, as discussed in Chapter 7 of this book.

The concept of multiplicity in the case when $\mathbb{K} = \mathbb{R}$ came later, and was introduced independently through four rather different devices. By using generalized Jordan chains (see Sarreither [116] and later Rabier [110]), through a finite-dimensional reduction of the family $\mathfrak{L}$ so that it can be defined as the multiplicity of the determinant of the reduced family (see Ize [61]), by means of a polynomial factorization of $\mathfrak{L}$ (see Gohberg [42], Gohberg & Sigal [54], and Magnus [92]), as discussed in Chapter 5 of this book, or through a change of variable giving rise to a *transversal* family (see Esquinas & López-Gómez [30], Esquinas [29], and López-Gómez [82]), as discussed in Chapter 4.

According to Theorem 0.0.1, all these concepts of multiplicity must coincide as long as they satisfy Axioms A1 and A2, though the reader should be aware of the fact that obtaining the product formula may be fraught with higher technical difficulties than directly checking that they do coincide from their own definitions.

This book gives a complete description of all these (equivalent) constructions of the algebraic multiplicity. The reasons to bring them together in this book include:

a) all of them involve interesting constructions in their own right;
b) although equivalent, a number of multiplicity concepts depart from rather separated starting points and, consequently, their mutual connections are not clear;
c) some properties of the multiplicity are easier to prove through a particular definition than departing from another.

Nevertheless, the existence of a number of constructions of the multiplicity from different perspectives facilitates the derivation of its properties, both from the theoretical point of view and from the point of view of applications. Although there are some previous and partial comparisons between several different theories of algebraic multiplicity (see, for example, Esquinas [29], Fitzpatrick & Pejsachowicz [33], Rabier [110] and López-Gómez [82]), this book should provide the most complete monograph encompassing all those different constructions from the point of view of the axiomatization of Theorem 0.0.1.

These algebraic multiplicities have many applications in a number of separate areas of mathematics. Among them, the constructions of the algebraic multiplicity are used for accomplishing the following tasks:

- Finding optimal criteria for the change of the topological degree of $\mathfrak{L}(\lambda)$ as the parameter λ crosses an eigenvalue of $\mathfrak{L}$, within the context of local and global bifurcation theory (see Sarreither [116], Ize [61], Magnus [92], Esquinas & López-Gómez [31], Esquinas [29], López-Gómez [82], Rabier [110], Fitzpatrick & Pejsachowicz [34], López-Gomez & Mora-Corral [91]).

- Establishing local classification of operator families $\mathfrak{L}$ through the existence and uniqueness of the Smith canonical form (see Gohberg, Kaashoek & Lay [44]), as well as global spectral results for general operator families (see Gohberg, Kaashoek & Lay [44], Leiterer [78, 79], Gohberg, Lancaster & Rodman [50], López-Gómez [82]).
- Finding completeness criteria for eigenvectors and generalized eigenvectors of operators and operator families (see Keldyš [66, 67], Friedman & Shinbrot [35], Gohberg & Kreĭn [45, 46]).
- Analyzing the solution manifold of general classes of ordinary linear differential equations with constant coefficients, and boundary-value problems (see Gohberg, Lancaster & Rodman [50, 51], Rodman [114], Wloka, Rowley & Lawruk [127]), as well as difference equations with constant coefficients (see Gohberg, Lancaster & Rodman [50]).
- Studying interpolation of matrix families (see Rodman [114], Ball, Gohberg & Rodman [8]),
- Studying linear systems in control engineering (see Gohberg, Lancaster & Rodman [51], Rodman [114], Wyman, Sain, Conte & Perdon [129]).

In the sequel, we will outline the contents of each chapter of this book, with particular emphasis on the new findings originated by the authors' research in the field.

Chapter 1 is devoted to a proof of the Jordan theorem and a construction of the underlying canonical form. Chapter 2 provides a short self-contained introduction to operator calculus, yet in finite dimension. Then, the Dunford integral formula is introduced and the spectral mapping theorem is studied. Chapter 3 describes the construction of the Jordan spectral projections. The proofs are based upon the fact that the ascent of each eigenvalue equals its order as a pole of the resolvent operator. This methodology substantially differs from the standard one adopted in most classic textbooks on elementary finite-dimensional spectral theory, shortens the inherent difficulties for the general infinite-dimensional framework, and facilitates the teaching of these materials at an undergraduate level. Actually, this is the pivotal idea upon which the development of the subsequent abstract general theory of this book relies.

Part II and the general theory of multiplicities begin at Chapter 4, after introducing all the spaces and most common notation used in the rest of the book. Throughout Part II, $r \geq 1$ is an integer or infinity, $\mathbb{K}$ is either the real field or the complex field, $\Omega \subset \mathbb{K}$ is an open set containing λ_0, and $\mathfrak{L} \in \mathcal{C}^r(\Omega, \mathcal{L}(U,V))$ satisfies $\mathfrak{L}(\lambda_0) \in \mathrm{Fred}_0(U,V)$. Besides, the following notation for the derivatives at $\mathfrak{L}$ at λ_0

$$\mathfrak{L}_j := \frac{1}{j!}\mathfrak{L}^{j)}(\lambda_0), \qquad 0 \leq j < r+1,$$

is used everywhere, so that $\mathfrak{L}_j$ equals the jth coefficient of the Taylor development of $\mathfrak{L}$ at λ_0. The parameter value λ_0 is said to be an *eigenvalue* of $\mathfrak{L}$ when $\mathfrak{L}(\lambda_0)$ is not invertible. As $\mathfrak{L}(\lambda_0)$ is Fredholm of index zero, this amounts to

$$N[\mathfrak{L}(\lambda_0)] \neq \{0\}. \tag{2}$$

Let $\mathrm{Iso}(U, V)$ denote the set of linear topological isomorphisms from U to V. When $\mathfrak{L}(\lambda_0) \in \mathrm{Iso}(U, V)$, all results become trivial; thus, we will subsequently focus our attention on the case when (2) occurs.

The theory of algebraic multiplicity presented in Chapter 4 is based on the concept of transversality, which is a pivotal concept throughout this book; it goes back to Esquinas & López-Gómez [31].

Definition 0.0.2. Given an integer $1 \leq k \leq r$, it is said that λ_0 is a k-transversal eigenvalue of $\mathfrak{L}$ if

$$\bigoplus_{j=1}^{k} \mathfrak{L}_j(N[\mathfrak{L}_0] \cap \cdots \cap N[\mathfrak{L}_{j-1}]) \oplus R[\mathfrak{L}_0] = V,$$

and

$$\mathfrak{L}_k(N[\mathfrak{L}_0] \cap \cdots \cap N[\mathfrak{L}_{k-1}]) \neq \{0\}.$$

In such a case, the algebraic multiplicity of $\mathfrak{L}$ at λ_0, denoted by $\chi[\mathfrak{L}; \lambda_0]$, is defined through

$$\chi[\mathfrak{L}; \lambda_0] := \sum_{j=1}^{k} j \dim \mathfrak{L}_j(N[\mathfrak{L}_0] \cap \cdots \cap N[\mathfrak{L}_{j-1}]).$$

Although, at a first glance, Definition 0.0.2 might look shocking, it turns out that transversal eigenvalues enjoy a special structure that makes most of the involved operator calculations straightforward from an algorithmic point of view. The concept of multiplicity was originally introduced to capture the order of $\det \mathfrak{L}$ at λ_0 in the finite-dimensional setting, but this is far from being an elementary feature that can be briefly explained.

Another key concept of Part II, going back to López-Gómez [82], is the one introduced by the following definition.

Definition 0.0.3. Given an integer $k \geq 1$, it is said that λ_0 is a k-algebraic eigenvalue of $\mathfrak{L}$ when there exists an open neighborhood $\tilde{\Omega} \subset \Omega$ of λ_0 such that $\mathfrak{L}(\lambda) \in \mathrm{Iso}(U, V)$ for all $\lambda \in \tilde{\Omega} \setminus \{\lambda_0\}$, the values of the family

$$\lambda \mapsto (\lambda - \lambda_0)^k \mathfrak{L}(\lambda)^{-1}, \qquad \lambda \in \tilde{\Omega} \setminus \{\lambda_0\},$$

are bounded, but $(\lambda - \lambda_0)^{k-1}\mathfrak{L}(\lambda)^{-1}$ is unbounded for λ in every perforated neighborhood of λ_0. In such a case, we will write that $\lambda_0 \in \mathrm{Alg}_k(\mathfrak{L})$.

When $\mathfrak{L}$ is analytic, it turns out that $\lambda_0 \in \mathrm{Alg}_k(\mathfrak{L})$ if and only if λ_0 is a pole of $\mathfrak{L}^{-1}$ of order k. Consequently, this concept is well known in the analytic case. But, since there is no available concept of pole for non-meromorphic families, the concept of algebraic eigenvalue is new in the general $\mathcal{C}^r$ setting adopted in this book.

Chapter 9 shows that the concept of algebraic eigenvalue introduced by Definition 0.0.3 indeed provides us with a concept of pole for $\mathfrak{L}^{-1}$, thus considerably extending the classic concept of pole. Quite strikingly, it turns out that if $\lambda_0 \in \mathrm{Alg}_k(\mathfrak{L})$ and $r \geq 2k$, then the inverse map $\mathfrak{L}^{-1}$ possesses a Laurent asymptotic expansion at λ_0,

$$\mathfrak{L}(\lambda)^{-1} = \sum_{n=-k}^{-1} (\mathfrak{L}^{-1})_n (\lambda - \lambda_0)^n + o\left((\lambda - \lambda_0)^{-1}\right), \qquad \text{as} \quad \lambda \to \lambda_0,$$

as well as $\mathfrak{L}'\mathfrak{L}^{-1}$,

$$\mathfrak{L}'(\lambda)\mathfrak{L}(\lambda)^{-1} = \sum_{n=-k}^{-1} \mathfrak{M}_n (\lambda - \lambda_0)^n + o\left((\lambda - \lambda_0)^{-1}\right), \qquad \text{as} \quad \lambda \to \lambda_0,$$

and, in addition, according to the main theorem of López-Gómez & Mora-Corral [90],

$$\operatorname{tr} \mathfrak{M}_n = \begin{cases} 0 & \text{if} \quad -k \leq n \leq -2, \\ \chi[\mathfrak{L}; \lambda_0] & \text{if} \quad n = -1, \end{cases}$$

which is a deep generalization to the case when $\mathbb{K} = \mathbb{R}$ and $\mathfrak{L}$ is of class $\mathcal{C}^r$ for some $r \in \mathbb{N}$ of the following classic well-known result attributable to Gohberg [42], Markus & Sigal [95] and Gohberg & Sigal [54] (see also Gohberg, Goldberg & Kaashoek [43]). That generalization constitutes the bulk of Chapter 9.

Theorem 0.0.4. *Let $\Omega \subset \mathbb{C}$ be open and connected, and $\mathfrak{L} \in \mathcal{H}(\Omega, \mathcal{L}(U, V))$ such that $\mathfrak{L}(\Omega) \subset \mathrm{Fred}_0(U, V)$ and $\mathfrak{L}(\Omega) \cap \mathrm{Iso}(U, V) \neq \varnothing$. Then, the set*

$$\Sigma(\mathfrak{L}) := \{\lambda \in \Omega : N[\mathfrak{L}(\lambda)] \neq \{0\}\} \tag{3}$$

is discrete in Ω, and $\mathfrak{L}^{-1}$ is holomorphic in $\Omega \backslash \Sigma(\mathfrak{L})$. Moreover, for each $\lambda_0 \in \Sigma(\mathfrak{L})$, the meromorphic function $\mathfrak{L}^{-1}$ admits a Laurent development

$$\mathfrak{L}(\lambda)^{-1} = \sum_{n=-k}^{\infty} (\mathfrak{L}^{-1})_n (\lambda - \lambda_0)^n$$

for some integer $k \geq 1$, where $(\mathfrak{L}^{-1})_0 \in \mathrm{Fred}_0(V, U)$ and $(\mathfrak{L}^{-1})_n$ has finite rank for all $-k \leq n \leq -1$. Furthermore, the coefficients of the Laurent development of $\mathfrak{L}'\mathfrak{L}^{-1}$ at λ_0,

$$\mathfrak{L}'(\lambda)\mathfrak{L}(\lambda)^{-1} = \sum_{n=-k}^{\infty} \mathfrak{M}_n (\lambda - \lambda_0)^n,$$

have finite rank for every $n \in \{-k, \dots, -1\}$, and their traces are given through

$$\operatorname{tr} \mathfrak{M}_n = \begin{cases} 0 & \textit{if} \quad -k \leq n \leq -2, \\ \mathfrak{m}[\mathfrak{L}; \lambda_0] & \textit{if} \quad n = -1. \end{cases}$$

In order to obtain our real non-analytic counterparts of Theorem 0.0.4, we need the previous analysis carried out in Chapters 4, 5 and 7.

The transversalization theorem of Chapter 4 establishes that if $\lambda_0 \in \mathrm{Alg}_k(\mathfrak{L})$ for some $0 \le k < r+1$, then there exist a polynomial $\Phi : \mathbb{K} \to \mathcal{L}(U)$ with $\Phi(\lambda_0) = I_U$, and an integer $\nu = \nu(\Phi) \in \{1, \dots, k\}$, such that λ_0 is a ν-transversal eigenvalue of the product family

$$\mathfrak{L}^{\Phi} := \mathfrak{L}\Phi.$$

Moreover, $\nu(\Phi)$ and $\chi[\mathfrak{L}^{\Phi}; \lambda_0]$ are independent of Φ; therefore, one can extend the concept of multiplicity introduced in Definition 0.0.2 by setting

$$\chi[\mathfrak{L}; \lambda_0] := \chi[\mathfrak{L}^{\Phi}; \lambda_0]$$

if $\lambda_0 \in \mathrm{Alg}_k(\mathfrak{L})$, and

$$\chi[\mathfrak{L}; \lambda_0] := \infty$$

if $r = \infty$ and λ_0 is not an algebraic eigenvalue of $\mathfrak{L}$. Further, from the analysis of Chapter 5, it will become apparent that, in fact, if $\lambda_0 \in \mathrm{Alg}_k(\mathfrak{L})$, then $\nu(\Phi) = k$ for all transversalizing families of isomorphisms Φ, whether polynomial or not. When $\mathfrak{L}(\lambda_0)$ is an isomorphism, the multiplicity $\chi[\mathfrak{L}; \lambda_0]$ is defined as zero. Consequently, the multiplicity χ is finite if and only if λ_0 is an *algebraic value* of $\mathfrak{L}$, in the sense that either $\mathfrak{L}(\lambda_0) \in \mathrm{Iso}(U, V)$, or $\lambda_0 \in \mathrm{Alg}_k(\mathfrak{L})$ for some integer $1 \le k < r+1$.

The main goal of Chapter 5 is to prove that this concept of multiplicity actually satisfies Axioms A1 and A2 of Theorem 0.0.1, thus providing us with the existence of the multiplicity $\mathfrak{m}$. The analysis of this chapter is based upon another independent concept of multiplicity, denoted by $\mu[\mathfrak{L}; \lambda_0]$, going back to Magnus [92], and defined through a certain polynomial factorization of the family $\mathfrak{L}$. Once some hidden connections between the algebraic invariants invoked in the definitions of μ and χ are established, it will follow that $\chi = \mu$ and that the following result is satisfied.

Theorem 0.0.5. *For every integer $1 \le k \le r$ there exist k finite-rank projections $P_1, \dots, P_k \in \mathcal{L}(U)$ and $\mathfrak{M}_k \in \mathcal{C}^{r-k}(\Omega, \mathcal{L}(U, V))$ such that*

$$\mathfrak{L}(\lambda) = \mathfrak{M}_k(\lambda)\,[(\lambda - \lambda_0)P_1 + I_U - P_1] \cdots [(\lambda - \lambda_0)P_k + I_U - P_k] \tag{4}$$

for all $\lambda \in \Omega$. Moreover, some of the following mutually exclusive options occur:

O1. *$\mathfrak{L}(\lambda_0) \in \mathrm{Iso}(U, V)$, and, in such a case,*

$$\chi[\mathfrak{L}; \lambda_0] = 0 \quad \textit{and} \quad P_1 = \cdots = P_k = 0.$$

O2. *$\lambda_0 \in \mathrm{Alg}_\nu(\mathfrak{L})$ for some $1 \le \nu < r+1$, and, in such a case, $\mathfrak{M}_\nu(\lambda_0) \in \mathrm{Iso}(U, V)$, the projections $P_1, \dots, P_\nu$ are non-zero, and*

$$\chi[\mathfrak{L}; \lambda_0] = \sum_{i=1}^{\nu} \dim R[P_i].$$

O3. *There does not exist any integer* $1 \leq \nu \leq r$ *for which* $\lambda_0 \in \mathrm{Alg}_\nu(\mathfrak{L})$*, and, in such a case,* $\chi[\mathfrak{L};\lambda_0] = \infty$ *if* $r = \infty$*, whereas* $\chi[\mathfrak{L};\lambda_0]$ *remains undefined if* $r \in \mathbb{N}$*. Moreover, the projections* $P_1, \ldots, P_k$ *are non-zero.*

Based on the Magnus factorization (4) guaranteed by Theorem 0.0.5, Chapter 6 proves Theorem 0.0.1 and some generalizations. Chapter 6 concludes by inferring Properties (iv), (v) and (viii) listed above from Theorem 0.0.1 and a number of variants of it.

Chapter 7 studies the classic concepts of Jordan chains, partial multiplicities and local Smith form. These are very old concepts going back to Smith [121] and Frobenius [36, 37], and characterize the local behavior of $\mathfrak{L}$ at λ_0. The starting point of Chapter 7 is the following classic result going back to Gohberg [41] (see Gohberg, Goldberg & Kaashoek [43], Gohberg & Sigal [54], Gohberg, Kaashoek & Lay [44] for later generalizations). Hereafter, $\mathcal{H}$ means holomorphic.

Theorem 0.0.6. *Suppose* $\mathbb{K} = \mathbb{C}$*, let* Ω *be an open and connected subset of* $\mathbb{C}$*, and let* $\mathfrak{L} \in \mathcal{H}(\Omega, \mathcal{L}(U,V))$ *satisfy* $\mathfrak{L}(\lambda_0) \in \mathrm{Fred}_0(U,V)$ *and* $\mathfrak{L}(\Omega) \cap \mathrm{Iso}(U,V) \neq \varnothing$*. Then, for every* $\lambda_0 \in \Omega$*, there exist an open neighborhood* $\tilde{\Omega} \subset \Omega$ *of* λ_0*, a topological decomposition* $U = U_0 \oplus U_1$ *with* $n := \dim U_0 = \dim N[\mathfrak{L}(\lambda_0)] < \infty$*, and three maps*

$$\mathfrak{E} \in \mathcal{H}(\tilde{\Omega}, \mathcal{L}(U,V)), \quad \mathfrak{F} \in \mathcal{H}(\tilde{\Omega}, \mathcal{L}(U)), \quad \mathfrak{S} \in \mathcal{H}(\tilde{\Omega}, \mathcal{L}(U_0))$$

such that $\mathfrak{E}(\lambda_0)$ *and* $\mathfrak{F}(\lambda_0)$ *are isomorphisms and*

$$\mathfrak{L}(\lambda) = \mathfrak{E}(\lambda)\,[\mathfrak{S}(\lambda) \oplus I_{U_1}]\,\mathfrak{F}(\lambda), \qquad \lambda \in \tilde{\Omega}, \tag{5}$$

where I_{U_1} *stands for the identity of* U_1 *and*

$$\mathfrak{S}(\lambda) = \mathrm{diag}\left\{(\lambda - \lambda_0)^{\kappa_1}, \ldots, (\lambda - \lambda_0)^{\kappa_n}\right\}, \qquad \lambda \in \mathbb{K} \tag{6}$$

for some integers $\kappa_1 \geq \cdots \geq \kappa_n \geq 1$.

The factorization (5), with $\mathfrak{S}$ given through (6), is called the *local Smith form* of $\mathfrak{L}$ at λ_0. The integers $\kappa_1, \ldots, \kappa_n$ of (6) must be uniquely determined, and are referred to as the *partial multiplicities* of $\mathfrak{L}$ at λ_0. One of the main results of Chapter 7 uses the transversalization theorem of Chapter 4 to show that

$$\mathfrak{m}[\mathfrak{L};\lambda_0] = \chi[\mathfrak{L};\lambda_0] = \sum_{j=1}^{n} \kappa_j.$$

Essentially, the proof of Theorem 0.0.6 is based on a preliminary finite-dimensional reduction relying on the property that $\mathfrak{L}(\lambda_0) \in \mathrm{Fred}_0(U,V)$, and on a further diagonalization algorithm based on the Gaussian elimination method. At each step of the underlying scheme one must make sure that the invariants of the transformed families remain unchanged, and this can be easily accomplished through the generalized Jordan chains of $\mathfrak{L}$ at λ_0.

In this context, the analysis of Chapter 7 reveals that the most appropriate device for characterizing the existence of a local Smith form is the concept of algebraic eigenvalue introduced by Definition 0.0.3, since it is a manageable and versatile invariant, intrinsic to the family $\mathfrak{L}$, which does not involve any constructive scheme, as occurs when using Jordan chains. Actually, the following sharp version of Theorem 0.0.6, valid for $\mathbb{K} \in \{\mathbb{R}, \mathbb{C}\}$, is satisfied; it constitutes the bulk of Chapter 7.

Theorem 0.0.7. *Suppose $r \in \mathbb{N}$ and $\mathfrak{L} \in \mathcal{C}^r(\Omega, \mathcal{L}(U,V))$ with $\mathfrak{L}(\lambda_0) \in \mathrm{Fred}_0(U,V)$. Then, the following assertions are equivalent:*

i) *$\mathfrak{L}$ admits a local Smith form at λ_0, in the sense that there exist an open neighborhood $\tilde{\Omega} \subset \Omega$ of λ_0, two families*

$$\mathfrak{E} \in \mathcal{C}(\tilde{\Omega}, \mathcal{L}(U,V)), \quad \mathfrak{F} \in \mathcal{C}(\tilde{\Omega}, \mathcal{L}(U))$$

such that $\mathfrak{E}(\lambda_0)$ and $\mathfrak{F}(\lambda_0)$ are isomorphisms, and a topological decomposition $U = U_0 \oplus U_1$ with $\dim U_0 = \dim N[\mathfrak{L}(\lambda_0)] = n$, such that (5) and (6) hold true for some integers $\kappa_1 \geq \cdots \geq \kappa_n$ with $\kappa_1 \leq r$ and $\kappa_n \geq 1$.

ii) *$\lambda_0 \in \mathrm{Alg}_\nu(\mathfrak{L})$ for some integer $\nu \leq r$.*

iii) *The length of every Jordan chain of $\mathfrak{L}$ at λ_0 is bounded above by $\nu \leq r$.*

iv) *The multiplicity $\chi[\mathfrak{L}; \lambda_0]$ is well defined and finite.*

Although some of the implications between the conditions stated in Theorem 0.0.7 were proved by Rabier [110], this characterization theorem is attributable to the authors.

Chapter 7 concludes with a short introduction to the theory of equivalence of families, as discussed, for example, by Gohberg, Kaashoek & Lay [44]. The pivotal feature of the underlying theory is the fact that the local Smith form of a family $\mathfrak{L}$ is the simplest representation of $\mathfrak{L}$ through pre- and post-multiplication by families of isomorphisms. As a by-product of Theorem 0.0.7, it becomes apparent that $\mathfrak{L}$ and $\mathfrak{M}$ are equivalent if and only if they possess the same local Smith form at λ_0.

Chapter 8 focuses the attention on analytic maps $\mathfrak{L} \in \mathcal{C}^\omega(\Omega, \mathcal{L}(U,V))$, and families of the form

$$\mathfrak{L}(\lambda) := \lambda I_U - A, \qquad \lambda \in \mathbb{C}, \tag{7}$$

for a given $A \in \mathcal{L}(U)$, in order to prove the following result (an early version of its first part can be found in Trofimov [124]).

Theorem 0.0.8. *Let Ω be an open and connected subset of $\mathbb{K}$, and $\mathfrak{L} \in \mathcal{C}^\omega(\Omega, \mathcal{L}(U,V))$ such that $\mathfrak{L}(\Omega) \subset \mathrm{Fred}_0(U,V)$ and $\mathfrak{L}(\Omega) \cap \mathrm{Iso}(U,V) \neq \varnothing$. Then, the spectrum (3) of $\mathfrak{L}$ in Ω is discrete in Ω and consists of algebraic eigenvalues of $\mathfrak{L}$, as discussed by Definition 0.0.3. Moreover, the map $\mathfrak{L}^{-1}$ is analytic in $\Omega \setminus \Sigma(\mathfrak{L})$, and exhibits a pole of order $\nu \in \mathbb{N}$ at every $\lambda_0 \in \mathrm{Alg}_\nu(\mathfrak{L})$.*

If, in addition, $\mathfrak{L}$ has the form (7), and $\lambda_0 \in \mathrm{Alg}_\nu(\mathfrak{L})$, then ν equals the algebraic ascent of λ_0, as an eigenvalue of A, and

$$\mathfrak{m}[\mathfrak{L}; \lambda_0] = \chi[\mathfrak{L}; \lambda_0] = m_a(\lambda_0) = \dim N[(\lambda_0 I_U - A)^\nu].$$

Chapter 8 concludes by establishing the stability of the algebraic multiplicity for holomorphic families, as well as its homotopy invariance, within the spirit of the classic theorem of Rouché.

Chapter 10 shows that the Jordan theorem and the fundamental theorem of algebra are particular cases of the spectral theorem for monic matrix polynomials. By a matrix polynomial is meant a family $\mathfrak{L} \in \mathcal{H}(\mathbb{C}, \mathcal{L}(\mathbb{C}^N))$ of the form

$$\mathfrak{L}(\lambda) = \sum_{n=0}^{\ell} A_n \lambda^n, \qquad \lambda \in \mathbb{C},$$

where $A_0, \ldots, A_\ell \in \mathcal{L}(\mathbb{C}^N)$. It is said to be *monic* when $A_\ell = I_{\mathbb{C}^N}$. The results of this chapter give some new insights into the available results in the literature. Actually, they are intended as a first step to generalize the theory developed in this book to cover more abstract settings within the context of spectral theory.

Chapter 11 has an expository and bibliographical character. It introduces briefly some further developments of the multiplicity not previously treated in this book that are of interest in their own right.

Finally, Chapter 12 combines the linear theory developed in this book together with the topological degree to show that the algebraic multiplicity studied in this book provides an optimal algebraic invariant to characterize the nonlinear eigenvalues of a given family $\mathfrak{L}$. Thus, facilitating the transference of information from linear to nonlinear functional analysis.

It should be emphasized that eliminating the standard assumption that $\mathfrak{L}$ is holomorphic in Ω (through the concept of algebraic eigenvalue) gives rise to some deep versions of important classic results; for instance, the following striking version of the celebrated theorem of Riemann on removable singularities. Indeed, besides providing a particular case of the theorem of Riemann when $\mathbb{K} = \mathbb{C}$, it also covers the case $\mathbb{K} = \mathbb{R}$, where no previous result seems to be available. Its number of applications might be huge.

Theorem 0.0.9. *Let $r \in \mathbb{N} \cup \{\infty\}$, a point $\lambda_0 \in \Omega$, and $\mathfrak{L} \in \mathcal{C}^r(\Omega, \mathcal{L}(U,V))$ with $\mathfrak{L}(\lambda_0) \in \mathrm{Fred}_0(U,V)$. Suppose that, for some integer $0 \leq \nu \leq r$, the mapping*

$$\lambda \mapsto (\lambda - \lambda_0)^\nu \mathfrak{L}(\lambda)^{-1} \tag{8}$$

exists and is bounded in $\Omega \setminus \{\lambda_0\}$. Then there exists an open subset $\Omega' \subset \Omega$ with $\lambda_0 \in \Omega'$ for which the mapping

$$\lambda \mapsto (\lambda - \lambda_0)^\nu \mathfrak{L}(\lambda)^{-1}, \qquad \lambda \in \Omega' \setminus \{\lambda_0\},$$

admits an extension of class $\mathcal{C}^{r-\nu}$ to Ω'.

Consequently, to remove an isolated singularity for mappings of the form (8), it suffices to impose $\mathfrak{L}(\lambda_0) \in \mathrm{Fred}_0(U,V)$ and $\lambda_0 \in \mathrm{Alg}_\nu(\mathfrak{L})$ for some integer $\nu \leq r$; the holomorphy is not necessary. This result illustrates the strength of the general theory developed in this book.

Except Chapters 11 and 12, all chapters of this book end by proposing a list of exercises and some bibliographical comments. Some of the exercises are computational, but most of them ask the reader to provide alternative proofs of the results of the corresponding chapter, or to adapt the proofs already given in the chapter to obtain some related results.

Although we are indeed scientifically indebted to many people, in this occasion we are extremely honored to express our deepest gratitude to Professor I. Gohberg, the Editor of this series. It is not only that his pioneering works for meromorphic maps have influenced many chapters of this book, but also that his strong and lucid mathematical criticism during the final steps of its preparation have been tremendously encouraging for us.

Finally, we should thank the Ministry of Education and Science of Spain for research funding under grant CGL2006–00524/BOS (National Plan of Global Change and Biodiversity).

Madrid and Oxford, February 2007

The authors

Part I

Finite-dimensional Classic Spectral Theory

Part I: Summary

Part I of this book is devoted to the finite-dimensional classic spectral theory. The main topics covered in this part are the Jordan theorem and operator calculus.

We assume that the readers are familiar with the basic concepts of linear algebra. Among them, finite-dimensional vector spaces, linear dependence and independence, bases, matrices and their elementary operations, range and kernel of a matrix and of a linear operator, systems of linear equations, determinants, and the Cramer rule. Some basic properties of polynomials are also assumed to be known. In particular, elementary properties of the sum, product, factorization and division of polynomials, and the fundamental theorem of algebra. Finally, we also assume that the reader has an elementary knowledge of the theory of holomorphic functions of one complex variable, including the Cauchy integral formula, the Cauchy estimates, the Riemann theorem on removable singularities, and the Taylor and Laurent developments.

Throughout this book, $\mathbb{C}$ and $\mathbb{R}$ stand for the fields of complex and real numbers, and, for any integer number $N \geq 1$ and $\mathbb{K} \in \{\mathbb{R}, \mathbb{C}\}$, the set $\mathcal{M}_N(\mathbb{K})$ denotes the space of square matrices of order N over the field $\mathbb{K}$.

Part I is divided into three chapters. In Chapter 1 we prove the Jordan theorem. This theorem establishes that for every $A \in \mathcal{M}_N(\mathbb{C})$, the space $\mathbb{C}^N$ decomposes into the direct sum of the ascent generalized eigenspaces associated with each of the eigenvalues of the matrix A. Then, we use the Jordan theorem to construct the complex and the real Jordan canonical forms of the matrix A. Chapter 2 is an introduction to operator calculus. Given a holomorphic function f and a square matrix A, one of the main aims of operator calculus is to give an appropriate definition of $f(A)$ and to study its properties. As a paradigmatic application of operator calculus, Chapter 2 presents a detailed calculation of the exponential of a matrix. Finally, in Chapter 3 we use the results of the previous two chapters to construct the spectral projections associated with a square matrix A; the spectral projections are the projections of $\mathbb{C}^N$ onto each of the ascent generalized eigenspaces of the matrix A.

Chapter 1

The Jordan Theorem

In this chapter we prove the Jordan theorem, a pivotal result in mathematics, which establishes that, for every $A \in \mathcal{M}_N(\mathbb{C})$, the space $\mathbb{C}^N$ decomposes as the direct sum of the ascent generalized eigenspaces associated with each of the eigenvalues of A. Then, by choosing an appropriate basis in each of the ascent generalized eigenspaces, the Jordan canonical form of A is constructed. These bases are chosen in order to attain a similar matrix to A with a maximum number of zeros.

This chapter is organized as follows. In Section 1.1 we review some basic concepts in the theory of matrices and finite-dimensional operators; among them, the relationship between operators and matrices, and the concepts of eigenvalue, eigenvector, eigenspace, characteristic polynomial and multiplicity. Section 1.2 is devoted to the proof of the Jordan theorem. Some important intermediate results will also be proved, such as the existence of interpolating polynomials and the Cayley–Hamilton theorem. In Section 1.3 we construct the Jordan canonical form, while in Section 1.4 we construct the real Jordan canonical form. Finally, Section 1.5 shows an example of the calculation of the Jordan canonical form.

1.1 Basic concepts

From now on, we consider the space $\mathcal{M}_N(\mathbb{C})$ of the square matrices of order N with coefficients in $\mathbb{C}$, and the space $\mathcal{L}(\mathbb{C}^N)$ of the linear operators from $\mathbb{C}^N$ to $\mathbb{C}^N$. Both have dimension N^2 and, hence, are isomorphic. Once a basis

$$\mathcal{B} := \{e_1, \ldots, e_N\},$$

in $\mathbb{C}^N$ is fixed, every matrix

$$A = (a_{ij})_{1 \le i,j \le N} \in \mathcal{M}_N(\mathbb{C})$$

defines a linear operator $T_A \in \mathcal{L}(\mathbb{C}^N)$ through

$$T_A u = \sum_{i=1}^{N}\sum_{j=1}^{N} a_{ij} u_j e_i \qquad \text{for each } u = \sum_{j=1}^{N} u_j e_j \in \mathbb{C}^N.$$

As the coordinates of the N-tuple $T_A u$ with respect to the basis $\mathcal{B}$ are obtained by multiplying the matrix A by the N-tuple of the coordinates of u with respect to $\mathcal{B}$, it is usual to identify T_A with A. The resulting map

$$\begin{array}{ccc} \mathcal{M}_N(\mathbb{C}) & \to & \mathcal{L}(\mathbb{C}^N) \\ A & \mapsto & T_A \end{array}$$

is an isomorphism of vector spaces and of algebras. Consequently, there is no ambiguity in making that identification. In practice, A is referred to as the matrix associated with the operator T_A with respect to the basis $\mathcal{B}$.

Pick $T \in \mathcal{L}(\mathbb{C}^N)$, let

$$\mathcal{B} := \{e_1, \dots, e_N\}, \qquad \hat{\mathcal{B}} := \{\hat{e}_1, \dots, \hat{e}_N\},$$

be two bases of $\mathbb{C}^N$, and let $A_{\mathcal{B}}$ and $A_{\hat{\mathcal{B}}}$ denote their associated matrices with respect to the bases $\mathcal{B}$ and $\hat{\mathcal{B}}$, respectively. Then,

$$A_{\hat{\mathcal{B}}} = P^{-1} A_{\mathcal{B}} P, \tag{1.1}$$

where $P \in \mathcal{M}_N(\mathbb{C})$ is the matrix whose jth column consists of the coordinates of $\hat{e}_j$ with respect to the basis $\mathcal{B}$, that is, $P = (p_{ij})_{1\le i,j\le N}$, where

$$\hat{e}_j = \sum_{i=1}^{N} p_{ij} e_i, \qquad 1 \le j \le N.$$

The matrix P is known as the *matrix of the change of basis*. Identity (1.1) follows easily by taking into account that the fact

$$u = \sum_{j=1}^{N} u_j e_j = \sum_{j=1}^{N} \hat{u}_j \hat{e}_j$$

implies

$$\begin{pmatrix} u_1 \\ \vdots \\ u_N \end{pmatrix} = P \begin{pmatrix} \hat{u}_1 \\ \vdots \\ \hat{u}_N \end{pmatrix}.$$

Two matrices are said to be *similar* when they represent the same linear operator with respect to two certain bases. In other words, $A, B \in \mathcal{M}_N(\mathbb{C})$ are similar when there exists an invertible matrix $P \in \mathcal{M}_N(\mathbb{C})$ such that

$$AP = PB.$$

One of the main objectives in finite-dimensional spectral theory is the construction of bases so that a given linear operator possesses an associated matrix with a maximum number of zeros; the closer to a diagonal matrix, the better. To that aim, one has to find the *eigenvectors* of the operator (or matrix) A, which are the vectors $u \in \mathbb{C}^N \setminus \{0\}$ for which there exists some $\lambda \in \mathbb{C}$ such that

$$Au = \lambda u.$$

The direction of an eigenvector is therefore invariant under the action of the linear operator A. The corresponding λ is called an *eigenvalue* of A. The matrix associated with A with respect to any basis containing an eigenvector exhibits a column with the associated eigenvalue in the entry of the principal diagonal, while all remaining entries of that column equal zero.

Hereafter, given $A \in \mathcal{L}(\mathbb{C}^N)$, the sets $N[A]$ and $R[A]$ will stand for the kernel (null-space) and the range (image) of A, respectively. Also, I denotes the identity matrix in $\mathcal{M}_N(\mathbb{C})$, or the identity operator in $\mathcal{L}(\mathbb{C}^N)$. The following definition makes the concepts of eigenvalue and eigenvector precise.

Definition 1.1.1. It is said that $\lambda \in \mathbb{C}$ is an eigenvalue of the matrix $A \in \mathcal{M}_N(\mathbb{C})$ if there exists $u \in \mathbb{C}^N \setminus \{0\}$ such that

$$Au = \lambda u.$$

Every $u \in N[A - \lambda I]$ is called an eigenvector of A associated with the eigenvalue λ. The linear subspace $N[A - \lambda I]$ is called the eigenspace of A associated with λ.

Thanks to the Cramer rule, the homogeneous linear system

$$Au = \lambda u$$

admits a non-zero solution u if and only if

$$\det(A - \lambda I) = 0.$$

Consequently, the set of eigenvalues of A coincides with the set of roots of the polynomial P_A defined by

$$P_A(z) := \det(A - zI), \qquad z \in \mathbb{C},$$

which will be referred to as the *characteristic polynomial* of A. A *polynomial* is a mapping $P : \mathbb{C} \to \mathbb{C}$ of the form

$$P(z) = \sum_{j=0}^{n} a_j z_j, \qquad z \in \mathbb{C},$$

for some integer $n \geq 0$ and some coefficients $a_0, \ldots, a_n \in \mathbb{C}$, of course.

The characteristic polynomial is invariant under changes of basis, and, hence, so are the eigenvalues. This feature reveals the consistency of referring to the characteristic polynomial *of a linear operator*, and, hence, to the eigenvalues *of a linear operator*. Indeed, if $A, Q \in \mathcal{M}_N(\mathbb{C})$ and $\det Q \neq 0$, then, for all $z \in \mathbb{C}$,

$$P_{Q^{-1}AQ}(z) = \det(Q^{-1}(A - zI)Q) = \det(A - zI) = P_A(z);$$

however, the eigenvectors are not invariant under changes of basis.

The equation

$$P_A(z) = 0$$

is called the *characteristic equation* of A, and its roots are the eigenvalues of A. It is easy to check that the characteristic polynomial has degree N, and that its leading coefficient is $(-1)^N$. Therefore, thanks to the fundamental theorem of algebra, it possesses exactly N roots, counting according to their multiplicities. Consequently, we can express

$$P_A(z) = (-1)^N \prod_{j=1}^{p} (z - \lambda_j)^{m_a(\lambda_j)}, \qquad z \in \mathbb{C}, \tag{1.2}$$

for some complex numbers $\lambda_1, \dots, \lambda_p$ and integers

$$m_a(\lambda_1), \dots, m_a(\lambda_p) \geq 1$$

with the convention that

$$\lambda_i \neq \lambda_j \quad \text{if} \quad 1 \leq i < j \leq p,$$

which will be maintained throughout this book. Then, the eigenvalues of the matrix A are $\lambda_1, \dots, \lambda_p$. Moreover,

$$m_a(\lambda_1) + \cdots + m_a(\lambda_p) = N.$$

Definition 1.1.2. Let $A \in \mathcal{M}_N(\mathbb{C})$. Then, the set of eigenvalues of A, hereafter denoted by $\sigma(A)$, is called the spectrum of A. Moreover, if the characteristic polynomial P_A of A is expressed by (1.2), then $m_a(\lambda_j)$ is called the algebraic multiplicity of λ_j as an eigenvalue of A.

For each $\lambda \in \sigma(A)$, the geometric multiplicity of the eigenvalue λ, subsequently denoted by $m_g(\lambda)$, is defined through

$$m_g(\lambda) := \dim N[A - \lambda I].$$

In general,

$$m_g(\lambda) \leq m_a(\lambda) \qquad \text{for each} \quad \lambda \in \sigma(A),$$

but the algebraic and geometric multiplicities do not always coincide. For example, for each $\alpha \in \mathbb{C}$, the characteristic polynomial P_A of the matrix

$$A = \begin{pmatrix} \alpha & 0 \\ 1 & \alpha \end{pmatrix}$$

is given by

$$P_A(z) = (z - \alpha)^2, \qquad z \in \mathbb{C},$$

and, hence, α is the unique eigenvalue of A. Its algebraic multiplicity equals 2, whereas its geometric multiplicity equals 1, since

$$N[A - \alpha I] = \operatorname{span}\left[\begin{pmatrix} 0 \\ 1 \end{pmatrix}\right].$$

The eigenvalues $\lambda \in \sigma(A)$ for which

$$m_g(\lambda) = m_a(\lambda)$$

are called *semisimple*. An eigenvalue is semisimple if and only if its algebraic multiplicity equals the maximum number of linearly independent associated eigenvectors. We will see that a matrix is diagonalizable (that is, it is similar to a diagonal matrix) if and only if all its eigenvalues are semisimple. An eigenvalue $\lambda \in \sigma(A)$ is said to be *simple* when $m_a(\lambda) = 1$. If λ is simple, necessarily $m_g(\lambda) = 1$, and, hence, λ is semisimple.

Besides the eigenspace $N[A - \lambda I]$ itself, associated with each eigenvalue $\lambda \in \sigma(A)$ there is the following increasing chain of *generalized eigenspaces*

$$N[A - \lambda I] \subset N[(A - \lambda I)^2] \subset N[(A - \lambda I)^3] \subset \cdots, \tag{1.3}$$

which is indeed increasing because

$$N[C] \subset N[BC]$$

for every $B, C \in \mathcal{M}_N(\mathbb{C})$. Moreover, as $\mathbb{C}^N$ is finite dimensional, the chain (1.3) must stabilize, and, hence, there exists some integer $j \geq 1$ such that

$$N[(A - \lambda I)^k] = N[(A - \lambda I)^j], \qquad k \geq j. \tag{1.4}$$

Actually, for every $j \geq 1$ such that

$$N[(A - \lambda I)^j] = N[(A - \lambda I)^{j+1}]$$

the identities (1.4) hold. Indeed, for each $u \in N[(A - \lambda I)^{j+2}]$, we have that

$$(A - \lambda I)u \in N[(A - \lambda I)^{j+1}] = N[(A - \lambda I)^j],$$

and, hence, $u \in N[(A - \lambda I)^{j+1}]$. Consequently,

$$N[(A - \lambda I)^{j+2}] \subset N[(A - \lambda I)^{j+1}]$$

and, therefore, (1.4) holds, because

$$N[(A - \lambda I)^{j+1}] \subset N[(A - \lambda I)^{j+2}].$$

Now, we are ready to introduce the following concept.

Definition 1.1.3. For each $\lambda \in \sigma(A)$, the ascent of λ, throughout denoted by $\nu(\lambda)$, is the minimum integer number $k \geq 1$ for which

$$N[(A-\lambda I)^k] = N[(A-\lambda I)^{k+1}].$$

For any integer $j \geq 1$, the linear subspace $N[(A-\lambda I)^j]$ is said to be a generalized eigenspace of A associated with the eigenvalue λ, and any $u \in N[(A-\lambda I)^j]$ is said to be a generalized eigenvector of A (associated with λ). The linear manifold $N[(A-\lambda I)^{\nu(\lambda)}]$ is called the ascent generalized eigenspace of A associated with the eigenvalue λ.

According to Definition 1.1.3 and the previous discussion, for every $\lambda \in \sigma(A)$ and $0 \leq i < j \leq \nu(\lambda) \leq k$, one has that

$$N[(A-\lambda I)^i] \subsetneq N[(A-\lambda I)^j] \subset N[(A-\lambda I)^{\nu(\lambda)}] = N[(A-\lambda I)^k]. \tag{1.5}$$

1.2 The Jordan theorem

Hereafter, for any $A \in \mathcal{M}_N(\mathbb{C})$, when we write

$$\sigma(A) = \{\lambda_1, \dots, \lambda_p\},$$

we will suppose that

$$\lambda_i \neq \lambda_j \qquad \text{if} \quad 1 \leq i < j \leq p,$$

even though it is not explicitly stated. The following result, which collects the most important properties of the ascent generalized eigenspaces, is the central result of the finite-dimensional classic spectral theory.

Theorem 1.2.1 (Jordan). *Let $A \in \mathcal{M}_N(\mathbb{C})$ be with*

$$\sigma(A) = \{\lambda_1, \dots, \lambda_p\}.$$

Then

$$A\left(N[(A-\lambda_j I)^{\nu(\lambda_j)}]\right) \subset N[(A-\lambda_j I)^{\nu(\lambda_j)}], \qquad 1 \leq j \leq p.$$

Actually, if for any $1 \leq j \leq p$ we denote by A_j the restriction of A from $N[(A-\lambda_j I)^{\nu(\lambda_j)}]$ to itself, then

$$\sigma(A_j) = \{\lambda_j\}, \qquad 1 \leq j \leq p.$$

Moreover,

$$\mathbb{C}^N = \bigoplus_{j=1}^{p} N[(A-\lambda_j I)^{\nu(\lambda_j)}] \tag{1.6}$$

and

$$m_a(\lambda_j) = \dim N[(A-\lambda_j I)^{\nu(\lambda_j)}], \qquad 1 \leq j \leq p.$$

The proof of Theorem 1.2.1 will be decomposed in a number of independent results of great intrinsic interest in their own right. Hereafter, given a matrix $A \in \mathcal{M}_N(\mathbb{C})$ and the polynomial P defined through

$$P(z) = \sum_{j=0}^{n} a_j z^j, \qquad z \in \mathbb{C},$$

we will denote by $P(A)$ the matrix

$$P(A) := \sum_{j=0}^{n} a_j A^j.$$

Note that if P is expressed as

$$P(z) = a_n \prod_{i=1}^{q} (z - \mu_i)^{m_i}, \qquad z \in \mathbb{C},$$

for some $q, m_1, \dots, m_q \in \mathbb{N}$, and $\mu_1, \dots, \mu_q \in \mathbb{C}$, then

$$P(A) = a_n \prod_{i=1}^{q} (A - \mu_i I)^{m_i};$$

the product being independent of the order of the factors, because A commutes with itself and with the identity operator I.

The next result shows that $P(A) = 0$ if $\sigma(A) = \{\lambda_1, \dots, \lambda_p\}$ and P is divided by the polynomial

$$z \mapsto \prod_{j=1}^{p} (z - \lambda_j)^{\nu(\lambda_j)}, \qquad z \in \mathbb{C}.$$

Every polynomial P satisfying $P(A) = 0$ will be called an *annihilator polynomial* of the matrix A.

Lemma 1.2.2. *Let $A \in \mathcal{M}_N(\mathbb{C})$ be with $\sigma(A) = \{\lambda_1, \dots, \lambda_p\}$. Then*

$$\prod_{j=1}^{p} (A - \lambda_j I)^{\nu(\lambda_j)} = 0.$$

Proof. Let $\mathcal{B} = \{e_1, \dots, e_N\}$ be a basis of $\mathbb{C}^N$. Then, for each $1 \le i \le N$, the $N+1$ vectors

$$A^j e_i, \qquad 0 \le j \le N,$$

must be linearly dependent, and, hence, there exist

$$(\alpha_{i0}, \dots, \alpha_{iN}) \in \mathbb{C}^{N+1} \setminus \{0\}$$

such that

$$\sum_{j=0}^{N} \alpha_{ij} A^j e_i = 0.$$

Now, consider the polynomials $Q_1, \dots, Q_N$ defined by

$$Q_i(z) := \sum_{j=0}^{N} \alpha_{ij} z^j, \qquad z \in \mathbb{C}, \qquad 1 \leq i \leq N.$$

Each of them is non-zero and satisfies

$$Q_i(A)e_i = 0.$$

Thus, the product polynomial

$$Q := Q_1 \cdots Q_N$$

satisfies

$$Q \neq 0 \quad \text{and} \quad Q(A)e_i = 0, \qquad 1 \leq i \leq N,$$

since $Q_1(A), \dots, Q_N(A)$ commute with one another. As the matrix $Q(A)$ vanishes over the basis $\mathcal{B}$, necessarily

$$Q(A) = 0.$$

We divide Q by a constant, so that its leading coefficient can be taken as the unity, and, hence, Q can be expressed in the form

$$Q(z) = \prod_{j=1}^{q} (z - \mu_j)^{m_j}, \qquad z \in \mathbb{C},$$

for some natural numbers $q, m_1, \dots, m_q \geq 1$, and certain $\mu_1, \dots, \mu_q \in \mathbb{C}$ mutually different. Now, we consider the polynomial R defined through

$$R(z) := \prod_{\substack{j=1 \\ \mu_j \in \sigma(A)}}^{q} (z - \mu_j)^{m_j}, \qquad z \in \mathbb{C},$$

if

$$\sigma(A) \cap \{\mu_1, \dots, \mu_q\} \neq \varnothing,$$

and by $R := 1$ if

$$\sigma(A) \cap \{\mu_1, \dots, \mu_q\} = \varnothing.$$

For these choices, Q admits the factorization

$$Q(z) = R(z) \prod_{\substack{j=1 \\ \mu_j \notin \sigma(A)}}^{q} (z - \mu_j)^{m_j}, \qquad z \in \mathbb{C}.$$

On the other hand, we obtain from the definition of eigenvalue that

$$\det \prod_{\substack{j=1 \\ \mu_j \notin \sigma(A)}}^{q} (A - \mu_j I)^{m_j} = \prod_{\substack{j=1 \\ \mu_j \notin \sigma(A)}}^{q} [\det(A - \mu_j I)]^{m_j} \neq 0,$$

and, hence, the matrix

$$\prod_{\substack{j=1 \\ \mu_j \notin \sigma(A)}}^{q} (A - \mu_j I)^{m_j}$$

is invertible. Consequently, $R(A) = 0$ since $Q(A) = 0$; therefore, there exists $i \in \{1, \dots, q\}$ such that $\mu_i \in \sigma(A)$. Let us check that the polynomial P defined by

$$P(z) := \prod_{\substack{j=1 \\ \mu_j \in \sigma(A)}}^{q} (z - \mu_j)^{\min\{\nu(\mu_j), m_j\}}, \qquad z \in \mathbb{C}, \tag{1.7}$$

also satisfies $P(A) = 0$. Indeed, if $m_j \leq \nu(\mu_j)$ for each $\mu_j \in \sigma(A)$, then $P = R$ and we already know that $R(A) = 0$. So suppose that there exists $\mu_j \in \sigma(A)$ such that $\nu(\mu_j) < m_j$. Without loss of generality, we can assume that

$$\nu(\mu_1) < m_1.$$

Then,

$$0 = R(A) = (A - \mu_1 I)^{m_1} \prod_{\substack{j=2 \\ \mu_j \in \sigma(A)}}^{q} (A - \mu_j I)^{m_j}$$

and, therefore,

$$R\Big[\prod_{\substack{j=2 \\ \mu_j \in \sigma(A)}}^{q} (A - \mu_j I)^{m_j} \Big] \subset N[(A - \mu_1 I)^{m_1}] = N[(A - \mu_1 I)^{\nu(\mu_1)}],$$

whence

$$(A - \mu_1 I)^{\nu(\mu_1)} \prod_{\substack{j=2 \\ \mu_j \in \sigma(A)}}^{q} (A - \mu_j I)^{m_j} = 0.$$

By repeating this process for each $\mu_j \in \sigma(A)$ with $\nu(\mu_j) < m_j$, it is apparent that the polynomial P defined by (1.7) satisfies $P(A) = 0$. This concludes the proof. □

The next result characterizes the annihilator polynomials of a matrix.

Theorem 1.2.3 (Cayley–Hamilton). *Let $A \in \mathcal{M}_N(\mathbb{C})$ be with*

$$\sigma(A) = \{\lambda_1, \ldots, \lambda_p\}.$$

Then, a polynomial P satisfies

$$P(A) = 0$$

if and only if

$$P(z) = Q(z) \prod_{j=1}^{p} (z - \lambda_j)^{\nu(\lambda_j)}, \qquad z \in \mathbb{C}, \tag{1.8}$$

for some polynomial Q.

Proof. Suppose that P admits the factorization (1.8). Then, thanks to Lemma 1.2.2,

$$P(A) = Q(A) \prod_{j=1}^{p} (A - \lambda_j I)^{\nu(\lambda_j)} = 0.$$

To show the converse suppose that a polynomial P satisfies $P(A) = 0$. Then, for each $1 \leq j \leq p$ and $x \in N[A - \lambda_j I] \setminus \{0\}$, one has that

$$0 = P(A)x = P(\lambda_j)x$$

and, hence, $P(\lambda) = 0$ for every $\lambda \in \sigma(A)$. Thus, there exist a polynomial Q and p natural numbers $m_1, \ldots, m_p \geq 1$, such that

$$P(z) = Q(z) \prod_{j=1}^{p} (z - \lambda_j)^{m_j}, \qquad z \in \mathbb{C}.$$

Moreover, by choosing appropriate exponents $m_1, \ldots, m_p \geq 1$, we can suppose that no eigenvalue of A is a root of the polynomial Q. To complete the proof of the theorem it suffices to check that

$$m_j \geq \nu(\lambda_j) \qquad \text{for each} \quad 1 \leq j \leq p. \tag{1.9}$$

The proof of (1.9) proceeds by contradiction. Suppose $m_j < \nu(\lambda_j)$ for some $1 \leq j \leq p$. Changing the order of the eigenvalues, if necessary, we can assume that

$$m_1 < \nu(\lambda_1).$$

Then, by (1.5), $N[(A - \lambda_1 I)^{m_1}]$ is a proper subspace of $N[(A - \lambda_1 I)^{m_1+1}]$. Let

$$x \in N[(A - \lambda_1 I)^{m_1+1}] \setminus N[(A - \lambda_1 I)^{m_1}]$$

and consider

$$y := (A - \lambda_1 I)^{m_1} x.$$

By construction, we have that

$$y \neq 0 \quad \text{and} \quad Ay = \lambda_1 y.$$

Therefore,

$$P(A)x = Q(A)\prod_{j=2}^{p}(A - \lambda_j I)^{m_j} y = Q(\lambda_1)\prod_{j=2}^{p}(\lambda_1 - \lambda_j)^{m_j} y \neq 0,$$

because no eigenvalue of A is a root of Q, the numbers $\lambda_1, \dots, \lambda_p$ are distinct, and $y \neq 0$. This contradicts the assumption $P(A) = 0$, which shows (1.9) and concludes the proof. □

The next result shows the existence of a class of interpolating polynomials.

Lemma 1.2.4. *Suppose $h \geq 1$ is an integer, and $r_1, \dots, r_h \in \mathbb{C}$ satisfy $r_i \neq r_j$ if $1 \leq i < j \leq h$. For each $j \in \{1, \dots, h\}$ consider any integer number $m_j \geq 0$ and any complex numbers $\alpha_{j,0}, \dots, \alpha_{j,m_j}$. Then there exists a polynomial P such that*

$$P^{k)}(r_j) = \alpha_{j,k}, \qquad 1 \leq j \leq h, \quad 0 \leq k \leq m_j.$$

Proof. We will argue by induction on h. When $h = 1$, the polynomial P defined by

$$P(z) := \sum_{k=0}^{m_1} \frac{\alpha_{1,k}}{k!}(z - r_1)^k, \qquad z \in \mathbb{C},$$

satisfies the required conditions. Suppose the result is true for $h = n$, and let P_n be a polynomial such that

$$P_n^{k)}(r_j) = \alpha_{j,k}, \qquad 1 \leq j \leq n, \quad 0 \leq k \leq m_j.$$

To prove the result for $h = n + 1$ it suffices to check that there exists a polynomial Q for which the polynomial P_{n+1} defined by

$$P_{n+1}(z) := P_n(z) + Q(z)\prod_{j=1}^{n}(z - r_j)^{m_j+1}, \quad z \in \mathbb{C}, \tag{1.10}$$

satisfies the required conditions. The auxiliary polynomial Π_n defined by

$$\Pi_n(z) := \prod_{j=1}^{n}(z - r_j)^{m_j+1}, \qquad z \in \mathbb{C},$$

satisfies

$$\Pi_n^{k)}(r_j) = 0, \qquad 1 \leq j \leq n, \quad 0 \leq k \leq m_j. \tag{1.11}$$

Moreover, for every polynomial Q satisfying (1.10), and $k \geq 0$, differentiating (1.10) gives

$$P_{n+1}^{k)} = P_n^{k)} + \sum_{\ell=0}^{k} \binom{k}{\ell} Q^{k-\ell)} \Pi_n^{\ell)}. \tag{1.12}$$

Suppose $1 \leq j \leq n$ and $0 \leq k \leq m_j$. Then, from (1.11), (1.12) and the induction hypothesis we obtain that

$$P_{n+1}^{k)}(r_j) = P_n^{k)}(r_j) = \alpha_{j,k}.$$

Thus, to conclude the proof it suffices to construct Q satisfying

$$\alpha_{n+1,k} = P_n^{k)}(r_{n+1}) + \sum_{\ell=0}^{k} \binom{k}{\ell} Q^{k-\ell)}(r_{n+1}) \Pi_n^{\ell)}(r_{n+1}) \tag{1.13}$$

for each $0 \leq k \leq m_{n+1}$. For $k = 0$, (1.13) becomes

$$\alpha_{n+1,0} = P_n(r_{n+1}) + Q(r_{n+1})\Pi_n(r_{n+1}),$$

and, since $\Pi_n(r_{n+1}) \neq 0$, the polynomial Q must satisfy $Q(r_{n+1}) = \beta_0$, where

$$\beta_0 := \frac{\alpha_{n+1,0} - P_n(r_{n+1})}{\Pi_n(r_{n+1})}. \tag{1.14}$$

In general, according to (1.13), for each $1 \leq k \leq m_{n+1}$ the polynomial Q must satisfy $Q^{k)}(r_{n+1}) = \beta_k$, where

$$\beta_k := \frac{1}{\Pi_n(r_{n+1})} \left[\alpha_{n+1,k} - P_n^{k)}(r_{n+1}) - \sum_{\ell=1}^{k} \binom{k}{\ell} \beta_{k-\ell} \Pi_n^{\ell)}(r_{n+1}) \right].$$

As this is a recurrence formula whose initialization is given by (1.14), substituting the polynomial Q defined through

$$Q(z) := \sum_{k=0}^{m_{n+1}} \frac{\beta_k}{k!} (z - r_{n+1})^k, \qquad z \in \mathbb{C},$$

into (1.10) provides us with a P satisfying all required conditions. □

Now, given $A \in \mathcal{M}_N(\mathbb{C})$, we use Lemma 1.2.4 to construct an auxiliary family of polynomials associated with A that will enable us to prove the Jordan theorem. From now on, δ_{ij} stands for the Kronecker delta, that is to say, $\delta_{ij} = 1$ if $i = j$, while $\delta_{ij} = 0$ if $i \neq j$.

Proposition 1.2.5. *Let $A \in \mathcal{M}_N(\mathbb{C})$ be a matrix with exactly $p \geq 1$ eigenvalues. Then there exist p polynomials $P_1, \ldots, P_p$ satisfying the following properties:*

- $P_j(A) \neq 0$ *for each* $1 \leq j \leq p$.
- $P_i(A)P_j(A) = \delta_{ij} P_i(A)$ *for each* $1 \leq i, j \leq p$.
- $I = \sum_{j=1}^{p} P_j(A)$.

Proof. Suppose
$$\sigma(A) = \{\lambda_1, \dots, \lambda_p\}.$$
By Lemma 1.2.4, for each $1 \leq j \leq p$ there exists a polynomial P_j such that
$$P_j(\lambda_j) = 1, \quad P_j^{k)}(\lambda_j) = 0, \qquad 1 \leq k \leq \nu(\lambda_j) - 1,$$
and
$$P_j^{k)}(\lambda_i) = 0, \qquad 0 \leq k \leq \nu(\lambda_i) - 1, \quad i \in \{1, \dots, p\} \setminus \{j\}.$$
These polynomials $P_1, \dots, P_p$ satisfy the conditions of the statement. Indeed, fix $1 \leq j \leq p$. By construction, the polynomial $P_j - 1$ has a zero of order greater than or equal to $\nu(\lambda_j)$ at λ_j and, consequently, there exists a polynomial Q_j such that
$$P_j(z) - 1 = (z - \lambda_j)^{\nu(\lambda_j)} Q_j(z), \qquad z \in \mathbb{C}. \tag{1.15}$$
Similarly, P_j has a zero of order greater than or equal to $\nu(\lambda_i)$ at each $\lambda_i \in \sigma(A) \setminus \{\lambda_j\}$, and, hence, there exists a polynomial R_j such that
$$P_j(z) = R_j(z) \prod_{\substack{i=1 \\ i \neq j}}^{p} (z - \lambda_i)^{\nu(\lambda_i)}, \qquad z \in \mathbb{C}. \tag{1.16}$$
As $P_j(\lambda_j) = 1 \neq 0$, the polynomial $z \mapsto (z - \lambda_j)^{\nu(\lambda_j)}$ cannot divide P_j and, hence, by Theorem 1.2.3, $P_j(A) \neq 0$. Moreover, by (1.15) and (1.16), we find that
$$P_j(z)[P_j(z) - 1] = Q_j(z) R_j(z) \prod_{i=1}^{p} (z - \lambda_i)^{\nu(\lambda_i)}, \quad z \in \mathbb{C},$$
and, hence, according to Theorem 1.2.3,
$$P_j(A)[P_j(A) - I] = 0.$$
Similarly, for any $1 \leq i \leq p$ with $i \neq j$, it is easy to see that
$$P_i(A)P_j(A) = 0.$$
Again by (1.15) and (1.16), the polynomial $z \mapsto (z - \lambda_j)^{\nu(\lambda_j)}$ divides $P_j - 1$ and each P_i with $i \neq j$. Consequently, it also divides their sum
$$\sum_{i=1}^{p} P_i - 1.$$

Actually, as this property is valid for each $1 \leq j \leq p$, the product polynomial

$$z \mapsto \prod_{j=1}^{p} (z - \lambda_j)^{\nu(\lambda_j)}$$

also divides $\sum_{i=1}^{p} P_i - 1$, since its factors are mutually prime. Consequently, thanks again to Theorem 1.2.3,

$$I = \sum_{j=1}^{p} P_j(A),$$

which completes the proof. □

Now, we have all necessary background to prove the Jordan theorem.

Proof of Theorem 1.2.1. Consider the polynomials $P_1, \ldots, P_p$, whose existence was guaranteed by Proposition 1.2.5, and the vector subspaces

$$N_j := R[P_j(A)], \qquad 1 \leq j \leq p.$$

We claim that

$$A(N_j) \subset N_j, \qquad 1 \leq j \leq p, \tag{1.17}$$

and

$$\mathbb{C}^N = N_1 + \cdots + N_p. \tag{1.18}$$

Indeed, (1.17) follows readily from the fact that

$$AP_j(A) = P_j(A)A\,, \qquad 1 \leq j \leq p,$$

and Property (1.18) follows straight away from the identity

$$I = \sum_{j=1}^{p} P_j(A).$$

Now, we will prove that

$$N_j = N[(A - \lambda_j I)^{\nu(\lambda_j)}], \qquad 1 \leq j \leq p. \tag{1.19}$$

Thanks to (1.16), for each $1 \leq i \leq p$, the polynomial

$$z \mapsto \prod_{j=1}^{p} (z - \lambda_j)^{\nu(\lambda_j)}$$

divides the polynomial

$$z \mapsto (z - \lambda_i)^{\nu(\lambda_i)} P_i(z)$$

and, hence, by Theorem 1.2.3,

$$(A - \lambda_i I)^{\nu(\lambda_i)} P_i(A) = 0.$$

Consequently,

$$N_i = R[P_i(A)] \subset N[(A - \lambda_i I)^{\nu(\lambda_i)}]. \tag{1.20}$$

Combining these inclusions with (1.18) shows that

$$\mathbb{C}^N = N[(A - \lambda_1 I)^{\nu(\lambda_1)}] + \cdots + N[(A - \lambda_p I)^{\nu(\lambda_p)}].$$

Therefore, in order to prove (1.6) it suffices to show that the facts

$$u_j \in N[(A - \lambda_j I)^{\nu(\lambda_j)}], \qquad 1 \leq j \leq p, \tag{1.21}$$

and

$$\sum_{j=1}^{p} u_j = 0 \tag{1.22}$$

imply

$$u_j = 0 \quad \text{for each} \quad 1 \leq j \leq p.$$

This can be obtained by contradiction. Suppose that $u_j \neq 0$ for some $j \in \{1, \dots, p\}$. By rearranging the eigenvalues, if necessary, we can assume that $u_1 \neq 0$. Then, by construction,

$$(A - \lambda_1 I)^{\nu(\lambda_1)} u_1 = 0.$$

Let m denote the largest $k \in \mathbb{N}$ for which

$$(A - \lambda_1 I)^k u_1 \neq 0;$$

this m exists because $u_1 \neq 0$ and $(A - \lambda_1 I)^0 = I$. Moreover, $0 \leq m < \nu(\lambda_1)$. Set

$$v_1 := (A - \lambda_1 I)^m u_1.$$

Then, $v_1 \neq 0$,

$$(A - \lambda_1 I) v_1 = (A - \lambda_1 I)^{m+1} u_1 = 0$$

and, hence,

$$A v_1 = \lambda_1 v_1. \tag{1.23}$$

Now, consider the auxiliary polynomial P defined by

$$P(z) := (z - \lambda_1)^m \prod_{j=2}^{p} (z - \lambda_j)^{\nu(\lambda_j)}, \qquad z \in \mathbb{C}.$$

By the definition of v_1, (1.23) implies that

$$P(A) u_1 = \prod_{j=2}^{p} (A - \lambda_j I)^{\nu(\lambda_j)} v_1 = \prod_{j=2}^{p} (\lambda_1 - \lambda_j)^{\nu(\lambda_j)} v_1 \neq 0.$$

On the other hand, using identity (1.22) gives

$$\begin{aligned} P(A)u_1 &= -\sum_{k=2}^{p} P(A)u_k \\ &= -\sum_{k=2}^{p} (A-\lambda_1 I)^m \prod_{\substack{j=2\\ j\neq k}}^{p} (A-\lambda_j I)^{\nu(\lambda_j)} (A-\lambda_k I)^{\nu(\lambda_k)} u_k \end{aligned}$$

and, hence, by (1.21), we find that $P(A)u_1 = 0$. This contradiction concludes the proof of (1.6). Identity (1.19) follows straight away from (1.6) and (1.20).

Thanks to (1.17) and (1.19), A leaves $N[(A-\lambda_j I)^{\nu(\lambda_j)}]$ invariant for each $1 \leq j \leq p$. Thus, we can define the operators

$$A_j \in \mathcal{L}(N[(A-\lambda_j I)^{\nu(\lambda_j)}]), \qquad 1 \leq j \leq p,$$

as the restriction of A from $N[(A-\lambda_j I)^{\nu(\lambda_j)}]$ to itself. Consider $1 \leq j \leq p$, a complex number μ, and

$$u \in N[(A-\lambda_j I)^{\nu(\lambda_j)}] \setminus \{0\}$$

such that $A_j u = \mu u$. Then

$$0 = (A-\lambda_j I)^{\nu(\lambda_j)} u = (\mu - \lambda_j)^{\nu(\lambda_j)} u$$

and, therefore, $\mu = \lambda_j$. Consequently,

$$\sigma(A_j) = \{\lambda_j\} \quad \text{for each} \quad 1 \leq j \leq p.$$

Now we set

$$d_j := \dim N[(A-\lambda_j I)^{\nu(\lambda_j)}], \qquad 1 \leq j \leq p.$$

For each $1 \leq j \leq p$, let $\mathcal{B}_j$ be a basis of $N[(A-\lambda_j I)^{\nu(\lambda_j)}]$. By (1.6), the (ordered) union of these bases

$$\mathcal{B} := \mathcal{B}_1 \cup \cdots \cup \mathcal{B}_p$$

provides us with a basis of $\mathbb{C}^N$. Moreover, thanks to the invariance property (1.17), with respect to $\mathcal{B}$, the linear operator A has the associated matrix

$$A_{\mathcal{B}} := \begin{pmatrix} A_{1,\mathcal{B}_1} & 0 & \cdots & 0 \\ 0 & A_{2,\mathcal{B}_2} & \cdots & 0 \\ \vdots & \vdots & \ddots & \vdots \\ 0 & 0 & \cdots & A_{p,\mathcal{B}_p} \end{pmatrix}, \tag{1.24}$$

where, for each $1 \leq j \leq p$, the matrix $A_{j,\mathcal{B}_j}$ denotes the associated matrix with A_j with respect to the basis $\mathcal{B}_j$. Note that $A_{j,\mathcal{B}_j} \in \mathcal{M}_{d_j}(\mathbb{C})$, for each $1 \leq j \leq p$, and that

$$N = \sum_{j=1}^{p} d_j.$$

Moreover, if $P_{\mathcal{B}}$ denotes the matrix whose columns consist of the coordinates of the vectors of $\mathcal{B}$ with respect to the original basis, then

$$A = P_{\mathcal{B}} A_{\mathcal{B}} P_{\mathcal{B}}^{-1}. \tag{1.25}$$

From (1.24) and (1.25), we find that

$$\det(A - zI) = \det(A_{\mathcal{B}} - zI) = \prod_{j=1}^{p} \det(A_{j,\mathcal{B}_j} - zI)$$

for all $z \in \mathbb{C}$. Moreover, as $\sigma(A_{j,\mathcal{B}_j}) = \{\lambda_j\}$ for all $1 \leq j \leq p$, it follows that

$$\det(A_{j,\mathcal{B}_j} - zI) = (-1)^{d_j}(z - \lambda_j)^{d_j}, \qquad z \in \mathbb{C}, \quad 1 \leq j \leq p,$$

and, therefore,

$$\det(A - zI) = (-1)^N \prod_{j=1}^{p} (z - \lambda_j)^{d_j}, \qquad z \in \mathbb{C}.$$

On the other hand, by definition,

$$\det(A - zI) = (-1)^N \prod_{j=1}^{p} (z - \lambda_j)^{m_a(\lambda_j)}, \qquad z \in \mathbb{C},$$

and, consequently,

$$d_j = m_a(\lambda_j) \quad \text{for each} \quad 1 \leq j \leq p.$$

This concludes the proof. □

1.3 The Jordan canonical form

Pick a natural number $n \geq 1$, and, for each integer $1 \leq j \leq n$, let $k_j \geq 1$ be an integer number, and $B_j \in \mathcal{M}_{k_j}(\mathbb{C})$. Then, one can define

$$\operatorname{diag}\{B_1, \ldots, B_n\} := \begin{pmatrix} B_1 & 0 & \cdots & 0 \\ 0 & B_2 & \cdots & 0 \\ \vdots & \vdots & \ddots & \vdots \\ 0 & 0 & \cdots & B_n \end{pmatrix} \in \mathcal{M}_N(\mathbb{C}),$$

where

$$N := k_1 + \cdots + k_n.$$

We consider a matrix $A \in \mathcal{M}_N(\mathbb{C})$ with spectrum $\sigma(A) = \{\lambda_1, \ldots, \lambda_p\}$. The main goal of this section consists in, for each $1 \leq j \leq p$, constructing a basis

$$\mathcal{B}_j := \{e_1^j, \ldots, e_{m_a(\lambda_j)}^j\}$$

of the ascent generalized eigenspace $N[(A-\lambda_j I)^{\nu(\lambda_j)}]$ so that the matrix $A_{j,\mathcal{B}_j}$ of the operator from $N[(A-\lambda_j I)^{\nu(\lambda_j)}]$ to itself defined by restriction of A is block-diagonal

$$A_{j,\mathcal{B}_j} = \operatorname{diag}\{J_{j,i_1}, J_{j,i_2}, \dots, J_{j,i_{n_j}}\}$$

for some integer number $n_j \geq 1$, where $i_h \in \{1, \dots, \nu(\lambda_j)\}$ for $1 \leq h \leq n_j$, and for each natural $n \geq 1$ the matrix $J_{j,n} \in \mathcal{M}_n(\mathbb{C})$ stands for the *Jordan block of order n associated with the eigenvalue* λ_j, which is defined by

$$J_{j,1} := \lambda_j, \quad J_{j,2} := \begin{pmatrix} \lambda_j & 0 \\ 1 & \lambda_j \end{pmatrix}, \quad J_{j,3} := \begin{pmatrix} \lambda_j & 0 & 0 \\ 1 & \lambda_j & 0 \\ 0 & 1 & \lambda_j \end{pmatrix}, \dots,$$

$$J_{j,n} := \begin{pmatrix} \lambda_j & 0 & 0 & \cdots & 0 & 0 \\ 1 & \lambda_j & 0 & \cdots & 0 & 0 \\ 0 & 1 & \lambda_j & \cdots & 0 & 0 \\ \vdots & \vdots & \vdots & \ddots & \vdots & \vdots \\ 0 & 0 & 0 & \cdots & \lambda_j & 0 \\ 0 & 0 & 0 & \cdots & 1 & \lambda_j \end{pmatrix}. \tag{1.26}$$

Fix $j \in \{1, \dots, p\}$. When $\nu(\lambda_j) = 1$, then the basis $\mathcal{B}_j$ consists of eigenvectors, and, hence,

$$A_j e_k^j = \lambda_j e_k^j, \qquad 1 \leq k \leq m_a(\lambda_j) = m_g(\lambda_j).$$

Consequently, in such a case the matrix $A_{j,\mathcal{B}_j}$ is diagonal and consists of $m_a(\lambda_j)$ Jordan blocks $J_{j,1}$ of order 1.

Suppose $\nu(\lambda_j) \geq 2$, and for each $1 \leq k \leq \nu(\lambda_j) - 1$ let C_k^j be a linear supplement of $N[(A-\lambda_j I)^k]$ in $N[(A-\lambda_j I)^{k+1}]$, that is,

$$N[(A-\lambda_j I)^{k+1}] = N[(A-\lambda_j I)^k] \oplus C_k^j, \qquad 1 \leq k \leq \nu(\lambda_j) - 1. \tag{1.27}$$

Then, setting

$$C_0^j := N[A - \lambda_j I],$$

we have that

$$N[(A-\lambda_j I)^{\nu(\lambda_j)}] = \bigoplus_{k=0}^{\nu(\lambda_j)-1} C_k^j$$

and, hence, in order to construct a basis of $N[(A-\lambda_j I)^{\nu(\lambda_j)}]$ it suffices to construct a basis of C_k^j for each $0 \leq k \leq \nu(\lambda_j) - 1$.

Now, for every $0 \leq k \leq \nu(\lambda_j) - 1$, we set

$$h_k := \dim C_k^j.$$

Then,

$$h_k = \dim N[(A-\lambda_j I)^{k+1}] - \dim N[(A-\lambda_j I)^k] \geq 1.$$

Let $\{e_1^j, \dots, e_{h_{\nu(\lambda_j)-1}}^j\}$ be a basis of $C_{\nu(\lambda_j)-1}^j$, and consider the images of the elements of this basis under the action of the operator $A - \lambda_j I$,

$$\{(A-\lambda_j I)e_i^j : 1 \leq i \leq h_{\nu(\lambda_j)-1}\}. \tag{1.28}$$

These images form a linearly independent system of $C_{\nu(\lambda_j)-2}^j$. Indeed, suppose there exist $c_i \in \mathbb{C}$, for $1 \leq i \leq h_{\nu(\lambda_j)-1}$, for which

$$\sum_{i=1}^{h_{\nu(\lambda_j)-1}} c_i (A-\lambda_j I)e_i^j = 0.$$

Then, since $\nu(\lambda_j) - 1 \geq 1$, we have that

$$\sum_{i=1}^{h_{\nu(\lambda_j)-1}} c_i e_i^j \in N[A-\lambda_j I] \subset N[(A-\lambda_j I)^{\nu(\lambda_j)-1}]$$

and, hence, since $e_1^j, \dots, e_{h_{\nu(\lambda_j)-1}}^j \in C_{\nu(\lambda_j)-1}^j$,

$$\sum_{i=1}^{h_{\nu(\lambda_j)-1}} c_i e_i^j \in N[(A-\lambda_j I)^{\nu(\lambda_j)-1}] \cap C_{\nu(\lambda_j)-1}^j.$$

Thus, owing to (1.27), this gives

$$\sum_{i=1}^{h_{\nu(\lambda_j)-1}} c_i e_i^j = 0.$$

As the set $\{e_i^j : 1 \leq i \leq h_{\nu(\lambda_j)-1}\}$ is linearly independent, necessarily $c_i = 0$ for each $1 \leq i \leq h_{\nu(\lambda_j)-1}$ and, therefore, the system (1.28) is linearly independent and contained in $C_{\nu(\lambda_j)-2}^j$, as claimed above. Consequently, if $h_{\nu(\lambda_j)-2} = h_{\nu(\lambda_j)-1}$, then

$$C_{\nu(\lambda_j)-2}^j = \operatorname{span}\left[(A-\lambda_j I)e_1^j, \dots, (A-\lambda_j I)e_{h_{\nu(\lambda_j)-1}}^j\right],$$

whereas in the case $h_{\nu(\lambda_j)-2} > h_{\nu(\lambda_j)-1}$ one can extend the system (1.28) to complete a basis of $C_{\nu(\lambda_j)-2}^j$, say

$$\{(A-\lambda_j I)e_1^j, \dots, (A-\lambda_j I)e_{h_{\nu(\lambda_j)-1}}^j, e_{h_{\nu(\lambda_j)-1}+1}^j, \dots, e_{h_{\nu(\lambda_j)-2}}^j\}.$$

Suppose $\nu(\lambda_j) = 2$. Then

$$C_{\nu(\lambda_j)-2}^j = N[A-\lambda_j I]$$

and the set

$$\mathcal{B}_j := \{e_1^j, (A-\lambda_j I)e_1^j, e_2^j, (A-\lambda_j I)e_2^j, \dots, \\ e^j_{h_{\nu(\lambda_j)-1}}, (A-\lambda_j I)e^j_{h_{\nu(\lambda_j)-1}}, e^j_{h_{\nu(\lambda_j)-1}+1}, \dots, e^j_{h_{\nu(\lambda_j)-2}}\}$$

provides us with a basis of $N[(A-\lambda_j I)^2]$. Moreover, since

$$\begin{aligned} Ae_i^j &= \lambda_j e_i^j + (A-\lambda_j I)e_i^j, && 1 \le i \le h_{\nu(\lambda_j)-1}, \\ A(A-\lambda_j I)e_i^j &= \lambda_j (A-\lambda_j I)e_i^j, && 1 \le i \le h_{\nu(\lambda_j)-1}, \\ Ae_i^j &= \lambda_j e_i^j, && h_{\nu(\lambda_j)-1}+1 \le i \le h_{\nu(\lambda_j)-2}, \end{aligned}$$

the matrix of A_j with respect to the basis $\mathcal{B}_j$ is the following:

$$A_{j,\mathcal{B}_j} = \operatorname{diag}\{\underbrace{J_{j,2}, J_{j,2}, \dots, J_{j,2}}_{h_{\nu(\lambda_j)-1}}, \underbrace{J_{j,1}, J_{j,1}, \dots, J_{j,1}}_{h_{\nu(\lambda_j)-2}-h_{\nu(\lambda_j)-1}}\}.$$

When $\nu(\lambda_j) \ge 3$, as in the previous step, we consider the system formed by all images under the action of $A-\lambda_j I$ of the vectors of the basis of $C^j_{\nu(\lambda_j)-2}$, that is,

$$\begin{aligned} &\left\{(A-\lambda_j I)^2 e_i^j : 1 \le i \le h_{\nu(\lambda_j)-1}\right\} \\ &\cup \left\{(A-\lambda_j I)e_i^j : h_{\nu(\lambda_j)-1}+1 \le i \le h_{\nu(\lambda_j)-2}\right\}. \end{aligned} \tag{1.29}$$

This system is linearly independent in $C^j_{\nu(\lambda_j)-3}$. Indeed, suppose $c_i \in \mathbb{C}$, for $1 \le i \le h_{\nu(\lambda_j)-2}$, satisfy

$$\sum_{i=1}^{h_{\nu(\lambda_j)-1}} c_i (A-\lambda_j I)^2 e_i^j + \sum_{i=h_{\nu(\lambda_j)-1}+1}^{h_{\nu(\lambda_j)-2}} c_i (A-\lambda_j I)e_i^j = 0.$$

Then, arguing as above, it follows from $\nu(\lambda_j)-2 \ge 1$ that

$$\sum_{i=1}^{h_{\nu(\lambda_j)-1}} c_i (A-\lambda_j I)e_i^j + \sum_{i=h_{\nu(\lambda_j)-1}+1}^{h_{\nu(\lambda_j)-2}} c_i e_i^j \in N[(A-\lambda_j I)^{\nu(\lambda_j)-2}] \cap C^j_{\nu(\lambda_j)-2}$$

and, therefore, due to (1.27), $c_i = 0$ for each $1 \le i \le h_{\nu(\lambda_j)-2}$. Similarly, (1.29) provides us with a basis of $C^j_{\nu(\lambda_j)-3}$ if $h_{\nu(\lambda_j)-3} = h_{\nu(\lambda_j)-2}$, whereas it can be extended to complete a basis of $C^j_{\nu(\lambda_j)-3}$, by adding the appropriate system

$$\{e_i^j : h_{\nu(\lambda_j)-2}+1 \le i \le h_{\nu(\lambda_j)-3}\},$$

if $h_{\nu(\lambda_j)-3} > h_{\nu(\lambda_j)-2}$. When $\nu(\lambda_j) = 3$, we are done. In the general case, after $\nu(\lambda_j)$ steps we will have constructed an ordered basis of the ascent generalized eigenspace of the form

$$\mathcal{B}_j := \bigcup_{i=1}^{\nu(\lambda_j)} \bigcup_{\ell=h_{\nu(\lambda_j)-i+1}+1}^{h_{\nu(\lambda_j)-i}} \bigcup_{k=0}^{\nu(\lambda_j)-i} \left\{(A-\lambda_j I)^k e_\ell^j\right\}.$$

The matrix of A_j with respect to $\mathcal{B}_j$ has exactly $h_{\nu(\lambda_j)-1}$ Jordan blocks of order $\nu(\lambda_j)$, exactly $h_{\nu(\lambda_j)-2} - h_{\nu(\lambda_j)-1}$ Jordan blocks of order $\nu(\lambda_j)-1$ and, in general, $h_{\nu(\lambda_j)-k} - h_{\nu(\lambda_j)-k+1}$ Jordan blocks of order $\nu(\lambda_j)-k+1$, for each $1 \leq k \leq \nu(\lambda_j)$. The ordered set of vectors of $\mathcal{B}_j$ is often called a *chain of generalized eigenvectors*.

Note that the operator A_j can be expressed as

$$A_j = \lambda_j I + (A_j - \lambda_j I)$$

and, consequently, A_j is a sum of the diagonal operator $\lambda_j I$ plus the nilpotent operator $A_j - \lambda_j I$, since

$$(A-\lambda_j I)^{\nu(\lambda_j)}(N[(A-\lambda_j)^{\nu(\lambda_j)}]) = \{0\}.$$

Moreover, the basis $\mathcal{B}_j$ that we have just constructed provides us with a matrix of $A_j - \lambda_j I$ with the maximum possible number of zero entries.

Let P be the matrix whose columns consist of the coordinates of the vectors of the bases $\mathcal{B}_1, \dots, \mathcal{B}_p$ that we have just constructed. The matrix

$$J := PAP^{-1}$$

is called the *Jordan canonical form* of the matrix A. In this section we have proved that

$$J = \operatorname{diag}\{J_1, J_2, \dots, J_p\},$$

where, for each $1 \leq j \leq p$, the matrix J_j consists of a finite diagonal union of Jordan blocks of the type $J_{j,k}$, for $1 \leq k \leq \nu(\lambda_j)$.

1.4 The real canonical form

Throughout this section we suppose that

$$A \in \mathcal{M}_N(\mathbb{R}).$$

In such a case, the characteristic polynomial of A has real coefficients, and, hence,

$$\sigma(A) = \{\bar{\lambda} : \lambda \in \sigma(A)\}$$

where the complex conjugate of any $z \in \mathbb{C}$ is denoted by $\bar{z}$. Consequently, the corresponding ascent generalized eigenspaces in the decomposition (1.6) also appear in pairs. Suppose

$$\sigma(A) = \{\lambda_1, \dots, \lambda_p\}$$

and pick $1 \leq j \leq p$. It is easy to see that the vectors of the basis $\mathcal{B}_j$ constructed in Section 1.3 all have real components if $\lambda_j \in \mathbb{R}$, whereas if

$$\lambda_j = \alpha_j + i\beta_j, \quad \text{with} \quad \alpha_j, \beta_j \in \mathbb{R}, \quad \beta_j \neq 0,$$

then, their components are complex. Suppose this is the case, and reorder the eigenvalues, if necessary, so that

$$\lambda_{j+1} = \alpha_j - i\beta_j = \bar{\lambda}_j.$$

Then, for each $k \geq 0$ and $e \in \mathbb{C}^N$,

$$(A - \lambda_j I)^k e = 0 \ \text{ if and only if } \ (A - \lambda_{j+1} I)^k \bar{e} = 0, \tag{1.30}$$

where $\bar{e}$ is the complex conjugate vector of e, that is, the vector whose components are the complex conjugate components of e. Therefore, for each $k \geq 0$,

$$\dim N[(A - \lambda_j I)^k] = \dim N[(A - \lambda_{j+1} I)^k]$$

and, in particular,

$$\nu(\lambda_j) = \nu(\lambda_{j+1}) \quad \text{and} \quad m_a(\lambda_j) = m_a(\lambda_{j+1}).$$

Let

$$\mathcal{B}_j := \{e_1^j, \dots, e_{m_a(\lambda_j)}^j\}$$

be the basis of $N[(A - \lambda_j I)^{\nu(\lambda_j)}]$ constructed in Section 1.3. Thanks to (1.30), it is easy to see that

$$\mathcal{B}_{j+1} := \{\bar{e}_1^j, \dots, \bar{e}_{m_a(\lambda_j)}^j\}$$

provides us with the Jordan basis of $N[(A - \lambda_{j+1} I)^{\nu(\lambda_{j+1})}]$ constructed in Section 1.3. In other words, we are choosing

$$e_k^{j+1} := \bar{e}_k^j, \qquad 1 \leq k \leq m_a(\lambda_j) = m_a(\lambda_{j+1}).$$

Actually, by construction, if a system

$$\{e_{k+i}^j : 0 \leq i \leq h-1\},$$

of $h \geq 1$ vectors of the basis $\mathcal{B}_j$, where $1 \leq k \leq m_a(\lambda_j)$, provides a Jordan block $J_{j,h}$ of order h, then, their conjugate vectors

$$e_{k+i}^{j+1} = \bar{e}_{k+i}^j, \qquad 0 \leq i \leq h-1,$$

also provide a Jordan block of type $J_{j+1,h}$.

From now on, for any integer $n \geq 1$ and $z \in \mathbb{C}^n$, the vectors $\operatorname{Re} z \in \mathbb{R}^n$ and $\operatorname{Im} z \in \mathbb{R}^n$ will stand for the real and the imaginary parts of z, respectively. According to this notation, when $h = 1$, we have that

$$Ae_k^j = \lambda_j e_k^j$$

and, hence,

$$\begin{aligned} A \operatorname{Re} e_k^j &= \alpha_j \operatorname{Re} e_k^j - \beta_j \operatorname{Im} e_k^j, \\ A \operatorname{Im} e_k^j &= \alpha_j \operatorname{Im} e_k^j + \beta_j \operatorname{Re} e_k^j. \end{aligned}$$

Consequently, A leaves the real linear manifold generated by $\operatorname{Re} e_k^j$ and $\operatorname{Im} e_k^j$ invariant, and the matrix of its restriction to this manifold with respect to the basis $\{\operatorname{Re} e_k^j, \operatorname{Im} e_k^j\}$ is

$$J_{j,2}^{\mathbb{R}} := \begin{pmatrix} \alpha_j & \beta_j \\ -\beta_j & \alpha_j \end{pmatrix}. \tag{1.31}$$

Note that the system $\{\operatorname{Re} e_k^j, \operatorname{Im} e_k^j\}$ is linearly independent, since the determinant of $J_{j,2}^{\mathbb{R}}$ is non-zero. Similarly, when $h \geq 2$, the identities

$$\begin{aligned} Ae_{k+i}^j &= \lambda_j e_{k+i}^j + e_{k+i+1}^j, \qquad 0 \leq i \leq h-2, \\ Ae_{k+h-1}^j &= \lambda_j e_{k+h-1}^j, \end{aligned}$$

hold and, hence, identifying real and imaginary parts in these identities shows that, for each $0 \leq i \leq h-2$,

$$\begin{aligned} A \operatorname{Re} e_{k+i}^j &= \operatorname{Re} \lambda_j \operatorname{Re} e_{k+i}^j - \operatorname{Im} \lambda_j \operatorname{Im} e_{k+i}^j + \operatorname{Re} e_{k+i+1}^j, \\ A \operatorname{Im} e_{k+i}^j &= \operatorname{Re} \lambda_j \operatorname{Im} e_{k+i}^j + \operatorname{Im} \lambda_j \operatorname{Re} e_{k+i}^j + \operatorname{Im} e_{k+i+1}^j, \end{aligned}$$

and

$$\begin{aligned} A \operatorname{Re} e_{k+h-1}^j &= \operatorname{Re} \lambda_j \operatorname{Re} e_{k+h-1}^j - \operatorname{Im} \lambda_j \operatorname{Im} e_{k+h-1}^j, \\ A \operatorname{Im} e_{k+h-1}^j &= \operatorname{Re} \lambda_j \operatorname{Im} e_{k+h-1}^j + \operatorname{Im} \lambda_j \operatorname{Re} e_{k+h-1}^j. \end{aligned}$$

Therefore, A leaves the real linear manifold

$$\operatorname{span}\left[\operatorname{Re} e_k^j, \operatorname{Im} e_k^j, \ldots, \operatorname{Re} e_{k+h-1}^j, \operatorname{Im} e_{k+h-1}^j\right] \tag{1.32}$$

invariant, and the matrix of its restriction to this linear manifold is

$$J_{j,2h}^{\mathbb{R}} = \begin{pmatrix} J_{j,2}^{\mathbb{R}} & 0 & \cdots & 0 & 0 \\ I_2 & J_{j,2}^{\mathbb{R}} & \cdots & 0 & 0 \\ \vdots & \vdots & \ddots & \vdots & \vdots \\ 0 & 0 & \cdots & J_{j,2}^{\mathbb{R}} & 0 \\ 0 & 0 & \cdots & I_2 & J_{j,2}^{\mathbb{R}} \end{pmatrix},$$

where $J^{\mathbb{R}}_{j,2}$ is the Jordan block (1.31), and I_2 is the identity matrix of order 2. Note that, since the determinant of $J^{\mathbb{R}}_{j,2h}$ is non-zero, the manifold (1.32) has dimension $2h$. These features show that choosing the basis

$$\mathcal{B}^{\mathbb{R}}_j := \{\operatorname{Re} e^j_1, \operatorname{Im} e^j_1, \ldots, \operatorname{Re} e^j_{m_a(\lambda_j)}, \operatorname{Im} e^j_{m_a(\lambda_j)}\},$$

instead of $\mathcal{B}_j \cup \mathcal{B}_{j+1}$, the endomorphism A leaves the real linear manifold generated by $\mathcal{B}^{\mathbb{R}}_j$ invariant, and that its associated matrix with respect to this basis is block-diagonal, each of its blocks being of type

$$J^{\mathbb{R}}_{j,2k}, \qquad 1 \leq k \leq \nu(\lambda_j).$$

Iterating the same process for each pair of complex conjugate eigenvalues, one is naturally driven to the construction of the so-called *real Jordan canonical form* of the matrix A. The block $J^{\mathbb{R}}_{j,2k}$ is often called the *real Jordan block* of order $2k$ associated with the complex eigenvalue λ_j.

1.5 An example

Throughout this section we make the particular choice

$$A := \begin{pmatrix} 2 & 2 & -1 & -1 \\ -1 & 3 & -1 & 1 \\ 1 & 1 & 1 & 0 \\ 0 & 1 & -1 & 2 \end{pmatrix} \in \mathfrak{M}_4(\mathbb{R}). \tag{1.33}$$

After some straightforward, but tedious, calculations, we arrive at the identity

$$\det(A - zI) = \begin{vmatrix} 2-z & 2 & -1 & -1 \\ -1 & 3-z & -1 & 1 \\ 1 & 1 & 1-z & 0 \\ 0 & 1 & -1 & 2-z \end{vmatrix} = [(z-2)^2+1]^2, \qquad z \in \mathbb{C}.$$

Thus,

$$\sigma(A) = \{2-i, 2+i\}$$

with algebraic multiplicities

$$m_a(2-i) = m_a(2+i) = 2.$$

We have that

$$A - (2-i)I = \begin{pmatrix} i & 2 & -1 & -1 \\ -1 & 1+i & -1 & 1 \\ 1 & 1 & -1+i & 0 \\ 0 & 1 & -1 & i \end{pmatrix}$$

and the minor of order 3 formed by the last three entries of the first, third and fourth columns, does not vanish; indeed,

$$\begin{vmatrix} -1 & -1 & 1 \\ 1 & -1+i & 0 \\ 0 & -1 & i \end{vmatrix} = -\begin{vmatrix} -1+i & 0 \\ -1 & i \end{vmatrix} - \begin{vmatrix} -1 & 1 \\ -1 & i \end{vmatrix} = 2i \neq 0.$$

Hence,

$$\dim R[A-(2-i)I] = 3, \qquad \dim N[A-(2-i)I] = 4-3 = 1,$$

and, therefore, since A is a real matrix, Theorem 1.2.1 implies that

$$\dim N[(A-(2-i)I)^2] = 2, \tag{1.34}$$

$$\begin{cases} N[A-(2+i)I] = \overline{N[A-(2-i)I]}, \\ N[(A-(2+i)I)^2] = \overline{N[(A-(2-i)I)^2]}, \end{cases} \tag{1.35}$$

and

$$\nu(2-i) = \nu(2+i) = 2,$$

where we have denoted by $\bar{L}$ the conjugate of the set $L \subset \mathbb{R}^4$, i.e.,

$$\bar{L} := \{\bar{u} \; : \; u \in L\}.$$

Clearly,

$$(x, y, z, t) \in N[A-(2-i)I]$$

if and only if

$$\begin{pmatrix} -1 & -1 & 1 \\ 1 & -1+i & 0 \\ 0 & -1 & i \end{pmatrix} \begin{pmatrix} x \\ z \\ t \end{pmatrix} = -\begin{pmatrix} 1+i \\ 1 \\ 1 \end{pmatrix} y, \qquad y \in \mathbb{C}.$$

By solving this system for the choice $y = 1$, we are led to

$$(x, z, t) = (1, 1+i, 1).$$

Therefore,

$$N[A-(2-i)I] = \operatorname{span}\left[\begin{pmatrix} 1 \\ 1 \\ 1+i \\ 1 \end{pmatrix}\right], \qquad N[A-(2+i)I] = \operatorname{span}\left[\begin{pmatrix} 1 \\ 1 \\ 1-i \\ 1 \end{pmatrix}\right].$$

Now, we will determine $N[(A-(2-i)I)^2]$. Multiplying $A-(2-i)I$ by itself we find that

$$(A-(2-i)I)^2 = \begin{pmatrix} -4 & 4i & -2i & 2-2i \\ -2-2i & -2+2i & -2i & 2+2i \\ -2+2i & 2+2i & -2-2i & 0 \\ -2 & 2i & -2i & 0 \end{pmatrix}.$$

According to (1.34),

$$\dim R[(A-(2-i)I)^2] = 2.$$

Moreover, a non-vanishing minor of order 2 of $(A-(2-i)I)^2$ is the one formed by the last two entries of the first and fourth rows; indeed,

$$\begin{vmatrix} -2i & 2-2i \\ -2i & 0 \end{vmatrix} = 2i(2-2i) \neq 0.$$

Thus, the second and third rows must depend linearly on the first and fourth ones; consequently,

$$(x,y,z,t) \in N[(A-(2-2i)I)^2]$$

if and only if

$$\begin{pmatrix} -i & 1-i \\ -i & 0 \end{pmatrix} \begin{pmatrix} z \\ t \end{pmatrix} = \begin{pmatrix} 2 \\ 1 \end{pmatrix} x - \begin{pmatrix} 2i \\ i \end{pmatrix} y. \tag{1.36}$$

For every $(x,y) \in \mathbb{C}^2$, the system (1.36) has a unique solution, and making the choices $(x,y) \in \{(2,0),(0,2)\}$, we are led to

$$N[(A-(2-i)I)^2] = \operatorname{span}\left[\begin{pmatrix} 2 \\ 0 \\ 2i \\ 1+i \end{pmatrix}, \begin{pmatrix} 0 \\ 2 \\ 2 \\ 1-i \end{pmatrix}\right].$$

Thus, owing to (1.35), we also obtain that

$$N[(A-(2+i)I)^2] = \operatorname{span}\left[\begin{pmatrix} 2 \\ 0 \\ -2i \\ 1-i \end{pmatrix}, \begin{pmatrix} 0 \\ 2 \\ 2 \\ 1+i \end{pmatrix}\right].$$

1.5.1 The complex Jordan form of A

As the first vector of the Jordan basis of A we consider

$$\overset{\mathbb{C}}{e}_1 := \begin{pmatrix} 2 \\ 0 \\ 2i \\ 1+i \end{pmatrix}.$$

It is easy to see that

$$\overset{\mathbb{C}}{e}_1 \in N[(A-(2-i)I)^2] \setminus N[A-(2-i)I]$$

and, hence,

$$\overset{\mathbb{C}}{e}_2 := (A-(2-i)I)\,\overset{\mathbb{C}}{e}_1 \in N[A-(2-i)I] \setminus \{0\}.$$

Clearly,

$$\overset{\mathbb{C}}{e}_2 = \begin{pmatrix} -1-i \\ -1-i \\ -2i \\ -1-i \end{pmatrix} = -(1+i) \begin{pmatrix} 1 \\ 1 \\ 1+i \\ 1 \end{pmatrix}$$

and, hence, since A is real, we also find that

$$\overset{\mathbb{C}}{e}_3 := \bar{\overset{\mathbb{C}}{e}}_1 = \begin{pmatrix} 2 \\ 0 \\ -2i \\ 1-i \end{pmatrix} \in N[(A-(2+i)I)^2] \setminus N[A-(2+i)I]$$

and

$$\overset{\mathbb{C}}{e}_4 := \bar{\overset{\mathbb{C}}{e}}_2 = \begin{pmatrix} -1+i \\ -1+i \\ 2i \\ -1+i \end{pmatrix} \in N[A-(2+i)I] \setminus \{0\}.$$

By construction,

$$\begin{aligned} A\overset{\mathbb{C}}{e}_1 &= (2-i)\overset{\mathbb{C}}{e}_1 + \overset{\mathbb{C}}{e}_2, \qquad & A\overset{\mathbb{C}}{e}_2 &= (2-i)\overset{\mathbb{C}}{e}_2, \\ A\overset{\mathbb{C}}{e}_3 &= (2+i)\overset{\mathbb{C}}{e}_3 + \overset{\mathbb{C}}{e}_4, \qquad & A\overset{\mathbb{C}}{e}_4 &= (2+i)\overset{\mathbb{C}}{e}_4. \end{aligned}$$

Therefore, with respect to the basis

$$\mathcal{B}_{\mathbb{C}} := \{\overset{\mathbb{C}}{e}_1, \overset{\mathbb{C}}{e}_2, \overset{\mathbb{C}}{e}_3, \overset{\mathbb{C}}{e}_4\},$$

the associated matrix to A equals its complex Jordan form

$$J_{\mathbb{C}} := \begin{pmatrix} 2-i & 0 & 0 & 0 \\ 1 & 2-i & 0 & 0 \\ 0 & 0 & 2+i & 0 \\ 0 & 0 & 1 & 2+i \end{pmatrix} = P_{\mathbb{C}}^{-1} A P_{\mathbb{C}} \tag{1.37}$$

where $P_{\mathbb{C}}$ is the matrix of change of basis

$$P_{\mathbb{C}} := \begin{pmatrix} 2 & -1-i & 2 & -1+i \\ 0 & -1-i & 0 & -1+i \\ 2i & -2i & -2i & 2i \\ 1+i & -1-i & 1-i & -1+i \end{pmatrix}. \tag{1.38}$$

1.5.2 The real Jordan form of A

The real Jordan form is obtained by considering the new basis

$$\mathcal{B}_{\mathbb{R}} := \{\overset{\mathbb{R}}{e_1}, \overset{\mathbb{R}}{e_2}, \overset{\mathbb{R}}{e_3}, \overset{\mathbb{R}}{e_4}\},$$

where

$$\overset{\mathbb{R}}{e_1} := \operatorname{Re} \overset{\mathbb{C}}{e_1} = \begin{pmatrix} 2 \\ 0 \\ 0 \\ 1 \end{pmatrix}, \qquad \overset{\mathbb{R}}{e_2} := \operatorname{Im} \overset{\mathbb{C}}{e_1} = \begin{pmatrix} 0 \\ 0 \\ 2 \\ 1 \end{pmatrix},$$

$$\overset{\mathbb{R}}{e_3} := \operatorname{Re} \overset{\mathbb{C}}{e_2} = \begin{pmatrix} -1 \\ -1 \\ 0 \\ -1 \end{pmatrix}, \qquad \overset{\mathbb{R}}{e_4} := \operatorname{Im} \overset{\mathbb{C}}{e_2} = \begin{pmatrix} -1 \\ -1 \\ -2 \\ -1 \end{pmatrix}.$$

Since

$$\begin{aligned} A\overset{\mathbb{R}}{e_1} &= 2\overset{\mathbb{R}}{e_1} + \overset{\mathbb{R}}{e_2} + \overset{\mathbb{R}}{e_3}, \\ A\overset{\mathbb{R}}{e_2} &= -\overset{\mathbb{R}}{e_1} + 2\overset{\mathbb{R}}{e_2} + \overset{\mathbb{R}}{e_4}, \\ A\overset{\mathbb{R}}{e_3} &= 2\overset{\mathbb{R}}{e_3} + \overset{\mathbb{R}}{e_4}, \\ A\overset{\mathbb{R}}{e_4} &= -\overset{\mathbb{R}}{e_3} + 2\overset{\mathbb{R}}{e_4}, \end{aligned}$$

the real Jordan form of A is given through

$$J_{\mathbb{R}} = \begin{pmatrix} 2 & -1 & 0 & 0 \\ 1 & 2 & 0 & 0 \\ 1 & 0 & 2 & -1 \\ 0 & 1 & 1 & 2 \end{pmatrix} = P_{\mathbb{R}}^{-1} A P_{\mathbb{R}}, \tag{1.39}$$

where

$$P_{\mathbb{R}} = \begin{pmatrix} 2 & 0 & -1 & -1 \\ 0 & 0 & -1 & -1 \\ 0 & 2 & 0 & -2 \\ 1 & 1 & -1 & -1 \end{pmatrix}. \tag{1.40}$$

1.6 Exercises

1. Determine the Jordan canonical forms (both complex and real) of each of the following real matrices:

$$\begin{pmatrix} 3 & 4 \\ -1 & 7 \end{pmatrix}, \begin{pmatrix} -1 & 2 \\ -7 & 8 \end{pmatrix}, \begin{pmatrix} 1 & -1 \\ 5 & -3 \end{pmatrix}, \begin{pmatrix} -1 & -1 \\ 2 & -1 \end{pmatrix}, \begin{pmatrix} 1 & 2 \\ -5 & -1 \end{pmatrix},$$

$$\begin{pmatrix} 10 & 4 & 13 \\ 5 & 3 & 7 \\ -9 & -4 & -12 \end{pmatrix}, \begin{pmatrix} 1 & 1 & 2 \\ 1 & 2 & 1 \\ 2 & 1 & 1 \end{pmatrix}, \begin{pmatrix} 1 & 1 & 1 \\ 2 & 1 & -1 \\ -3 & 2 & 4 \end{pmatrix}, \begin{pmatrix} 5 & -3 & -2 \\ 8 & -5 & -4 \\ -4 & 3 & 3 \end{pmatrix},$$

$$\begin{pmatrix} 0 & 1 & 1 \\ 1 & 0 & 1 \\ 1 & 1 & 0 \end{pmatrix}, \begin{pmatrix} 3 & 2 & 4 \\ 2 & 0 & 2 \\ 4 & 2 & 3 \end{pmatrix}, \begin{pmatrix} 1 & 1 & 1 & 1 \\ 2 & 2 & 2 & 2 \\ 3 & 3 & 3 & 3 \\ 4 & 4 & 4 & 4 \end{pmatrix}.$$

2. Let A be a square matrix such that $\sigma(A) = \{\alpha, \beta, \gamma\}$ and

$$\begin{aligned} \dim N[A-\alpha I] = 3, \quad & \dim N[(A-\alpha I)^2] = 3, \quad \dim N[A-\beta I] = 3, \\ \dim N[(A-\beta I)^2] = 4, \quad & \dim N[(A-\beta I)^3] = 5, \quad \dim N[(A-\beta I)^4] = 5, \\ \dim N[A-\gamma I] = 2, \quad & \dim N[(A-\gamma I)^2] = 3, \quad \dim N[(A-\gamma I)^3] = 3. \end{aligned}$$

Determine its order and its Jordan canonical form.

3. Given $A \in \mathcal{M}_N(\mathbb{C})$, the adjoint of A, denoted by A^*, is defined as the conjugate matrix of the transposed matrix of A, that is,

$$A^* := \bar{A}^T.$$

The matrix $A \in \mathcal{M}_N(\mathbb{C})$ is said to be *Hermitian* if $A = A^*$. Let $\langle \cdot, \cdot \rangle$ be the skew-linear product in $\mathbb{C}^N$,

$$\langle u, v \rangle = \sum_{j=1}^{N} u_j \, \bar{v}_j,$$

for each $u, v \in \mathbb{C}^N$ with $u = (u_1, \dots, u_N)$ and $v = (v_1, \dots, v_N)$. Prove the following:

- The matrix A is Hermitian if and only if

$$\langle Au, v \rangle = \langle u, Av \rangle \qquad \text{for each} \quad u, v \in \mathbb{C}^N.$$

- The spectrum of every Hermitian matrix consists of real semisimple eigenvalues.

- Every Hermitian matrix is diagonalizable.
- The eigenvectors associated with different eigenvalues of a Hermitian matrix must be orthogonal. By *orthogonal* vectors it is meant that their skew-linear product equals zero.
- The matrix P of the change of basis from any Hermitian matrix to its Jordan canonical form can be chosen to satisfy

$$P^* P = I.$$

Such matrices P are said to be *orthogonal.*

4. Suppose $\Omega \subset \mathbb{R}$ is an open interval containing zero, $k \geq 1$ is an integer number, and $a_{ij} \in \mathcal{C}^k(\Omega, \mathbb{C})$ for each $i, j \in \{1, \ldots, N\}$. Consider, for each $s \in \Omega$, the matrix

$$A(s) = (a_{ij}(s))_{1 \leq i,j \leq N} \in \mathcal{M}_N(\mathbb{C}).$$

Suppose $\lambda_0 \in \mathbb{C}$ is a simple eigenvalue of $A_0 := A(0)$ and

$$\varphi_0 \in N[A_0 - \lambda_0 I] \setminus \{0\}.$$

We ask the following:

- Check that

$$\mathbb{C}^N = N[A_0 - \lambda_0 I] \oplus R[A_0 - \lambda_0 I].$$

- Given a linear subspace $Y \subset \mathbb{C}^N$ such that

$$\mathbb{C}^N = N[A_0 - \lambda_0 I] \oplus Y,$$

use the implicit function theorem to prove that there exist a real number $\delta > 0$ and two unique functions $\Lambda \in \mathcal{C}^k((-\delta, \delta), \mathbb{C})$ and $\Phi \in \mathcal{C}^k((-\delta, \delta), Y)$ such that

$$\Lambda(0) = \lambda_0, \qquad \Phi(0) = \varphi_0,$$

and for all $s \in (-\delta, \delta)$,

$$\Phi(s) - \varphi_0 \in Y, \qquad A(s)\Phi(s) = \Lambda(s)\Phi(s).$$

- Calculate the jth derivative of Λ at 0 for each $1 \leq j \leq k$, and prove that

$$\Lambda'(0) = 0 \text{ if and only if } A'(0)\varphi_0 \in R[A_0 - \lambda_0 I].$$

5. Suppose $\Omega \subset \mathbb{R}$ is an open interval containing zero, $a_{ij} \in \mathcal{C}(\Omega, \mathbb{C})$ for each $i, j \in \{1, \ldots, N\}$, and consider the matrices

$$A(s) = (a_{ij}(s))_{1 \leq i,j \leq N} \in \mathcal{M}_N(\mathbb{C}), \qquad s \in \Omega.$$

Suppose $\sigma(A(0)) = \{\lambda_1, \ldots, \lambda_p\}$, and let

$$L := \min_{1 \leq i < j \leq N} \{|\lambda_i - \lambda_j|\} > 0.$$

Use a celebrated theorem on complex variable theory to prove that for each $R \in (0, L)$ there exists $\delta > 0$ with $(-\delta, \delta) \subset \Omega$ such that for every $s \in (-\delta, \delta)$ and $1 \leq j \leq p$ the matrix $A(s)$ has exactly $m_a(\lambda_j)$ eigenvalues in the open disk of radius R centered at λ_j, counting according to their algebraic multiplicities.

6. Let
$$A = (a_{ij})_{1\le i,j\le N} \in \mathcal{M}_N(\mathbb{C}).$$
Prove that there exists a sequence $\{A_n\}_{n\in\mathbb{N}}$ of matrices
$$A_n = (a_{ij}^n)_{1\le i,j\le N} \in \mathcal{M}_N(\mathbb{C}), \qquad n \ge 1,$$
such that, for each $n \in \mathbb{N}$, all the eigenvalues of A_n are simple, and
$$\lim_{n\to\infty} a_{ij}^n = a_{ij} \quad \text{for all} \quad i,j \in \{1,\dots,N\}.$$

7. Let $\alpha \in \mathbb{C}\setminus\{0\}$ and $A \in \mathcal{M}_N(\mathbb{C})$ be such that $\sigma(A) = \{\lambda_1,\dots,\lambda_p\}$. Construct a basis in $\mathbb{C}^N$ so that the associated matrix of A is block-diagonal with blocks of order n of the form
$$\begin{pmatrix} \lambda_j & 0 & 0 & \cdots & 0 & 0 \\ \alpha & \lambda_j & 0 & \cdots & 0 & 0 \\ 0 & \alpha & \lambda_j & \cdots & 0 & 0 \\ \vdots & \vdots & \vdots & \ddots & \vdots & \vdots \\ 0 & 0 & 0 & \cdots & \lambda_j & 0 \\ 0 & 0 & 0 & \cdots & \alpha & \lambda_j \end{pmatrix},$$
where $1 \le n \le \nu(\lambda_j)$ and $1 \le j \le p$.

8. Prove that for each invertible matrix $A \in \mathcal{M}_N(\mathbb{C})$ there exists a polynomial P such that
$$P(A)^2 = A.$$
Prove also that for all $B \in \mathcal{M}_2(\mathbb{C})$ and every polynomial Q,
$$Q(B) \ne \begin{pmatrix} 0 & 0 \\ 1 & 0 \end{pmatrix}.$$

9. Let $A \in \mathcal{M}_N(\mathbb{C})$ be with $\nu(0) \ge 2$. Prove that $B^2 \ne A$ if $B \in \mathcal{M}_N(\mathbb{C})$.

10. For any matrix
$$A = (a_{ij})_{1\le i,j\le N} \in \mathcal{M}_N(\mathbb{C}),$$
the trace of A, hereafter denoted by $\operatorname{tr} A$, is defined through
$$\operatorname{tr} A := \sum_{j=1}^N a_{jj}.$$
Prove the validity of the following assertions:

- For each $B \in \mathcal{M}_N(\mathbb{C})$,
$$\operatorname{tr}(AB) = \operatorname{tr}(BA).$$
- Two similar matrices have the same trace.
- Let $\lambda_1,\dots,\lambda_N$ be the eigenvalues of A, where each eigenvalue is repeated according to its multiplicity. Then,
$$\operatorname{tr} A = \sum_{j=1}^N \lambda_j \qquad \text{and} \qquad \det A = \prod_{j=1}^N \lambda_j.$$

11. Let $A \in \mathcal{M}_N(\mathbb{R})$ be a diagonal matrix such that

$$(A+I)^2 = 0.$$

Determine its Jordan canonical form.

12. Determine all matrices $A \in \mathcal{M}_3(\mathbb{C})$ satisfying the equation

$$A^3 - 2A^2 + A = 0.$$

13. Let $A \in \mathcal{M}_N(\mathbb{C})$ be such that

$$\sigma(A) = \{\alpha, \beta\}, \quad \nu(\alpha) = 2, \quad \nu(\beta) = 1.$$

Construct a polynomial P such that $P(A) = 0$.

14. Let $A \in \mathcal{M}_N(\mathbb{C})$ be a Hermitian matrix (see Exercise 3). The Rayleigh quotient of A is the mapping

$$\mathfrak{R} : \mathbb{C}^N \setminus \{0\} \to \mathbb{C}$$

defined by

$$\mathfrak{R}(u) := \frac{\langle Au, u \rangle}{\langle u, u \rangle}, \qquad u \in \mathbb{C}^N \setminus \{0\},$$

where $\langle \cdot, \cdot \rangle$ is the skew-linear product of $\mathbb{C}^N$. Let

$$\lambda_1 \leq \lambda_2 \leq \cdots \leq \lambda_N$$

be the eigenvalues of A, where each eigenvalue is repeated according to its algebraic multiplicity. Thanks to the result of Exercise 3, we already know that there exist N eigenvectors

$$p_j \in N[A - \lambda_j I], \qquad 1 \leq j \leq N,$$

such that the basis $\{p_1, \ldots, p_N\}$ is orthonormal in $(\mathbb{C}^N, \langle \cdot, \cdot \rangle)$, that is,

$$\langle p_i, p_j \rangle = \delta_{ij} \quad \text{for all} \quad 1 \leq i, j \leq N.$$

For each $1 \leq k \leq N$ set

$$V_k := \operatorname{span}[p_1, \ldots, p_k],$$

and let $\mathcal{G}_k$ denote the set of vector subspaces of $\mathbb{C}^N$ of dimension k. Finally, define $V_0 := \{0\}$ and $\mathcal{G}_0 := \{V_0\}$. Prove the following properties:

- For each $1 \leq k \leq N$,

$$\lambda_k = \mathfrak{R}(p_k) = \max_{u \in V_k \setminus \{0\}} \mathfrak{R}(u).$$

- For each $1 \leq k \leq N$,

$$\lambda_k = \min_{u \in V_{k-1}^\perp \setminus \{0\}} \mathfrak{R}(u),$$

  where, given a linear subspace V, we define by $V^\perp$ the orthogonal subspace of V with respect to the skew-linear product $\langle \cdot, \cdot \rangle$, that is to say,

$$V^\perp := \{u \in \mathbb{C}^N : \langle u, v \rangle = 0 \ \text{ for each } \ v \in V\}.$$

- For each $1 \le k \le N$,

$$\lambda_k = \min_{V \in \mathcal{G}_k} \max_{u \in V \setminus \{0\}} \mathfrak{R}(u) = \max_{V \in \mathcal{G}_{k-1}} \min_{u \in V^{\perp} \setminus \{0\}} \mathfrak{R}(u).$$

- $\{\mathfrak{R}(u) : u \in \mathbb{C}^N \setminus \{0\}\} = [\lambda_1, \lambda_N]$.

15. A Hermitian matrix $A \in \mathcal{M}_N(\mathbb{C})$ is said to be *positive* if

$$\langle Au, u \rangle \ge 0 \quad \text{for each} \quad u \in \mathbb{C}^N.$$

Prove that A is positive if and only if $\lambda \ge 0$ for all $\lambda \in \sigma(A)$.

16. A Hermitian matrix $A \in \mathcal{M}_N(\mathbb{C})$ is said to be *positive definite* if

$$\langle Au, u \rangle > 0 \quad \text{for each} \quad u \in \mathbb{C}^N \setminus \{0\}.$$

Prove that A is positive definite if and only if $\lambda > 0$ for all $\lambda \in \sigma(A)$.

17. Let $A \in \mathcal{M}_N(\mathbb{C})$ be invertible, and consider its characteristic polynomial P_A. Prove that the characteristic polynomial $P_{A^{-1}}$ of A^{-1} satisfies

$$P_{A^{-1}}(z) = \frac{1}{\det A}(-z)^n P_A(z^{-1}), \quad z \in \mathbb{C} \setminus \{0\}.$$

1.7 Comments on Chapter 1

The material of this chapter is standard and can be found in most advanced text-books on linear algebra. Classic books covering most of the material are Jacobson [63], Burton [17], Hirsch & Smale [59], Blyth & Robertson [13], Lang [76], and Lax [77]. In addition, Gantmacher [39] is an outstanding book in matrix theory covering both theory and applications; it counts among the most paradigmatic books in matrix theory of the 20th Century.

There are a number of different approaches to the proof of the Jordan theorem, of course. In this chapter we have followed the one based on polynomial interpolation, which is based on the text-book López-Gómez [83], and was inspired by Guzmán [55]. An interesting approach to the Jordan theorem with emphasis on invariant subspaces can be found in Gohberg, Lancaster & Rodman [51].

Exercise 4 establishes a classic very celebrated result on perturbation from simple eigenvalues, and Exercise 5 establishes the continuous dependence of the spectrum. Both are pivotal results in the theory of perturbation of linear operators. The review paper by Gohberg & Kreĭn [45] and the comprehensive book by Kato [65] are two of the most paradigmatic references on these topics; for a different perspective, see the book by Gohberg, Lancaster & Rodman [51].

Exercise 14 establishes the classic Courant–Fischer theorem. It plays a crucial role in the spectral theory of self-adjoint (or Hermitian) operators, and, in particular, in classic potential theory.

Chapter 2

Operator Calculus

In Section 1.2 we defined $P(A)$ when P is a polynomial and $A \in \mathcal{M}_N(\mathbb{K})$ with $\mathbb{K} \in \{\mathbb{R}, \mathbb{C}\}$. More generally, one of the main goals of operator calculus consists in giving an appropriate definition of $f(A)$ when $f : \mathbb{K} \to \mathbb{K}$ is an arbitrary function, as well as in studying the most important analytical properties of $f(A)$. This chapter covers these issues for the special, but important, case when f is a certain holomorphic function and $A \in \mathcal{M}_N(\mathbb{C})$.

This chapter begins by introducing the concept of norm of a linear operator, which will be very useful to get norm estimates for linear operators. Then, it will describe the construction of $f(A)$ for the special case when f is holomorphic at zero and the square matrix A has sufficiently small norm. As in the case when f is a polynomial, $f(A)$ can be defined as the matrix obtained by substituting the variable z by A in the Taylor series of $f(z)$ at $z = 0$. As a paradigmatic example, for each $A \in \mathcal{M}_N(\mathbb{C})$ the exponential of the matrix A can be defined through

$$e^A := \sum_{n=0}^{\infty} \frac{A^n}{n!}.$$

Then, we will obtain the Dunford integral formula, which is an extension of the Cauchy integral formula to the class of matrix-valued holomorphic functions. To obtain the Dunford integral formula we will have to study some fundamental properties of the *resolvent operator*

$$\mathcal{R}(z; A) := (zI - A)^{-1}, \qquad z \in \mathbb{C} \setminus \sigma(A),$$

of a matrix $A \in \mathcal{M}_N(\mathbb{C})$. Finally, we will determine the exponential of A through its Jordan canonical form.

More precisely, Section 2.1 introduces the concept of norm equivalence, and proves that all norms in a finite-dimensional space are equivalent. It also presents the subordinated operator norm and establishes some of its main properties. Section 2.2 defines $f(A)$ for holomorphic functions f at the origin, and matrices A

with sufficiently small norm. This section provides the basic elements of operator calculus. Section 2.3 studies the properties of the resolvent operator and introduces the Dunford integral formula. Section 2.4 shows that, under the previous assumptions,

$$\sigma\left(f(A)\right) = f\left(\sigma(A)\right).$$

Finally, Section 2.5 provides the algorithms to determine e^A from the Jordan canonical form of A.

2.1 Norm of a linear operator

Suppose $\mathbb{K} \in \{\mathbb{R}, \mathbb{C}\}$. Then, the space $\mathcal{M}_N(\mathbb{K})$ of the square matrices of order N with coefficients in $\mathbb{K}$ is isomorphic to $\mathbb{K}^{N^2}$ and, hence, it can be equipped with the structure of Banach space from any norm of $\mathbb{K}^{N^2}$. For example, for each $p \in [1, \infty)$ we can consider the norm $\|\cdot\|_p$ defined by

$$\|A\|_p := \left(\sum_{i=1}^{N}\sum_{j=1}^{N} |a_{ij}|^p\right)^{1/p}, \qquad A \in \mathcal{M}_N(\mathbb{K}),$$

or the norm $\|\cdot\|_\infty$ defined by

$$\|A\|_\infty := \max_{1\leq i,j\leq N} \{|a_{ij}|\}, \qquad A \in \mathcal{M}_N(\mathbb{K}).$$

But none of these norms can be comfortably handled when the matrices are considered as endomorphisms of $\mathbb{K}^N$ and, consequently, we ought to introduce a specific norm for linear operators. Since every norm defines a topology, we need to ascertain whether or not two norms define the same topology.

Definition 2.1.1. Two norms of a vector space U are said to be equivalent if they induce in U the same topology.

The next result provides a necessary and sufficient condition for two norms to be equivalent.

Lemma 2.1.2. *Let $\|\cdot\|$ and $\|\cdot\|'$ be two norms of the vector space U. Then, they are equivalent if and only if there exist constants $\alpha, \beta > 0$ such that*

$$\alpha\|u\| \leq \|u\|' \leq \beta\|u\| \quad \text{for each} \quad u \in U. \tag{2.1}$$

Proof. Let $u_0 \in U$ and $r > 0$. If there exist $\alpha, \beta > 0$ satisfying (2.1), then

$$\begin{aligned}\{u \in U : \|u - u_0\| < r/\beta\} &\subset \{u \in U : \|u - u_0\|' < r\} \\ &\subset \{u \in U : \|u - u_0\| < r/\alpha\},\end{aligned}$$

and, hence, the norms $\|\cdot\|$ and $\|\cdot\|'$ induce the same topology in U.

Reciprocally, if the two induced topologies coincide, then the unit ball

$$\{u \in U : \|u\|' < 1\}$$

is open in the normed vector space $(U, \|\cdot\|)$, and, therefore, there exists $m > 0$ such that

$$\{u \in U : \|u\| < m\} \subset \{u \in U : \|u\|' < 1\}.$$

By symmetry, there exists $M > 0$ such that

$$\{u \in U : \|u\|' < M\} \subset \{u \in U : \|u\| < 1\}.$$

From these inclusions, (2.1) can be easily obtained. □

From the topological point of view, it is inessential to consider one norm or another if U is finite dimensional, since the next result establishes that, for each integer $d \geq 1$, all norms in $\mathbb{K}^d$ are equivalent; in other words, $\mathbb{K}^d$ possesses a unique topology of normed vector space. The key property for this result to hold is the compactness of the unit ball in any finite-dimensional space. It is well known that every closed and bounded set in $\mathbb{K}$ is compact. Also, the space $\mathbb{K}^d$ is equipped with the product topology, which coincides with the topology induced by $\|\cdot\|_\infty$. Moreover, the product of compact sets is compact. Consequently, every closed and bounded set of $\mathbb{K}^d$ equipped with the $\|\cdot\|_\infty$-topology is compact.

Theorem 2.1.3. *For every $d \geq 1$, all norms of $\mathbb{K}^d$ are equivalent.*

Proof. It suffices to prove that the norm $\|\cdot\|_\infty$ defined by

$$\|u\|_\infty := \max_{1\leq j\leq d} |u_j|, \qquad u = (u_1, \dots, u_d) \in \mathbb{K}^d,$$

is equivalent to every norm of $\mathbb{K}^d$. So let $\|\cdot\|$ be a norm in $\mathbb{K}^d$, and let $e_1, \dots, e_d$ denote the vectors of the canonical basis of $\mathbb{K}^d$. Then,

$$\|u\| = \|\sum_{j=1}^d u_j e_j\| \leq \sum_{j=1}^d \|u_j e_j\| = \sum_{j=1}^d |u_j| \cdot \|e_j\| \leq \|u\|_\infty \sum_{j=1}^d \|e_j\|.$$

In particular, the function

$$\|\cdot\| \; : \; (\mathbb{K}^d, \|\cdot\|_\infty) \to [0, \infty) \tag{2.2}$$

is continuous. The boundary

$$\partial B_1 := \{u \in \mathbb{K}^d : \|u\|_\infty = 1\}$$

of the unit ball, being closed and bounded, is compact in $(\mathbb{K}^d, \|\cdot\|_\infty)$. As the function (2.2) is continuous and ∂B_1 is compact, there exists $u^* \in \partial B_1$ such that

$$\|u\| \geq \|u^*\| \quad \text{for each} \quad u \in \partial B_1.$$

Therefore, for each $u \in \mathbb{K}^d$,

$$\|u\| \geq \|u^*\| \cdot \|u\|_\infty.$$

By Lemma 2.1.2, the norms $\|\cdot\|$ and $\|\cdot\|_\infty$ are equivalent. □

According to Theorem 2.1.3, there is a unique topological structure of finite-dimensional Banach space.

Now, fixing an integer $N \geq 1$ and an arbitrary norm $\|\cdot\|_{\mathbb{K}^N}$ in $\mathbb{K}^N$, we define, for each $A \in \mathcal{M}_N(\mathbb{K})$,

$$\|A\|_{\mathcal{L}(\mathbb{K}^N)} := \sup\left\{\|Au\|_{\mathbb{K}^N} \;:\; u \in \mathbb{K}^N,\ \|u\|_{\mathbb{K}^N} = 1\right\}. \tag{2.3}$$

As every linear operator in $\mathbb{K}^N$ is continuous, and the boundary of the unit ball is a compact subset of $\mathbb{K}^N$, the supremum of (2.3) is attained, and the magnitude $\|A\|_{\mathcal{L}(\mathbb{K}^N)}$ is finite. The number $\|A\|_{\mathcal{L}(\mathbb{K}^N)}$ is called the *norm of the operator* A subordinated to the norm $\|\cdot\|_{\mathbb{K}^N}$. The function

$$\begin{array}{ccc} \mathcal{M}_N(\mathbb{K}) & \to & [0,\infty) \\ A & \mapsto & \|A\|_{\mathcal{L}(\mathbb{K}^N)} \end{array}$$

is called the *operator norm* subordinated to $\|\cdot\|_{\mathbb{K}^N}$. The following result establishes its main properties.

Lemma 2.1.4. *Let $\|\cdot\|_{\mathbb{K}^N}$ be a norm in $\mathbb{K}^N$. Then, the map*

$$\|\cdot\|_{\mathcal{L}(\mathbb{K}^N)} : \mathcal{M}_N(\mathbb{K}) \to [0,\infty)$$

defined by (2.3) is a norm. Moreover, for each $u \in \mathbb{K}^N$ and $A \in \mathcal{M}_N(\mathbb{K})$,

$$\|Au\|_{\mathbb{K}^N} \leq \|A\|_{\mathcal{L}(\mathbb{K}^N)} \|u\|_{\mathbb{K}^N} \tag{2.4}$$

and, hence, for each $A, B \in \mathcal{M}_N(\mathbb{K})$,

$$\|AB\|_{\mathcal{L}(\mathbb{K}^N)} \leq \|A\|_{\mathcal{L}(\mathbb{K}^N)} \|B\|_{\mathcal{L}(\mathbb{K}^N)}. \tag{2.5}$$

Therefore, for each $A \in \mathcal{M}_N(\mathbb{K})$ and $n \in \mathbb{N}$, the following estimate holds,

$$\|A^n\|_{\mathcal{L}(\mathbb{K}^N)} \leq \|A\|^n_{\mathcal{L}(\mathbb{K}^N)}. \tag{2.6}$$

Proof. Pick $A \in \mathcal{M}_N(\mathbb{K})$. By definition, $\|A\|_{\mathcal{L}(\mathbb{K}^N)} \geq 0$. Moreover, if $\|A\|_{\mathcal{L}(\mathbb{K}^N)} = 0$, then $\|Au\|_{\mathbb{K}^N} = 0$ for each $u \in \mathbb{K}^N$ with $\|u\|_{\mathbb{K}^N} = 1$ and, hence, A is zero over any basis, whence $A = 0$.

For each $\lambda \in \mathbb{K}$ and $u \in \mathbb{K}^N$ with $\|u\|_{\mathbb{K}^N} = 1$, one has that

$$\|\lambda A u\|_{\mathbb{K}^N} = |\lambda| \cdot \|Au\|_{\mathbb{K}^N}$$

and, therefore,
$$\|\lambda A\|_{\mathcal{L}(\mathbb{K}^N)} = |\lambda| \cdot \|A\|_{\mathcal{L}(\mathbb{K}^N)}.$$
Finally, for each $B \in \mathcal{M}_N(\mathbb{K}^N)$ and $u \in \mathbb{K}^N$ with $\|u\|_{\mathbb{K}^N} = 1$, the following estimates hold,
$$\|(A+B)u\|_{\mathbb{K}^N} \leq \|Au\|_{\mathbb{K}^N} + \|Bu\|_{\mathbb{K}^N} \leq \|A\|_{\mathcal{L}(\mathbb{K}^N)} + \|B\|_{\mathcal{L}(\mathbb{K}^N)},$$
and, hence,
$$\|A+B\|_{\mathcal{L}(\mathbb{K}^N)} \leq \|A\|_{\mathcal{L}(\mathbb{K}^N)} + \|B\|_{\mathcal{L}(\mathbb{K}^N)}.$$
This shows that (2.3) indeed is a norm in $\mathcal{M}_N(\mathbb{K})$.

Now we check (2.4). Obviously, (2.4) is satisfied at $u = 0$. Suppose $u \neq 0$. Then,
$$\frac{\|Au\|_{\mathbb{K}^N}}{\|u\|_{\mathbb{K}^N}} = \|A\frac{u}{\|u\|_{\mathbb{K}^N}}\|_{\mathbb{K}^N} \leq \|A\|_{\mathcal{L}(\mathbb{K}^N)},$$
whence (2.4).

Finally, we will verify (2.5). By (2.4), for each $u \in \mathbb{K}^N$ with $\|u\|_{\mathbb{K}^N} = 1$, we have that
$$\|ABu\|_{\mathbb{K}^N} \leq \|A\|_{\mathcal{L}(\mathbb{K}^N)} \|Bu\|_{\mathbb{K}^N} \leq \|A\|_{\mathcal{L}(\mathbb{K}^N)} \|B\|_{\mathcal{L}(\mathbb{K}^N)}.$$
This implies (2.5); inequality (2.6) is an immediate consequence of (2.5). □

Lemma 2.1.4 and Theorem 2.1.3 prove that (2.3) equips $\mathcal{M}_N(\mathbb{K})$ with its unique structure of normed space; actually, by Exercise 1, of Banach space. Property (2.6) is the main reason for having introduced the norm (2.3).

2.2 Introduction to operator calculus

The next result explains how to generate operators from holomorphic functions at the origin. From now on, we will work in the complex field, so we make the choice $\mathbb{K} = \mathbb{C}$. The set D_R stands for the open disk in $\mathbb{C}$ of radius R and center 0. Also, for any open subset $\Omega \subset \mathbb{C}$, the set $\mathcal{H}(\Omega)$ denotes the space of holomorphic functions from Ω to $\mathbb{C}$. Throughout the remaining of this chapter, we consider fixed a norm $\|\cdot\|_{\mathbb{C}^N}$ of $\mathbb{C}^N$, and its associated subordinated operator norm $\|\cdot\|_{\mathcal{L}(\mathbb{C}^N)}$. When no ambiguity can arise, we will simply write $\|\cdot\|$ to denote either norm.

Theorem 2.2.1. *Let $R > 0$. Consider a function $f \in \mathcal{H}(D_R)$ and a matrix $A \in \mathcal{M}_N(\mathbb{C})$ satisfying*
$$\|A\|_{\mathcal{L}(\mathbb{C}^N)} < R.$$
Then the series
$$\sum_{j=0}^{\infty} \frac{f^{j)}(0)}{j!} A^j \tag{2.7}$$

converges absolutely in $\mathcal{L}(\mathbb{C}^N)$ to a matrix that will be denoted by $f(A)$. Moreover, for any invertible matrix $P \in \mathcal{M}_N(\mathbb{C})$, one has that

$$f(P^{-1}AP) = P^{-1}f(A)P. \tag{2.8}$$

Proof. Since $\|A\| < R$ and $f \in \mathcal{H}(D_R)$, by (2.6) and the Cauchy estimates, it is apparent that

$$\sum_{j=0}^{\infty} \| \frac{f^{j)}(0)}{j!} A^j \| \leq \sum_{j=0}^{\infty} \frac{|f^{j)}(0)|}{j!} R^j < \infty.$$

From the convergence of these series, it is easy to see that the sequence of partial sums

$$S_n := \sum_{j=0}^{n} \frac{f^{j)}(0)}{j!} A^j \in \mathcal{M}_N(\mathbb{C}), \qquad n \geq 1,$$

is a Cauchy sequence in $\mathcal{M}_N(\mathbb{C})$. By completeness, the limit

$$f(A) := \lim_{n\to\infty} S_n$$

is well defined and, in fact, the series (2.7) converges absolutely to $f(A)$.

Now, suppose

$$B = P^{-1}AP.$$

Then, a straightforward induction argument shows that

$$B^j = P^{-1}A^jP, \qquad j \geq 0.$$

Consequently, for each $n \geq 1$,

$$\sum_{j=0}^{n} \frac{f^{j)}(0)}{j!} B^j = P^{-1} \sum_{j=0}^{n} \frac{f^{j)}(0)}{j!} A^j P$$

and, taking limits as $n \to \infty$, we conclude (2.8). □

Property (2.8) is crucial from the point of view of the applications, since it establishes that in order to determine $f(A)$ it suffices to determine $f(J)$, where J is the Jordan canonical form of A. In general, calculating $f(J)$ tends to be substantially easier than calculating $f(A)$.

In the next definition, we recall the concept of vector-valued holomorphic function.

Definition 2.2.2. Consider an open set $\Omega \subset \mathbb{C}$, an integer $d \geq 1$ and a function

$$F : \Omega \to \mathbb{C}^d.$$

It is said that F is holomorphic in Ω when

$$\lim_{w\to z} \frac{F(w) - F(z)}{w - z}$$

exists in $\mathbb{C}^d$ for each $z \in \Omega$. In such a case, this limit is denoted by $F'(z)$, and called the derivative of F at z. The set of holomorphic functions from Ω to $\mathbb{C}^d$ will be hereafter denoted by $\mathcal{H}(\Omega, \mathbb{C}^d)$.

By expressing F through its components

$$F := (F_1, \dots, F_d),$$

it is apparent $F \in \mathcal{H}(\Omega, \mathbb{C}^d)$ if and only if $F_j \in \mathcal{H}(\Omega)$ for each $j \in \{1, \dots, d\}$. Moreover, in such a case,

$$F' = (F_1', \dots, F_d').$$

From this feature, it is easy to see that all algebraic properties of $\mathbb{C}$-valued holomorphic functions remain true for vector-valued holomorphic functions. For instance, the derivative of a sum equals the sum of the derivatives, and the product formula for the derivative of a product of a scalar-valued function by a vector-valued function holds. More importantly, defining the path integral of a continuous vector-valued function as the vector formed by the integrals of each of the components of the function, makes the Cauchy integral formula remain valid for vector-valued functions.

The proof of the following elementary lemma is proposed in Exercise 4.

Lemma 2.2.3. *Suppose α, β are two distinct complex numbers, and $n \geq 2$ is an integer. Then*

$$\frac{\alpha^n - \beta^n}{\alpha - \beta} - n\beta^{n-1} = (\alpha - \beta) \sum_{k=1}^{n-1} k\beta^{k-1}\alpha^{n-k-1}.$$

Lemma 2.2.3 will provide us with a proof of a special, but important, case of the chain rule. Although there are more general versions of the chain rule, the next one suffices for the purposes of this book.

Theorem 2.2.4. *Suppose $R > 0$, consider $f \in \mathcal{H}(D_R)$ and $A \in \mathcal{M}_N(\mathbb{C})$, and set*

$$\rho := R/\|A\|_{\mathcal{L}(\mathbb{C}^N)}.$$

Then, the matrix-valued function $F : D_\rho \to \mathcal{M}_N(\mathbb{C})$ defined by

$$F(z) := f(zA), \quad z \in D_\rho,$$

is holomorphic in D_ρ and

$$F'(z) = Af'(zA), \qquad z \in D_\rho.$$

Proof. Throughout this proof, we denote

$$a_n := \frac{f^{n)}(0)}{n!}, \qquad n \geq 0,$$

and fix $z \in D_\rho$. Let $\varepsilon > 0$ be such that

$$(1+\varepsilon)|z| < \rho.$$

By Theorem 2.2.1 and our choice of ρ, both $f(zA)$ and $f'(zA)$ are well defined. By definition, for each $h \sim 0$ with $h \neq 0$, we have that

$$\frac{F(z+h)-F(z)}{h} - Af'(zA) = \sum_{n=2}^{\infty} a_n \left[\frac{(z+h)^n - z^n}{h} - nz^{n-1} \right] A^n.$$

Moreover, due to Lemma 2.2.3,

$$\frac{(z+h)^n - z^n}{h} - nz^{n-1} = h \sum_{k=1}^{n-1} kz^{k-1}(z+h)^{n-k-1}$$

for each $n \geq 2$ and, hence,

$$\frac{F(z+h)-F(z)}{h} - Af'(zA) = h \sum_{n=2}^{\infty} a_n \sum_{k=1}^{n-1} kz^{k-1}(z+h)^{n-k-1} A^n. \tag{2.9}$$

On the other hand, for all $h \in \bar{D}_{\varepsilon|z|}$ the following inequalities hold,

$$\begin{aligned} |\sum_{k=1}^{n-1} kz^{k-1}(z+h)^{n-k-1}| &\leq \sum_{k=1}^{n-1} k|z|^{k-1}(|z|+|h|)^{n-k-1} \\ &\leq \sum_{k=1}^{n-1} k(|z|+|h|)^{n-2} \leq n^2[(1+\varepsilon)|z|]^{n-2} \end{aligned}$$

and, hence, for each $n \geq 2$,

$$\|a_n \sum_{k=1}^{n-1} kz^{k-1}(z+h)^{n-k-1} A^n\|_{\mathcal{L}(\mathbb{C}^N)} \leq |a_n| n^2 [(1+\varepsilon)|z|]^{n-2} \|A\|^n_{\mathcal{L}(\mathbb{C}^N)}. \tag{2.10}$$

Moreover, by the Cauchy estimates, we get

$$\begin{aligned} \limsup_{n\to\infty} \left(|a_n| n^2 [(1+\varepsilon)|z|]^{n-2} \|A\|^n_{\mathcal{L}(\mathbb{C}^N)} \right)^{1/n} &\leq \frac{(1+\varepsilon)|z|}{R} \|A\|_{\mathcal{L}(\mathbb{C}^N)} \\ &< \frac{\rho}{R} \|A\|_{\mathcal{L}(\mathbb{C}^N)} = 1. \end{aligned}$$

Therefore, by (2.10) and the root test for the convergence of series, it is apparent that the matrix series of the right-hand side of (2.9) is convergent and that, actually, its sum is uniformly bounded for $h \in \bar{D}_{\varepsilon|z|}$. Finally, taking limits as $h \to 0$ in (2.9) completes the proof. □

The following property provides us with the series of the product of the matrix functions generated through Theorem 2.2.1 by means of the Cauchy product of the corresponding matrix series.

Lemma 2.2.5. *Suppose $R > 0$, take $f, g \in \mathcal{H}(D_R)$ and let $A, B \in \mathcal{M}_N(\mathbb{C})$ satisfy*

$$\|A\|_{\mathcal{L}(\mathbb{C}^N)} < R, \qquad \|B\|_{\mathcal{L}(\mathbb{C}^N)} < R.$$

Then

$$f(A)g(B) = \sum_{n=0}^{\infty} \sum_{j=0}^{n} \frac{f^{n-j)}(0)}{(n-j)!} \frac{g^{j)}(0)}{j!} A^{n-j} B^j.$$

Proof. Clearly, for each $z, w \in D_R$ one has that

$$f(z)g(w) = \sum_{n=0}^{\infty} \sum_{j=0}^{n} \frac{f^{n-j)}(0)}{(n-j)!} \frac{g^{j)}(0)}{j!} w^j.$$

Consequently, arguing as is the proof of Theorem 2.2.1, one can easily complete all technical details of the proof. □

2.3 Resolvent operator. Dunford's integral formula

In this section we will obtain a generalized version of the Cauchy integral formula valid for matrix-valued functions; namely, the Dunford integral formula. We begin by introducing the concept of resolvent and by studying some of its main properties.

Definition 2.3.1. Let $A \in \mathcal{M}_N(\mathbb{C})$. The resolvent set of A, denoted by $\varrho(A)$, is defined as the complement of the spectrum of A in $\mathbb{C}$,

$$\varrho(A) := \mathbb{C} \setminus \sigma(A).$$

The resolvent operator of A is the matrix function $\mathcal{R}(\cdot; A)$ defined by

$$\mathcal{R}(z; A) := (zI - A)^{-1}, \qquad z \in \varrho(A).$$

It turns out that, for each $z \in \varrho(A)$ and $v \in \mathbb{C}^N$, the vector

$$u := \mathcal{R}(z; A)v$$

provides us with the unique solution of the linear system

$$(zI - A)u = v,$$

hence the name of *resolvent operator*.

The following classic result provides a sufficient condition for an operator to be invertible. For each $R > 0$ and $z_0 \in \mathbb{C}$, the subset $D_R(z_0) \subset \mathbb{C}$ denotes the open disk of radius R centered at z_0. We have already defined D_R as $D_R(0)$.

Lemma 2.3.2. *Let* $T \in \mathcal{L}(\mathbb{C}^N)$ *be such that*

$$\|T\|_{\mathcal{L}(\mathbb{C}^N)} < 1. \tag{2.11}$$

Then $I - T$ *is invertible and its inverse is given through the series*

$$(I - T)^{-1} = \sum_{n=0}^{\infty} T^n. \tag{2.12}$$

Moreover,

$$\|(I - T)^{-1}\|_{\mathcal{L}(\mathbb{C}^N)} \leq \frac{1}{1 - \|T\|_{\mathcal{L}(\mathbb{C}^N)}}. \tag{2.13}$$

Proof. The function $f : D_1 \to \mathbb{C}$ defined by

$$f(z) := \frac{1}{1 - z} = \sum_{n=0}^{\infty} z^n, \qquad z \in D_1,$$

is holomorphic, and, hence, according to Theorem 2.2.1, the matrix

$$f(T) = \sum_{j=0}^{\infty} T^j$$

is well defined. To prove (2.12) it suffices to show that $f(T)$ is the inverse of $I - T$. Indeed, for each $n \geq 1$ one has that

$$\sum_{j=0}^{n} T^j (I - T) = (I - T) \sum_{j=0}^{n} T^j = I - T^{n+1}. \tag{2.14}$$

On the other hand, by (2.6) and (2.11),

$$\lim_{n \to \infty} T^{n+1} = 0.$$

Consequently, passing to the limit as $n \to \infty$ in (2.14) gives

$$f(T)(I - T) = (I - T) f(T) = I.$$

Finally, the estimate (2.13) is easily obtained from the next one,

$$\|\sum_{j=0}^{n} T^j\|_{\mathcal{L}(\mathbb{C}^N)} \leq \sum_{j=0}^{n} \|T\|^j_{\mathcal{L}(\mathbb{C}^N)}, \qquad n \geq 1,$$

by taking limits as $n \to \infty$. □

The next result, although straightforward, is extremely useful.

Lemma 2.3.3. *For each $A \in \mathcal{M}_N(\mathbb{C})$ and $z, w \in \varrho(A)$, the following identity holds*

$$\mathcal{R}(z;A) - \mathcal{R}(w;A) = (w-z)\,\mathcal{R}(z;A)\mathcal{R}(w;A). \tag{2.15}$$

Proof. By definition,

$$\begin{aligned}\mathcal{R}(z;A) &= \mathcal{R}(z;A)(wI-A)\mathcal{R}(w;A)\\ &= \mathcal{R}(z;A)[(w-z)I+(zI-A)]\mathcal{R}(w;A)\\ &= (w-z)\mathcal{R}(z;A)\mathcal{R}(w;A)+\mathcal{R}(w;A),\end{aligned}$$

whence (2.15) follows. □

Identity (2.15) is usually referred to as the *resolvent formula*. The next result establishes that the resolvent operator is holomorphic.

Theorem 2.3.4. *For each $A \in \mathcal{M}_N(\mathbb{C})$ the operator-valued function*

$$\mathcal{R}(\cdot;A) : \varrho(A) \to \mathcal{L}(\mathbb{C}^N)$$

is holomorphic, and its derivative at each $z \in \varrho(A)$ equals

$$-[\mathcal{R}(z;A)]^2.$$

Moreover, $\mathcal{R}(\cdot;A)$ admits the following Laurent development at zero:

$$\mathcal{R}(z;A) = \sum_{n=0}^{\infty} \frac{1}{z^{n+1}} A^n, \qquad z \in \Omega := \mathbb{C} \setminus \bar{D}_{\|A\|_{\mathcal{L}(\mathbb{C}^N)}}, \tag{2.16}$$

uniformly in compact subsets of Ω.

Proof. First, we show the continuity of the resolvent in $\varrho(A)$. By (2.15), for each $z, w \in \varrho(A)$,

$$\mathcal{R}(z;A) = [I+(w-z)\mathcal{R}(z;A)]\mathcal{R}(w;A).$$

Fix $z \in \varrho(A)$. Then, there exists $\varepsilon > 0$ such that for each $w \in \bar{D}_\varepsilon(z)$,

$$\|(w-z)\mathcal{R}(z;A)\|_{\mathcal{L}(\mathbb{C}^N)} < 1$$

and, hence, by Lemma 2.3.2, the operator $I+(w-z)\mathcal{R}(z;A)$ is invertible. Therefore, for each $w \in \bar{D}_\varepsilon(z)$ the following identity holds

$$\mathcal{R}(w;A) = [I+(w-z)\mathcal{R}(z;A)]^{-1}\mathcal{R}(z;A). \tag{2.17}$$

Applying again Lemma 2.3.2 yields

$$[I+(w-z)\mathcal{R}(z;A)]^{-1} = \sum_{n=0}^{\infty} (z-w)^n [\mathcal{R}(z;A)]^n,$$

whence

$$\lim_{w\to z}[I+(w-z)\mathcal{R}(z;A)]^{-1}=I.$$

Consequently, taking limits in (2.17) gives

$$\lim_{w\to z}\mathcal{R}(w;A)=\mathcal{R}(z;A),$$

which proves the continuity of the function $\mathcal{R}(\cdot;A)$ in $\varrho(A)$. Now, its holomorphy is an immediate consequence of (2.15). Indeed, for all $z,w\in\varrho(A)$ with $w\neq z$, we have that

$$\frac{\mathcal{R}(w;A)-\mathcal{R}(z;A)}{w-z}=-\mathcal{R}(w;A)\mathcal{R}(z;A)$$

and, hence,

$$\lim_{w\to z}\frac{\mathcal{R}(w;A)-\mathcal{R}(z;A)}{w-z}=-[\mathcal{R}(z;A)]^2.$$

Finally, let K be a compact subset of the open set Ω. Then, there exists $\varepsilon>0$ such that

$$|z|>\|A\|_{\mathcal{L}(\mathbb{C}^N)}+\varepsilon \quad \text{for each} \quad z\in K. \tag{2.18}$$

According to Lemma 2.3.2 applied to $z^{-1}A$, it is apparent that, for each $z\in K$,

$$\mathcal{R}(z;A)=\frac{1}{z}(I-\frac{A}{z})^{-1}=\sum_{n=0}^{\infty}\frac{1}{z^{n+1}}A^n.$$

Moreover, this series is uniformly convergent in K, because, due to (2.18), we have that, for all $n\geq 1$ and $z\in K$,

$$\begin{aligned}\|\mathcal{R}(z;A)-\sum_{j=0}^{n}\frac{1}{z^{j+1}}A^j\|_{\mathcal{L}(\mathbb{C}^N)}&=\|\sum_{j=n+1}^{\infty}\frac{1}{z^{j+1}}A^j\|_{\mathcal{L}(\mathbb{C}^N)}\\&\leq\frac{1}{\|A\|_{\mathcal{L}(\mathbb{C}^N)}+\varepsilon}\sum_{j=n+1}^{\infty}\left(\frac{\|A\|_{\mathcal{L}(\mathbb{C}^N)}}{\|A\|_{\mathcal{L}(\mathbb{C}^N)}+\varepsilon}\right)^j\\&=\frac{1}{\varepsilon}\left(\frac{\|A\|_{\mathcal{L}(\mathbb{C}^N)}}{\|A\|_{\mathcal{L}(\mathbb{C}^N)}+\varepsilon}\right)^{n+1}.\end{aligned}$$

This concludes the proof. □

Note that Theorem 2.3.4 actually proves the convergence of the series (2.16) to the resolvent operator uniformly in the open unbounded set $\mathbb{C}\setminus\bar{D}_\rho$ as long as

$$\rho>\|A\|_{\mathcal{L}(\mathbb{C}^N)}.$$

Now we introduce some preliminaries for the Dunford integral formula. That formula, besides being a substantial generalization of the Cauchy integral formula, plays a pivotal role in Functional Analysis; very especially, in the spectral theory of linear operators. Prior to this, we need to recall some basic concepts about closed path (or contour) integrals.

First, we should make precise the class of contours that will be used in the integrals. By definition, a *Cauchy contour* is the oriented boundary of a Cauchy domain in $\mathbb{C}$. By a *Cauchy domain* is meant any disjoint union in $\mathbb{C}$ of finitely many non-empty open, bounded and connected sets $\Delta_1, \dots, \Delta_r$ such that

$$\Delta_i \cap \Delta_j = \emptyset, \qquad 1 \leq i < j \leq r,$$

and, for each $1 \leq j \leq r$, the boundary of Δ_j consists of finitely many non-intersecting closed rectifiable Jordan curves, each of them C is oriented clockwise or counterclockwise according to whether the points of Δ_j near a point of C are outside or inside of C. In such a case, the inner domain of the Cauchy contour γ is defined through

$$D_\gamma := \bigcup_{i=1}^{r} \Delta_i.$$

For every non-empty compact subset σ of an open set $\Omega \subset \mathbb{C}$, there exists a Cauchy contour γ in Ω for which $\sigma \subset D_\gamma$. Indeed, construct in the complex plane a grid of congruent hexagons of diameter less than one third of the distance between σ and $\mathbb{C} \setminus \Omega$, and let Δ be the interior of the union of all closed hexagons of the grid that have non-empty intersection with σ. Clearly, the boundary of Δ provides us with a Cauchy contour in Ω whose inner domain contains σ.

Let $d \geq 1$ be an integer number, γ a Cauchy contour, and $f : \gamma \to \mathbb{C}^d$ a continuous function component-wise expressed by

$$f = (f_1, \dots, f_d).$$

Then, the integral of f along γ is defined through

$$\int_\gamma f(z)\, dz := \left(\int_\gamma f_1(z)\, dz, \dots, \int_\gamma f_d(z)\, dz \right).$$

Finally, we make precise the version of the Cauchy integral formula that will be subsequently used.

Theorem 2.3.5 (Cauchy's integral formula). *Suppose $\Omega \subset \mathbb{C}$ is open, $f \in \mathcal{H}(\Omega)$, and γ is a Cauchy contour whose inner domain D_γ satisfies $\bar{D}_\gamma \subset \Omega$. Then,*

$$f(z_0) = \frac{1}{2\pi i} \int_\gamma \frac{f(z)}{z - z_0}\, dz \quad \textit{for each} \quad z_0 \in D_\gamma. \tag{2.19}$$

The following result provides us with a substantial generalization of the Cauchy integral formula (2.19). It establishes that the operator $f(A)$ defined by Theorem 2.2.1 can be represented through a vector-valued contour integral within the spirit of (2.19). The formula, being attributable to Dunford, is usually referred to as the Dunford integral formula. Throughout the remaining of this book, C_R stands for the positively oriented boundary of D_R, for each $R > 0$.

Theorem 2.3.6. *Suppose $R > 0$, the matrix $A \in \mathcal{M}_N(\mathbb{C})$ satisfies $\|A\|_{\mathcal{L}(\mathbb{C}^N)} < R$, and $f \in \mathcal{H}(D_R)$. Let γ be a Cauchy contour in D_R whose inner domain contains the spectrum of A. Then*

$$f(A) = \frac{1}{2\pi i} \int_\gamma f(z)\mathcal{R}(z;A)\,dz. \tag{2.20}$$

Proof. For each $\lambda \in \sigma(A)$ there exists $u \in \mathbb{C}^N$ such that

$$\|u\|_{\mathbb{C}^N} = 1 \quad \text{and} \quad Au = \lambda u.$$

Thus,

$$|\lambda| = \|Au\|_{\mathbb{C}^N} \le \|A\|_{\mathcal{L}(\mathbb{C}^N)} < R$$

and, therefore,

$$\sigma(A) \subset \bar{D}_{\|A\|_{\mathcal{L}(\mathbb{C}^N)}} \subset D_R. \tag{2.21}$$

Now we consider r satisfying

$$\|A\|_{\mathcal{L}(\mathbb{C}^N)} < r < R.$$

By (2.21), the circle C_r is a Cauchy contour satisfying the hypothesis of the theorem. Moreover, according to Theorem 2.3.4,

$$\mathcal{R}(z;A) = \sum_{n=0}^{\infty} \frac{1}{z^{n+1}} A^n \quad \text{uniformly in} \quad z \in C_r.$$

Therefore,

$$\frac{1}{2\pi i}\int_{C_r} f(z)\mathcal{R}(z;A)\,dz = \sum_{n=0}^{\infty} \frac{1}{2\pi i}\int_{C_r} \frac{f(z)}{z^{n+1}}\,dz\; A^n = \sum_{n=0}^{\infty} \frac{f^{n)}(0)}{n!} A^n = f(A).$$

This proves (2.20) for the special case when $\gamma = C_r$.

To complete the proof in the general case, it suffices to note that, according to Theorem 2.3.5, by the definition of holomorphic vector-valued function,

$$\int_\gamma f(z)\mathcal{R}(z;A)\,dz = \int_{C_r} f(z)\mathcal{R}(z;A)\,dz.$$

This concludes the proof. □

Actually, Theorem 2.3.6 shows the consistence of the following concept, which provides us with a substantial and sharp generalization of the corresponding one introduced by Theorem 2.2.1.

Definition 2.3.7. Let $A \in \mathcal{M}_N(\mathbb{C})$, and denote by $\mathcal{F}(A)$ the family of complex functions of a complex variable that are holomorphic in an open neighborhood (not necessarily connected) of $\sigma(A)$. Let $f \in \mathcal{F}(A)$, and $\Omega \subset \mathbb{C}$ be an open set such that

$$\sigma(A) \subset \Omega \quad \text{and} \quad f \in \mathcal{H}(\Omega).$$

Then, $f(A) \in \mathcal{M}_N(\mathbb{C})$ is the linear operator defined through

$$f(A) := \frac{1}{2\pi i} \int_{\partial\Delta} f(z)\mathcal{R}(z; A)\, dz, \tag{2.22}$$

where Δ is a Cauchy domain in Ω such that

$$\sigma(A) \subset \Delta \subset \bar{\Delta} \subset \Omega.$$

Remark 2.3.8. By applying Theorem 2.3.5 to every coefficient of the matrix

$$f(z)\mathcal{R}(z; A), \qquad z \in \partial\Delta,$$

it is apparent that the Dunford integral of (2.22) does not depend on the choice of Ω, but exclusively on the function f and on the operator A. Moreover, under the assumptions of Theorem 2.3.6, it turns out that (2.22) equals the operator $f(A)$ as defined through its Taylor series centered at zero, by Theorems 2.2.1 and 2.3.6.

It should be remembered that the existence of Cauchy domains Δ fulfilling the requirements of Definition 2.3.7 has already been established.

2.4 The spectral mapping theorem

We begin by establishing a very natural and useful product formula. The notation introduced in Section 2.3 and, in particular, in Definition 2.3.7, will be maintained through this section.

Proposition 2.4.1. *If $A \in \mathcal{M}_N(\mathbb{C})$ and $f, g \in \mathcal{F}(A)$, then $fg \in \mathcal{F}(A)$ and*

$$(fg)(A) = f(A)g(A).$$

Proof. Since fg is defined in the intersection of the domains of f and g, and fg is holomorphic therein, necessarily $fg \in \mathcal{F}(A)$. Let Δ_1 and Δ_2 be two Cauchy domains such that

$$\sigma(A) \subset \Delta_1 \subset \bar{\Delta}_1 \subset \Delta_2$$

for which f and g are holomorphic in a neighborhood of $\bar{\Delta}_2$. Then, by Remark 2.3.8, we have that

$$\begin{aligned} f(A)g(A) &= \left(\frac{1}{2\pi i}\right)^2 \int_{\partial\Delta_1} f(z)\mathcal{R}(z; A)\, dz \int_{\partial\Delta_2} g(w)\mathcal{R}(w; A)\, dw \\ &= \left(\frac{1}{2\pi i}\right)^2 \int_{\partial\Delta_1} \int_{\partial\Delta_2} f(z)g(w)\mathcal{R}(z; A)\mathcal{R}(w; A)\, dw\, dz \end{aligned}$$

and, hence, using Lemma 2.3.3 gives

$$\begin{aligned} f(A)g(A) &= \left(\frac{1}{2\pi i}\right)^2 \int_{\partial\Delta_1}\int_{\partial\Delta_2} f(z)g(w)\frac{\mathcal{R}(z;A)-\mathcal{R}(w;A)}{w-z}\,dw\,dz \\ &= \frac{1}{2\pi i}\int_{\partial\Delta_1} f(z)\mathcal{R}(z;A)\left[\frac{1}{2\pi i}\int_{\partial\Delta_2}\frac{g(w)}{w-z}\,dw\right]dz \\ &\quad - \frac{1}{2\pi i}\int_{\partial\Delta_2} g(w)\mathcal{R}(w;A)\left[\frac{1}{2\pi i}\int_{\partial\Delta_1}\frac{f(z)}{w-z}\,dz\right]dw. \end{aligned}$$

On the other hand, since $\bar{\Delta}_1 \subset \Delta_2$, according to Theorem 2.3.5, for each $z \in \partial\Delta_1$ and $w \in \partial\Delta_2$, we have that

$$\frac{1}{2\pi i}\int_{\partial\Delta_2}\frac{g(w)}{w-z}\,dw = g(z) \quad \text{and} \quad \frac{1}{2\pi i}\int_{\partial\Delta_1}\frac{f(z)}{w-z}\,dz = 0.$$

Therefore, inserting these identities in the previous one gives

$$f(A)g(A) = \frac{1}{2\pi i}\int_{\partial\Delta_1} f(z)g(z)\mathcal{R}(z;A)\,dz = (fg)(A),$$

as claimed in the statement. □

The main result of this section is the following. It describes how the spectrum of a matrix is transformed by holomorphic functions.

Theorem 2.4.2 (Spectral mapping theorem). *For every $A \in \mathcal{M}_N(\mathbb{C})$ and $f \in \mathcal{F}(A)$, the following identity of sets is satisfied,*

$$\sigma(f(A)) = f(\sigma(A)).$$

Proof. Pick $\lambda \in \sigma(A)$, and let $D \subset \mathbb{C}$ be an open set containing $\sigma(A)$ such that f is holomorphic in D. Now, let g denote the auxiliary function defined by

$$g(z) := \begin{cases} (f(z)-f(\lambda))(z-\lambda)^{-1} & \text{if } z \in D\setminus\{\lambda\}, \\ f'(\lambda) & \text{if } z = \lambda. \end{cases}$$

By the celebrated theorem of Riemann on removable singularities, we obtain that $g \in \mathcal{H}(D)$. Moreover, by construction,

$$g(z)(z-\lambda) = f(z) - f(\lambda), \qquad z \in D,$$

and, hence, owing to Proposition 2.4.1,

$$g(A)(A-\lambda I) = f(A) - f(\lambda)I.$$

Consequently, since $\lambda \in \sigma(A)$,

$$\det[f(A)-f(\lambda)I] = \det g(A)\cdot\det(A-\lambda I) = 0,$$

whence $f(\lambda) \in \sigma(f(A))$ and, therefore,

$$f(\sigma(A)) \subset \sigma(f(A)).$$

Now, suppose $\lambda \notin f(\sigma(A))$. Then, there exists an open subset $\Omega \subset \mathbb{C}$ such that $\sigma(A) \subset \Omega$ and

$$f(z) \neq \lambda \quad \text{for each} \quad z \in \Omega.$$

Consequently, the auxiliary function h defined by

$$h(z) := \frac{1}{f(z) - \lambda}, \qquad z \in \Omega,$$

is holomorphic in Ω. According to Proposition 2.4.1, we have that

$$h(A)(f(A) - \lambda I) = I$$

and, in particular,

$$1 = \det h(A) \cdot \det[f(A) - \lambda I].$$

Therefore $\lambda \notin \sigma(f(A))$. This ends the proof. □

2.5 The exponential matrix

Owing to Theorem 2.2.1, for each $A \in \mathcal{M}_N(\mathbb{C})$ we can define the exponential e^A of the matrix A through the identity

$$e^A := \sum_{n=0}^{\infty} \frac{A^n}{n!}. \tag{2.23}$$

The following result collects its main properties.

Proposition 2.5.1. *Let $A, B \in \mathcal{M}_N(\mathbb{C})$. Then,*

(a) $\|e^A\|_{\mathcal{L}(\mathbb{C}^N)} \leq e^{\|A\|_{\mathcal{L}(\mathbb{C}^N)}}$.

(b) $e^{A+B} = e^A e^B = e^B e^A$ *if* $AB = BA$.

(c) *The matrix* e^A *is invertible,* $e^0 = I$, *and* $(e^A)^{-1} = e^{-A}$.

(d) $e^B = P^{-1} e^A P$ *if* $B = P^{-1}AP$ *for some invertible* $P \in \mathcal{M}_N(\mathbb{C})$.

Proof. For each $n \geq 1$ the following is satisfied,

$$\|\sum_{j=0}^{n} \frac{A^j}{j!}\|_{\mathcal{L}(\mathbb{C}^N)} \leq \sum_{j=0}^{n} \frac{\|A\|^j_{\mathcal{L}(\mathbb{C}^N)}}{j!} \leq e^{\|A\|_{\mathcal{L}(\mathbb{C}^N)}}.$$

Taking limits as $n \to \infty$ shows Part (a).

Suppose

$$AB = BA.$$

Then, an easy induction argument proves that

$$(A+B)^k = \sum_{j=0}^{k} \binom{k}{j} A^j B^{k-j}$$

for every integer $k \geq 0$. Consequently,

$$e^{A+B} = \sum_{k=0}^{\infty} \frac{(A+B)^k}{k!} = \sum_{k=0}^{\infty}\sum_{j=0}^{k} \frac{A^j}{j!} \frac{B^{k-j}}{(k-j)!}.$$

As the second series is the Cauchy product of the series defining e^A and e^B, by Lemma 2.2.5, Part (b) follows.

The identity $e^0 = I$ follows from the definition of the exponential. Moreover, as A commutes with $-A$, Part (b) yields

$$I = e^0 = e^{A-A} = e^A e^{-A} = e^{-A} e^A$$

and, therefore, e^{-A} is the inverse matrix of e^A. This proves Part (c). Finally, Property (d) is a consequence of identity (2.8). □

In practice, calculating the matrix e^A from (2.23) may be extraordinarily involved, unless A is nilpotent or an appropriate basis of $\mathbb{C}^N$ has been chosen to get an easy algorithm to identify A^n in terms of A, for each $n \geq 2$. Nevertheless, according to Proposition 2.5.1(d), the problem of calculating e^A can be reduced to the problem of determining e^J, where J is the Jordan form of A. Consequently, the problem simplifies to the calculation of the exponential of the Jordan blocks, which can be accomplished very easily.

Indeed, for $A \in \mathcal{M}_N(\mathbb{C})$, let P be the matrix of the change of basis to its Jordan canonical form J. Then,

$$A = PJP^{-1}$$

and, hence, by Proposition 2.5.1(d),

$$e^{zA} = Pe^{zJ}P^{-1} \quad \text{for each} \quad z \in \mathbb{C}.$$

Remember that the columns of P consist of the generalized eigenvectors constructed in Section 1.3. Moreover, the canonical form J has the form

$$J = \operatorname{diag}\{J_1, J_2, \dots, J_{\ell-1}, J_\ell\}$$

for a certain family of Jordan blocks $J_1, \dots, J_\ell$, each of them having the form $J_{\lambda,h}$ (see (1.26)), for some $\lambda \in \sigma(A)$ and some integer $1 \leq h \leq \nu(\lambda)$. In other words, for each $1 \leq k \leq \ell$, the block J_k is a square matrix of order h whose main diagonal consists of λ's, whose subdiagonal consists of 1's, whereas all remaining coefficients vanish. This has already been discussed in Section 1.3. Multiplying by blocks shows that, for each integer $n \geq 0$,

$$J^n = \operatorname{diag}\left\{J_1^n, J_2^n, \ldots, J_{\ell-1}^n, J_\ell^n\right\},$$

and, consequently, for each $z \in \mathbb{C}$,

$$e^{zJ} = \operatorname{diag}\left\{e^{zJ_1}, e^{zJ_2}, \ldots, e^{zJ_{\ell-1}}, e^{zJ_\ell}\right\}.$$

Fix $k \in \{1, \ldots, \ell\}$ and suppose $J_k = J_{\lambda,h}$ (see (1.26)), for some $\lambda \in \sigma(A)$ and $1 \leq h \leq \nu(\lambda)$. Then

$$J_k = J_{0,h} + \lambda I_h,$$

where I_h stands for the identity matrix of order h. As the identity commutes with all matrices, by Proposition 2.5.1(b), it follows that

$$e^{zJ_k} = e^{\lambda z} e^{zJ_{0,h}}, \qquad z \in \mathbb{C}.$$

From now on, given two natural numbers $n, k \geq 1$ and $B_1, \ldots, B_n \in \mathcal{M}_k(\mathbb{C})$, we will denote

$$\operatorname{trng}\{B_1, B_2, \ldots, B_n\} := \begin{pmatrix} B_1 & 0 & \cdots & 0 \\ B_2 & B_1 & \cdots & 0 \\ \vdots & \vdots & \ddots & \vdots \\ B_n & B_{n-1} & \cdots & B_1 \end{pmatrix} \in \mathcal{M}_{nk}(\mathbb{C}). \tag{2.24}$$

Using this notation (with $n = h$ and $k = 1$), we find from $J_{0,h}^h = 0$ that

$$e^{zJ_{0,h}} = \sum_{n=0}^{h-1} \frac{z^n}{n!} J_{0,h}^n = \operatorname{trng}\left\{1, z, \frac{z^2}{2!}, \ldots, \frac{z^{h-2}}{(h-2)!}, \frac{z^{h-1}}{(h-1)!}\right\}$$

and, therefore,

$$e^{zJ_k} = \operatorname{trng}\left\{e^{\lambda z}, z\, e^{\lambda z}, \frac{z^2}{2!}\, e^{\lambda z}, \ldots, \frac{z^{h-2}}{(h-2)!}\, e^{\lambda z}, \frac{z^{h-1}}{(h-1)!}\, e^{\lambda z}\right\}.$$

Consequently, the exponential e^{zA} can be determined through a finite algorithm from the Jordan canonical form of A.

When A has real coefficients, then so does e^{tA} for each $t \in \mathbb{R}$, and, in a number of circumstances, it is imperative to determine e^{tA} from the real Jordan canonical form $J^{\mathbb{R}}$ of A. As we discussed earlier, that problem can be reduced to calculate e^{tJ_k} for each $1 \leq k \leq \ell$. When J_k is a Jordan block associated with a real eigenvalue of A we can proceed as described above. If $\alpha, \beta \in \mathbb{R}$ and

$$J_k = J^{\mathbb{R}}_{\alpha+i\beta,2} := \begin{pmatrix} \alpha & \beta \\ -\beta & \alpha \end{pmatrix},$$

then, the calculation of e^{tJ_k} should be carried out as follows. First, decompose

$$\begin{pmatrix} \alpha & \beta \\ -\beta & \alpha \end{pmatrix} = \alpha\, I_2 + \begin{pmatrix} 0 & \beta \\ -\beta & 0 \end{pmatrix}.$$

Then, according to Proposition 2.5.1(b) and using the definition of the exponential, one obtains

$$e^{tJ_k} = e^{\alpha t} \begin{pmatrix} \cos(\beta t) & \sin(\beta t) \\ -\sin(\beta t) & \cos(\beta t) \end{pmatrix},$$

because αI_2 commutes with all matrices. In general, when

$$J_k = J^{\mathbb{R}}_{\alpha+i\beta,2h} = \operatorname{trng}\left\{ J^{\mathbb{R}}_{\alpha+i\beta,2}, I_2, 0, \dots, 0 \right\} \quad \text{for some} \quad h \geq 2,$$

one first decomposes $J^{\mathbb{R}}_{\alpha+i\beta,2h}$ into the sum of two commuting matrices

$$J^{\mathbb{R}}_{\alpha+i\beta,2h} = \operatorname{trng}\{0, I_2, 0, \dots, 0\} + \operatorname{trng}\{J^{\mathbb{R}}_{\alpha+i\beta,2}, 0, 0, \dots, 0\},$$

in order to determine $e^{tJ^{\mathbb{R}}_{\alpha+i\beta,2h}}$ by means of Proposition 2.5.1(b). This shows that

$$e^{tJ^{\mathbb{R}}_{\alpha+i\beta,2h}} = \operatorname{trng}\left\{ e^{tJ^{\mathbb{R}}_{\alpha+i\beta,2}}, t\, e^{tJ^{\mathbb{R}}_{\alpha+i\beta,2}}, \dots, \frac{t^{h-1}}{(h-1)!}\, e^{tJ^{\mathbb{R}}_{\alpha+i\beta,2}} \right\},$$

where

$$e^{tJ^{\mathbb{R}}_{\alpha+i\beta,2}} = e^{\alpha t} \begin{pmatrix} \cos(\beta t) & \sin(\beta t) \\ -\sin(\beta t) & \cos(\beta t) \end{pmatrix}.$$

2.6 An example

In this section we shall determine e^{zA}, for each $z \in \mathbb{C}$, where A is the matrix introduced in (1.33). According to (1.37) and Proposition 2.5.1(d) we have that, for every $z \in \mathbb{C}$,

$$e^{zA} = P_{\mathbb{C}} e^{zJ_{\mathbb{C}}} P_{\mathbb{C}}^{-1},$$

where $J_{\mathbb{C}}$ and $P_{\mathbb{C}}$ are the matrices defined in (1.37) and (1.38), respectively. Moreover, owing to the results of Section 2.5,

$$e^{zJ_{\mathbb{C}}} = \operatorname{diag}\left\{e^{z\begin{pmatrix} 2-i & 0 \\ 1 & 2-i \end{pmatrix}}, e^{z\begin{pmatrix} 2+i & 0 \\ 1 & 2+i \end{pmatrix}}\right\}$$
$$= \begin{pmatrix} e^{(2-i)z} & 0 & 0 & 0 \\ ze^{(2-i)z} & e^{(2-i)z} & 0 & 0 \\ 0 & 0 & e^{(2+i)z} & 0 \\ 0 & 0 & ze^{(2+i)z} & e^{(2+i)z} \end{pmatrix}$$

and, therefore,

$$e^{zA} = P_{\mathbb{C}} \begin{pmatrix} e^{(2-i)z} & 0 & 0 & 0 \\ ze^{(2-i)z} & e^{(2-i)z} & 0 & 0 \\ 0 & 0 & e^{(2+i)z} & 0 \\ 0 & 0 & ze^{(2+i)z} & e^{(2+i)z} \end{pmatrix} P_{\mathbb{C}}^{-1} =$$
$$e^{2z} \begin{pmatrix} \cos z - z\sin z & z\cos z + \sin z & -\sin z & -z(\cos z - \sin z) \\ -(z+1)\sin z & (z+1)\cos z & -\sin z & -z\cos z + (z+2)\sin z \\ z(\cos z - \sin z) & z(\cos z + \sin z) & \cos z - \sin z & -2(z\cos z - \sin z) \\ -z\sin z & z\cos z & -\sin z & (1-z)\cos z + (z+1)\sin z \end{pmatrix}$$

for every $z \in \mathbb{C}$.

Alternatively, it follows from (1.39) that, for every $z \in \mathbb{C}$,

$$e^{zA} = P_{\mathbb{R}} e^{zJ_{\mathbb{R}}} P_{\mathbb{R}}^{-1} = P_{\mathbb{R}} \begin{pmatrix} e^{2z}\cos z & -e^{2z}\sin z & 0 & 0 \\ e^{2z}\sin z & e^{2z}\cos z & 0 & 0 \\ ze^{2z}\cos z & -ze^{2z}\sin z & e^{2z}\cos z & -e^{2z}\sin z \\ ze^{2z}\sin z & ze^{2z}\cos z & e^{2z}\sin z & e^{2z}\cos z \end{pmatrix} P_{\mathbb{R}}^{-1} =$$
$$e^{2z} \begin{pmatrix} \cos z - z\sin z & z\cos z + \sin z & -\sin z & -z(\cos z - \sin z) \\ -(z+1)\sin z & (z+1)\cos z & -\sin z & -z\cos z + (z+2)\sin z \\ z(\cos z - \sin z) & z(\cos z + \sin z) & \cos z - \sin z & -2(z\cos z - \sin z) \\ -z\sin z & z\cos z & -\sin z & (1-z)\cos z + (z+1)\sin z \end{pmatrix},$$

where $P_{\mathbb{R}}$ is the matrix defined in (1.40).

2.7 Exercises

1. Prove that every norm of $\mathbb{K}^d$ equips it with the structure of Banach space (complete normed space).

2. Prove that the norm (2.3) satisfies

$$\|A\|_{\mathcal{L}(\mathbb{K}^N)} = \max\{\|Au\|_{\mathbb{K}^N} \;:\; \|u\|_{\mathbb{K}^N} \le 1\} = \max\left\{\frac{\|Au\|_{\mathbb{K}^N}}{\|u\|_{\mathbb{K}^N}} \;:\; u \in \mathbb{K}^N \setminus \{0\}\right\}$$
$$= \min\{L \ge 0 \;:\; \|Au\|_{\mathbb{K}^N} \le L\|u\|_{\mathbb{K}^N}\}$$

for every $A \in \mathcal{M}_N(\mathbb{C})$.

3. Let $\|\cdot\|$ and $\|\cdot\|'$ be two norms in a given vector space U. Consider their subordinated norms $\|\cdot\|_{\mathcal{L}(U)}$ and $\|\cdot\|'_{\mathcal{L}(U)}$ in $\mathcal{L}(U)$, respectively. Show that $\|\cdot\|$ and $\|\cdot\|'$ are equivalent if and only if so are $\|\cdot\|_{\mathcal{L}(U)}$ and $\|\cdot\|'_{\mathcal{L}(U)}$.

4. Use induction on n to prove Lemma 2.2.3.

5. Use Lemma 2.3.2 to prove that the set of invertible matrices in $\mathcal{M}_N(\mathbb{K})$ is open, where $\mathbb{K} \in \{\mathbb{R}, \mathbb{C}\}$.

6. Calculate e^{tA}, for every $t \in \mathbb{R}$, where A is each one of the matrices of Exercise 1 of Chapter 1.

7. Calculate $\sin A$ and $\cos A$, where

$$A = \begin{pmatrix} 0 & 0 & 0 & 0 \\ 1 & 0 & 0 & 0 \\ 0 & 1 & 0 & 0 \\ 0 & 0 & 1 & 0 \end{pmatrix}.$$

8. Let

$$A = \begin{pmatrix} 0 & 0 & 1 \\ 1 & 0 & 0 \\ 0 & 1 & 0 \end{pmatrix}.$$

Calculate $\cos A$, $\sin A$ and e^A from the corresponding Taylor series. This is one of the few circumstances where the Jordan form does not simplify the necessary calculations.

9. Let $A \in \mathcal{M}_N(\mathbb{C})$. Prove that for each $\varepsilon > 0$ there exist an invertible matrix P, a diagonal matrix D and a nilpotent matrix R such that

$$P^{-1}AP = D + R \quad \text{and} \quad \|R\|_{\mathcal{L}(\mathbb{C}^N)} \le \varepsilon.$$

10. Let $A \in \mathcal{M}_N(\mathbb{C})$ and $\alpha, \beta \in \mathbb{R}$ be such that

$$\alpha < \operatorname{Re}\lambda < \beta \quad \text{for each} \quad \lambda \in \sigma(A).$$

Prove that there exists a basis $\{e_1, \dots, e_N\}$ of $\mathbb{C}^N$ such that

$$\alpha\, \|u\|^2 \le \operatorname{Re}\,\langle Au, u\rangle \le \beta\, \|u\|^2 \quad \text{for each} \quad u \in \mathbb{C}^N,$$

where, for each $u_j, v_j \in \mathbb{C}$ and $1 \le j \le N$, we have denoted

$$\left\langle \sum_{j=1}^{N} u_j e_j, \sum_{j=1}^{N} v_j e_j \right\rangle := \sum_{j=1}^{N} u_j \bar{v}_j \quad \text{and} \quad \| \sum_{j=1}^{N} u_j e_j \| := \left(\sum_{j=1}^{N} |u_j|^2 \right)^{1/2} .$$

11. For each $A \in \mathcal{M}_N(\mathbb{C})$, the *spectral radius* of A, usually denoted by $\operatorname{spr} A$, is the real number defined through

$$\operatorname{spr} A := \max \{ |\lambda| \; : \; \lambda \in \sigma(A) \} .$$

Prove the following properties:

- $\operatorname{spr} A \le \|A\|_{\mathcal{L}(\mathbb{C}^N)}$ for every norm $\| \cdot \|$ of $\mathbb{C}^N$.
- For each $\varepsilon > 0$ there exists a norm $\| \cdot \|$ of $\mathbb{C}^N$ such that

$$\|A\|_{\mathcal{L}(\mathbb{C}^N)} \le \operatorname{spr} A + \varepsilon .$$

- Let $\mathfrak{S}$ denote the set of norms $\| \cdot \|_{\mathcal{L}(\mathbb{C}^N)}$ of $\mathcal{L}(\mathbb{C}^N)$ that are subordinated to some norm of $\mathbb{C}^N$. Then,

$$\operatorname{spr} A = \inf \left\{ \|A\|_{\mathcal{L}(\mathbb{C}^N)} \; : \; \| \cdot \|_{\mathcal{L}(\mathbb{C}^N)} \in \mathfrak{S} \right\} .$$

12. Prove that, for all $A \in \mathcal{M}_N(\mathbb{C})$ and $z \in \mathbb{C}$,

$$\det e^{zA} = e^{z \operatorname{tr} A} .$$

The concept of trace was given in Exercise 10 of Chapter 1.

13. Pick $R > 0$ and $f \in \mathcal{H}(D_R)$; suppose $A \in \mathcal{M}_N(\mathbb{C})$ satisfies

$$\operatorname{spr} A < R .$$

Use the Jordan canonical form to prove that

$$\sigma(f(A)) = f(\sigma(A))$$

and that, actually, $f(\lambda)$ is an eigenvalue of $f(A)$ of multiplicity m if λ is an eigenvalue of A of multiplicity m. This result provides us with a sharp version of Theorem 2.4.2, though the proposed proof cannot be adapted to the infinite-dimensional setting.

14. Let $A \in \mathcal{M}_N(\mathbb{C})$ be an invertible matrix. Prove that

$$A = e^L$$

for some $L \in \mathcal{M}_N(\mathbb{C})$.

15. Suppose that $A \in \mathcal{M}_N(\mathbb{C})$ and $f \in \mathcal{F}(A)$ satisfy

$$f^{-1}(0) \cap \sigma(A) = \varnothing .$$

Prove that $f(A)$ is invertible.

16. Let $A \in \mathcal{M}_N(\mathbb{C})$ and an open set $\Omega \subset \mathbb{C}$ such that $\sigma(A) \subset \Omega$. Prove that there exists $\varepsilon > 0$ such that if $B \in \mathcal{M}_N(\mathbb{C})$ and $\|A - B\| \leq \varepsilon$, then $\sigma(B) \subset \Omega$.

17. Let $A \in \mathcal{M}_N(\mathbb{C})$ and $x \in \mathbb{C}^N$. Prove the following:

- The function $f : \mathbb{C} \to \mathcal{M}_N(\mathbb{C})$ defined by $f(z) = e^{zA}$ for $z \in \mathbb{C}$ is holomorphic, and $f'(z) = Af(z)$ for each $z \in \mathbb{C}$.
- For each $x \in \mathbb{C}^N$, the function $\varphi_x : \mathbb{C} \to \mathbb{C}^N$ defined by $\varphi_x(z) = e^{zA}x$ for $z \in \mathbb{C}$ provides us with the unique solution in $\mathcal{H}(\mathbb{C}, \mathbb{C}^N)$ of the initial-value problem

$$\begin{cases} u'(z) = Au(z), & z \in \mathbb{C}, \\ u(0) = x. \end{cases}$$

18. For each $1 \leq p \leq \infty$ and $A \in \mathcal{M}_N(\mathbb{C})$, let $\|A\|_p$ denote the norm of A subordinated to the norm $\|\cdot\|_p$ of $\mathbb{C}^N$. Set

$$A = (a_{ij})_{i,j=1}^N \in \mathcal{M}_N(\mathbb{C}).$$

The following properties are requested to be proved:

- $\|A\|_1 = \max_{1 \leq j \leq N} \sum_{i=1}^N |a_{ij}|$.
- $\|A\|_2 = [\mathrm{spr}(AA^*)]^{1/2}$, where A^* is the adjoint matrix of A, and spr denotes the spectral radius (see Exercise 3 of Chapter 1 and Exercise 11 above for the definitions, if necessary).
- $\|A\|_2 = \mathrm{spr}(A)$ if $A \in \mathcal{M}_N(\mathbb{C})$ is Hermitian.
- $\|A\|_\infty = \max_{1 \leq i \leq N} \sum_{j=1}^N |a_{ij}|$.
- $\|A\|_2 \leq \left(\sum_{i=1}^N \sum_{j=1}^N |a_{ij}|^2 \right)^{1/2} = \sqrt{\mathrm{tr}(AA^*)} \leq N \|A\|_2$.

19. Let $A \in \mathcal{M}_N(\mathbb{C}^N)$. Prove the equivalence of the following four conditions:

- $\lim_{n \to \infty} A^n = 0$.
- $\lim_{n \to \infty} A^n u = 0$ for each $u \in \mathbb{C}^N$.
- $\mathrm{spr}\, A < 1$ (see Exercise 11 above, if necessary).
- There exists a norm $\|\cdot\|$ in $\mathbb{C}^N$ for which

$$\|A\|_{\mathcal{L}(\mathbb{C}^N)} < 1,$$

where $\|\cdot\|_{\mathcal{L}(\mathbb{C}^N)}$ is the norm subordinated to $\|\cdot\|$.

Conclude that, for every norm $\|\cdot\|$ of $\mathbb{C}^N$,

$$\mathrm{spr}\, A = \lim_{n \to \infty} \|A^n\|_{\mathcal{L}(\mathbb{C}^N)}^{1/n}.$$

20. Let $b \in \mathbb{C}^N$ and $A \in \mathcal{M}_N(\mathbb{C})$ be such that $I - A$ is invertible, and, for every $u_0 \in \mathbb{C}$, let $\{u_n\}_{n\geq 0}$ denote the sequence of iterates defined by

$$u_{n+1} := Au_n + b, \qquad n \geq 0.$$

Prove the equivalence of the following conditions:

- The sequence $\{u_n\}_{n\geq 0}$ converges for all $u_0 \in \mathbb{C}$.
- For every $u_0 \in \mathbb{C}$,
$$\lim_{n\to\infty} u_n = (I - A)^{-1} b.$$
- $\operatorname{spr} A < 1$.

21. Let $A, B \in \mathcal{M}_N(\mathbb{C})$ be two Hermitian matrices, and let

$$\alpha_1 \leq \cdots \leq \alpha_N \qquad \text{and} \qquad \beta_1 \leq \cdots \leq \beta_N$$

be their respective systems of eigenvalues, counting according to their multiplicity. Prove that

$$\max_{1\leq n\leq N} |\alpha_n - \beta_n| \leq \|A - B\|_2,$$

where $\|\cdot\|_2$ denotes the operator norm subordinated to the norm $\|\cdot\|_2$ in $\mathbb{C}^N$.

22. Let $A \in \mathcal{M}_N(\mathbb{C})$, and consider

$$\omega := \max\{\operatorname{Re}\lambda \ : \ \lambda \in \sigma(A)\}.$$

This exercise suggests two different proofs of the fact that, for every $\varepsilon > 0$, there exists $M(\varepsilon) > 0$ such that

$$\|e^{tA}\|_{\mathcal{L}(\mathbb{C}^N)} \leq M(\varepsilon) e^{(\omega+\varepsilon)t} \quad \text{for all} \quad t > 0.$$

- Choose an appropriate Cauchy contour whose inner domain contains $\sigma(A)$, and use the Dunford integral formula to prove the result.
- Use, instead, the Jordan canonical form.

23. Let $A \in \mathcal{M}_N(\mathbb{C})$ and $\lambda \in \sigma(A)$. Use Exercise 5 of Chapter 1 to prove the existence of $\delta >$ and $R > 0$ with the property that for every $B \in \mathcal{M}_N(\mathbb{C})$ satisfying $\|A - B\| \leq \delta$, the matrix B has exactly $m_a(\lambda)$ eigenvalues in $D_R(\lambda)$, counting according to their algebraic multiplicities.

24. Show that assertion "The set of matrices whose spectrum consists of simple eigenvalues is dense in $\mathcal{M}_N(\mathbb{C})$" is nothing but a re-statement of the main result established by Exercise 6 of Chapter 1.

2.8 Comments on Chapter 2

This chapter contains standard material, though not really elementary, which can be found in a number of classic textbooks such as Taylor [122], Gantmacher [39], Dunford & Schwartz [27], Yosida [130], or Gohberg, Goldberg & Kaashoek [43].

Usually, operator calculus is introduced from the very beginning in the infinite-dimensional setting, which makes it difficult to teach it at an undergraduate level. In such a setting, f is a holomorphic function, A is a linear endomorphism of a Banach space, and the main goal is to analyze the properties of $f(A)$. Nevertheless, these general approaches do not differ substantially from the one adopted in this chapter, which goes back to the textbooks of López-Gómez [83, 84] and can be comfortably taught to undergraduate students. Other finite-dimensional approaches to operator calculus can be found in the books of Gohberg, Lancaster & Rodman [51], and Gantmacher [39]. In the latter, a very detailed and complete description is given.

Certainly, operator calculus is a very active field of research. So far, one of the major advances has been carried out for holomorphic functions f. Also well known is the operator calculus for continuous functions $f : \mathbb{R} \to \mathbb{R}$ and for selfadjoint operators A (see, for example, Gohberg, Goldberg & Kaashoek [43]). The analysis of the next chapters will give some answers to these and other related questions in the general case when $f \in \mathcal{C}^\infty(\mathbb{R}, \mathbb{R})$, not necessarily real analytic, and A is a Fredholm operator of index zero, not necessarily selfadjoint.

Lemma 2.3.2 is sometimes known as the Banach lemma. The series of (2.12) is called the Neumann series.

Exercise 17 is the starting point for the theory of linear differential equations. Classic books on this vast subject are Coddington & Levinson [21], Hartman [57], Hale [56], Arrowsmith & Place [7], and Amann [2], though this list is not exhaustive. Applications of the operator calculus to the theory of linear differential equations, from a viewpoint of matrix theory, can be found in Gantmacher [39].

Exercise 20 presents a paradigmatic iterative scheme for solving linear equations (see Ciarlet [20]).

A number of the proposed exercises were borrowed from Guzmán [55] and López-Gómez [83, 84].

Chapter 3

Spectral Projections

This chapter considers $A \in \mathcal{M}_N(\mathbb{C})$ and uses the Dunford integral formula to construct the spectral projections of $\mathbb{C}^N$ onto $N[(A-\lambda I)^{\nu(\lambda)}]$ for every $\lambda \in \sigma(A)$. It also shows that, for each $\lambda \in \sigma(A)$, the algebraic ascent $\nu(\lambda)$ equals the order of λ as a pole of the associated resolvent operator

$$\mathcal{R}(z;A) := (zI - A)^{-1}, \qquad z \in \varrho(A) := \mathbb{C} \setminus \sigma(A). \tag{3.1}$$

Precisely, this chapter is structured as follows. Section 3.1 gives a universal estimate for the norm of the inverse of a matrix in terms of its determinant and its norm. From this estimate it will become apparent that the eigenvalues of A are poles of the resolvent operator (3.1). The necessary analysis to show this feature will be carried out in Section 3.3. Section 3.2 gives a result on Laurent series valid for vector-valued holomorphic functions. Finally, Section 3.4 constructs the spectral projections associated with the direct sum decomposition (1.6), whose validity was established by the Jordan Theorem 1.2.1.

3.1 Estimating the inverse of a matrix

From now on, for any integer numbers $d, k \geq 1$ and $a_1, \dots, a_k \in \mathbb{C}^d$, we shall denote by

$$[a_1, \dots, a_k]$$

the matrix with k columns and d rows, whose k columns consist of the ordered column vectors $a_1, \dots, a_k$.

The next inequality estimates the norm of the inverse of a matrix in terms of the norm of the matrix and of its determinant.

Proposition 3.1.1. *Let $\|\cdot\|$ be a norm in $\mathcal{M}_N(\mathbb{C})$. Then there exists a constant $\gamma > 0$ such that, for every invertible matrix $A \in \mathcal{M}_N(\mathbb{C})$,*

$$\|A^{-1}\| \leq \gamma \frac{\|A\|^{N-1}}{|\det A|}.$$

Proof. Pick $v \in \mathbb{C}^N$, set $u := Av$, and let $a_1, \dots, a_N \in \mathbb{C}^N$ denote the columns of A. Then

$$A = [a_1, \dots, a_N].$$

Now we consider, for each $1 \leq j \leq N$, the value

$$b_j := \det[a_1, \dots, a_{j-1}, u, a_{j+1}, \dots, a_N].$$

By the Cramer rule, for each $1 \leq j \leq N$, the jth component v_j of the vector v satisfies

$$v_j = \frac{b_j}{\det A}.$$

Let $a_{j1}, \dots, a_{jN} \in \mathbb{C}$ denote the components of a_j for $1 \leq j \leq N$. By definition of determinant, for each $1 \leq j \leq N$, we have that

$$b_j = \sum_{\sigma \in \mathcal{P}_N} (-1)^{\operatorname{sign} \sigma} u_{\sigma(j)} \prod_{\substack{i=1 \\ i \neq j}}^{N} a_{i,\sigma(i)},$$

where $\mathcal{P}_N$ stands for the set of permutations of $\{1, \dots, N\}$, and the sign or parity of any $\sigma \in \mathcal{P}_N$ is denoted by $\operatorname{sign} \sigma$. Then, for each $1 \leq j \leq N$,

$$|b_j| \leq \sum_{\sigma \in \mathcal{P}_N} |u_{\sigma(j)}| \prod_{\substack{i=1 \\ i \neq j}}^{N} |a_{i,\sigma(i)}| \leq N! \, \|u\|_\infty \|A\|_\infty^{N-1}.$$

Therefore,

$$\|v\|_\infty \leq N! \frac{1}{|\det A|} \|u\|_\infty \|A\|_\infty^{N-1}. \tag{3.2}$$

Take $\|\cdot\|_{\mathbb{C}^N}$ any norm in $\mathbb{C}^N$, and let $\|\cdot\|_{\mathcal{L}(\mathbb{C}^N)}$ be its subordinated operator norm. By the equivalence of norms established by Theorem 2.1.3, it follows from (3.2) that there exists a constant $\gamma' > 0$ such that

$$\|A^{-1}u\|_{\mathbb{C}^N} \leq \gamma' \frac{1}{|\det A|} \|u\|_{\mathbb{C}^N} \|A\|_{\mathcal{L}(\mathbb{C}^N)}^{N-1},$$

whence

$$\|A^{-1}\|_{\mathcal{L}(\mathbb{C}^N)} \leq \gamma' \frac{1}{|\det A|} \|A\|_{\mathcal{L}(\mathbb{C}^N)}^{N-1}.$$

A further application of Theorem 2.1.3 concludes the proof. ☐

3.2 Vector-valued Laurent series

The theory of Laurent series is valid not only for complex-valued holomorphic functions, but also for complex vector-valued holomorphic functions. Although the general theory works in more general contexts, this chapter uses exclusively the following special case of it, valid for matrix-valued holomorphic functions on a perforated disk.

Theorem 3.2.1. *Suppose $N \geq 1$ is an integer, and*

$$f \in \mathcal{H}\left(D_R(z_0) \setminus \{z_0\}, \mathcal{M}_N(\mathbb{C})\right)$$

for some $R > 0$ and $z_0 \in \mathbb{C}$. Then there exists a unique sequence of matrices $\{L_n\}_{n\in\mathbb{Z}}$ in $\mathcal{M}_N(\mathbb{C})$ such that

$$f(z) = \sum_{n=-\infty}^{\infty} (z - z_0)^n L_n \quad \textit{for each} \quad z \in D_R(z_0) \setminus \{z_0\}, \tag{3.3}$$

uniformly for z in compact subsets of $D_R(z_0) \setminus \{z_0\}$. Moreover,

$$L_n = \frac{1}{2\pi i} \int_{C_r(z_0)} \frac{1}{(z - z_0)^{n+1}} f(z)\, dz$$

for each $r \in (0, R)$ and $n \in \mathbb{Z}$.

The series (3.3) is called the Laurent series of f at z_0, and its coefficient operators L_n, for $n \in \mathbb{Z}$, are the Laurent coefficients of f at z_0.

Proof. Define $\Omega := D_R(z_0) \setminus \{z_0\}$. For each $1 \leq i, j \leq N$ and $z \in \Omega$, let $f_{ij}(z)$ denote the (i, j)th entry of the matrix $f(z)$; fix such an (i, j). Then $f_{ij} \in \mathcal{H}(\Omega)$, and, hence, by the classic Laurent theorem, there exists a unique sequence of complex numbers $\{L_n^{ij}\}_{n\in\mathbb{Z}}$ such that

$$f_{ij}(z) = \sum_{n=-\infty}^{\infty} L_n^{ij} (z - z_0)^n \quad \text{for each} \quad z \in \Omega,$$

uniformly on compact subsets of Ω. Moreover, for each $r \in (0, R)$ and $n \in \mathbb{Z}$,

$$L_n^{ij} = \frac{1}{2\pi i} \int_{C_r(z_0)} \frac{1}{(z - z_0)^{n+1}} f_{ij}(z)\, dz.$$

For each $n \in \mathbb{Z}$, let $L_n \in \mathcal{M}_N(\mathbb{C})$ be the matrix whose (i, j)th entry equals L_n^{ij}. It is easy to see that the sequence $\{L_n\}_{n\in\mathbb{Z}}$ satisfies all the requirements of the theorem. The uniqueness is an immediate consequence of the uniqueness of the L_n^{ij}'s, once the entries in (3.3) are identified. □

Now, we can make the following concept precise.

Definition 3.2.2 (Pole). Suppose $N \geq 1$ is an integer, and

$$f \in \mathcal{H}(D_R(z_0) \setminus \{z_0\}, \mathcal{M}_N(\mathbb{C}))$$

for some $R > 0$ and $z_0 \in \mathbb{C}$. Then, it is said that z_0 is a pole of order $m \geq 1$ of f when the coefficients L_n, for $n \in \mathbb{Z}$, of the Laurent series (3.3) satisfy

$$L_{-m} \neq 0 \quad \text{and} \quad L_{-(m+k)} = 0 \quad \text{for each} \quad k \geq 1. \tag{3.4}$$

3.3 The eigenvalues are poles of the resolvent

The next result shows that the eigenvalues of any matrix $A \in \mathcal{M}_N(\mathbb{C})$ are poles of its resolvent operator (3.1), and it establishes some important relationships between the underlying Laurent coefficients. It should be remembered that, for any $\lambda \in \sigma(A)$, we denote by $m_a(\lambda)$ the algebraic multiplicity of λ.

Proposition 3.3.1. *Consider $A \in \mathcal{M}_N(\mathbb{C})$. Let $\lambda \in \sigma(A)$ and $R > 0$ be such that*

$$D_R(\lambda) \setminus \{\lambda\} \subset \varrho(A) = \mathbb{C} \setminus \sigma(A).$$

Let

$$L_n := \frac{1}{2\pi i} \int_{C_r(\lambda)} \frac{\mathcal{R}(z; A)}{(z-\lambda)^{n+1}}\, dz, \qquad n \in \mathbb{Z}, \quad 0 < r < R,$$

be the Laurent coefficients of $\mathcal{R}(\cdot; A)$ at λ, whose existence is guaranteed by Theorems 2.3.4 and 3.2.1. Then

$$L_{-(n+1)} = (A - \lambda I)^n L_{-1}, \qquad n \geq 0, \tag{3.5}$$

and there exists an integer $m = m(\lambda)$ with $1 \leq m(\lambda) \leq m_a(\lambda)$ for which (3.4) and

$$R[L_{-1}] \subset N[(A - \lambda I)^{m(\lambda)}]$$

are satisfied. In particular, λ is a pole of order $m(\lambda)$ of the resolvent operator $\mathcal{R}(\cdot; A)$.

Proof. By Proposition 3.1.1, there exists $\gamma > 0$ such that

$$\|\mathcal{R}(z; A)\| \leq \gamma \frac{\|zI - A\|^{N-1}}{|\det(zI - A)|} \qquad \text{for each} \quad z \in \varrho(A). \tag{3.6}$$

As the function

$$z \mapsto \|zI - A\|$$

is continuous and the set $\bar{D}_R(\lambda)$ is compact, there exists a constant $C_1 > 0$ such that

$$\gamma \|zI - A\|^{N-1} \leq C_1 \quad \text{for each} \quad z \in \bar{D}_R(\lambda).$$

Moreover, since λ is a root of order $m_a(\lambda)$ of the characteristic polynomial

$$z \mapsto \det(zI - A),$$

there exists a constant $C_2 > 0$ such that

$$|\det(zI - A)| \geq C_2 |z - \lambda|^{m_a(\lambda)} \quad \text{for each} \quad z \in \bar{D}_{\frac{R}{2}}(\lambda).$$

Thus, inserting these inequalities in (3.6) shows that there exists a constant $C_3 > 0$ for which

$$\|\mathcal{R}(z; A)\| \leq \frac{C_3}{|z - \lambda|^{m_a(\lambda)}} \qquad \text{for each} \quad z \in \bar{D}_{\frac{R}{2}}(\lambda) \setminus \{\lambda\}. \tag{3.7}$$

Therefore, for each $r \in (0, R/2)$ and $k \geq 1$, we find from (3.7) that

$$\begin{aligned} \|L_{-[m_a(\lambda)+k]}\| &\leq \frac{1}{2\pi} \int_{C_r(\lambda)} \frac{\|\mathcal{R}(z; A)\|}{|z - \lambda|^{-[m_a(\lambda)+k]+1}} dz \\ &\leq \frac{1}{2\pi} \int_{C_r(\lambda)} \frac{C_3}{|z - \lambda|^{-k+1}} dz = C_3 r^k \end{aligned}$$

and, consequently, passing to the limit as $r \searrow 0$ shows that

$$L_{-[m_a(\lambda)+k]} = 0 \quad \text{for all} \quad k \geq 1. \tag{3.8}$$

Now, we will check that

$$L_{-n} \neq 0 \quad \text{for some} \quad n \in \{1, \ldots, m_a(\lambda)\}. \tag{3.9}$$

The proof proceeds by contradiction. Suppose (3.9) does not hold. Then, using (3.8) shows that λ is a removable singularity of $\mathcal{R}(\cdot; A)$ and

$$\lim_{z \to \lambda} \mathcal{R}(z; A) = L_0.$$

Then

$$1 = \lim_{z \to \lambda} \det\left[\mathcal{R}(z; A)(zI - A)\right] = \det L_0 \det(\lambda I - A),$$

which shows that λ is not an eigenvalue of A. This contradiction proves (3.9). Obviously, combining (3.8) and (3.9) shows that λ is a pole of order m of the function $\mathcal{R}(\cdot; A)$ for some $1 \leq m \leq m_a(\lambda)$.

Now, by the definition of $\mathcal{R}(z; A)$, Theorem 3.2.1 implies that, for every $z \in D_R(\lambda) \setminus \{\lambda\}$,

$$I = (zI - A)\mathcal{R}(z; A) = [(z - \lambda)I + (\lambda I - A)] \sum_{n=-\infty}^{\infty} L_n (z - \lambda)^n$$

and, hence,

$$I = \sum_{n=-\infty}^{\infty} \left[L_{n-1} + (\lambda I - A) L_n \right] (z-\lambda)^n.$$

Therefore, by the uniqueness of the Laurent expansion as a consequence of Theorem 3.2.1, it is apparent that

$$0 = L_{n-1} + (\lambda I - A) L_n \quad \text{for all} \quad n \leq -1.$$

By induction, (3.5) is easily obtained. In particular,

$$(A - \lambda I)^m L_{-1} = L_{-(m+1)} = 0,$$

because λ is a pole of order m of $\mathcal{R}(\cdot; A)$. Equivalently,

$$R[L_{-1}] \subset N[(A - \lambda I)^m],$$

which concludes the proof. □

3.4 Spectral projections

This section constructs the *spectral projections* associated with each of the factors of the direct sum decomposition of $\mathbb{C}^N$ given by (1.6) in the Jordan Theorem 1.2.1. These projections are constructed through the Dunford integrals of the resolvent operator $\mathcal{R}(z; A)$ over a series of Cauchy contours surrounding each of the eigenvalues of the matrix, and no other eigenvalue. It should be remembered that δ_{ij} stands for the Kronecker delta, and that, given a Cauchy contour γ, we are denoting by D_γ its inner domain.

Theorem 3.4.1. *Let $A \in \mathcal{M}_N(\mathbb{C})$ be with $\sigma(A) = \{\lambda_1, \dots, \lambda_p\}$, and set*

$$L := \min_{1 \leq i < j \leq N} |\lambda_i - \lambda_j| > 0.$$

Then, the following properties are satisfied:

(a) *For every Cauchy contour γ such that $\sigma(A) \subset D_\gamma$,*

$$I = \frac{1}{2\pi i} \int_\gamma \mathcal{R}(z; A)\, dz.$$

(b) *For each $1 \leq j \leq p$, let $\mathcal{P}_j$ be the operator defined by*

$$\mathcal{P}_j := \frac{1}{2\pi i} \int_{C_r(\lambda_j)} \mathcal{R}(z; A)\, dz,$$

where $r \in (0, L)$ is arbitrary. Then,

$$\sum_{k=1}^{p} \mathcal{P}_k = I$$

and

$$\mathcal{P}_i \mathcal{P}_j = \delta_{ij} \mathcal{P}_j, \qquad i, j \in \{1, \dots, p\}.$$

(c) $\mathbb{C}^N = \bigoplus_{k=1}^{p} R[\mathcal{P}_k]$.

(d) *For each $1 \le j \le p$, the ascent $\nu(\lambda_j)$ of the eigenvalue λ_j equals the order of λ_j as a pole of $\mathcal{R}(\cdot; A)$. Moreover,*

$$R[\mathcal{P}_j] = N[(A - \lambda_j I)^{\nu(\lambda_j)}].$$

Remark 3.4.2. The operators $\mathcal{P}_1, \dots, \mathcal{P}_p$ are usually referred to as the *spectral projections* of the matrix A. According to Theorems 2.3.4 and 3.2.1, for each $1 \le j \le p$, the spectral projection $\mathcal{P}_j$ of $\mathbb{C}^N$ onto $N[(A - \lambda_j I)^{\nu(\lambda_j)}]$ equals the operator coefficient L_{-1} of the Laurent series of $\mathcal{R}(\cdot; A)$ at λ_j. As L_{-1} is usually referred to as the residue of the resolvent operator at λ_j, one might re-state the previous theorem by simply saying that *the spectral projections of a matrix are the residua of its resolvent operator at each of the eigenvalues of the matrix.*

Proof. Part (a) is a direct consequence of Theorem 2.3.6 applied to the constant polynomial 1.

Now we consider the operators $\mathcal{P}_1, \dots, \mathcal{P}_p$ defined in Part (b). According to Part (a) and Theorem 2.3.6, it is apparent that

$$\sum_{k=1}^{p} \mathcal{P}_k = \frac{1}{2\pi i} \int_{\gamma} \mathcal{R}(z; A)\, dz = I.$$

Now, fix $0 < \rho \le L/2$ and, for each $1 \le j \le p$, consider the function

$$f_j \; : \; \bigcup_{k=1}^{p} D_\rho(\lambda_k) \to \mathbb{C}$$

defined by

$$f_j(z) := \begin{cases} 1 & \text{if } \; z \in D_\rho(\lambda_j), \\ 0 & \text{if } \; z \in \bigcup_{k=1}^{p} D_\rho(\lambda_k) \setminus D_\rho(\lambda_j). \end{cases}$$

Clearly, $f_j \in \mathcal{F}(A)$ and

$$\mathcal{P}_j = \frac{1}{2\pi i} \int_{C_\rho(\lambda_j)} f_j(z) \mathcal{R}(z; A)\, dz = f_j(A), \qquad 1 \le j \le p.$$

In addition,

$$f_i f_j = \delta_{ij} f_i, \qquad 1 \le i, j \le p.$$

Thus, we find from Proposition 2.4.1 that $f_i f_j \in \mathcal{F}(A)$ and

$$\mathcal{P}_i \mathcal{P}_j = \frac{1}{2\pi i} \int_\gamma f_i(z) f_j(z) \mathcal{R}(z; A) dz = \delta_{ij} \mathcal{P}_i, \qquad 1 \le i, j \le p,$$

which concludes the proof of Part (b). So far, we have shown that the operators $\mathcal{P}_1, \dots, \mathcal{P}_p$ are disjoint projections whose sum equals the identity. In particular, this implies Part (c).

Finally, we will prove Part (d). For each $1 \le j \le p$, let m_j denote the order of λ_j as a pole of the resolvent $\mathcal{R}(\cdot; A)$. By Proposition 3.3.1 and Remark 3.4.2, we have that

$$1 \le m_j \le m_a(\lambda_j) \quad \text{and} \quad R[\mathcal{P}_j] \subset N[(\lambda_j I - A)^{m_j}], \qquad 1 \le j \le p,$$

and, hence,

$$R[\mathcal{P}_j] \subset N[(A - \lambda_j I)^{m_j}] \subset N[(A - \lambda_j I)^{\nu(\lambda_j)}], \quad 1 \le j \le p. \tag{3.10}$$

On the other hand, by Part (c), the Jordan Theorem 1.2.1 shows that

$$\mathbb{C}^N = \bigoplus_{j=1}^p R[\mathcal{P}_j] = \bigoplus_{j=1}^p N[(A - \lambda_j I)^{\nu(\lambda_j)}].$$

Consequently, the inclusions (3.10) must be equalities, and, therefore, according to the definition of algebraic ascent, it is apparent that

$$\nu(\lambda_j) = m_j \quad \text{for every} \quad 1 \le j \le p,$$

which concludes the proof. □

3.5 Exercises

1. Prove that, necessarily, $\alpha = N - 1$ if there exist $\gamma, \alpha \in \mathbb{R}$ satisfying

$$\|A^{-1}\| \le \gamma \frac{\|A\|^\alpha}{|\det A|}$$

for all invertible matrices $A \in \mathcal{M}_N(\mathbb{C})$.

2. The purpose of this exercise is to give an alternative proof of Theorem 3.4.1, based on the Jordan canonical form. Let $A \in \mathcal{M}_N(\mathbb{C})$ be with

$$\sigma(A) = \{\lambda_1, \dots, \lambda_p\},$$

and let J denote its Jordan canonical form constructed in Section 1.3. It is already known that $A = P^{-1}JP$ for some invertible matrix $P \in \mathcal{M}_N(\mathbb{C})$, and that J can be expressed in the form

$$J = \operatorname{diag}\{J_1, \dots, J_p\},$$

where, for each $1 \le j \le p$,

$$J_j = \operatorname{diag}\{J_{j,i_1}, \dots, J_{j,i_{n_j}}\},$$

with

$$\nu(\lambda_j) = \max\{i_h : 1 \le h \le n_j\},$$

and for each natural number $n \ge 1$, the matrix $J_{j,n}$ is a Jordan block of the form (1.26). Prove successively the following:

- For each $z \in \varrho(A)$,

$$\mathcal{R}(z; A) = P\mathcal{R}(z; J)P^{-1}, \quad \mathcal{R}(z; J) = \operatorname{diag}\{\mathcal{R}(z; J_1), \dots, \mathcal{R}(z; J_p)\},$$

  and

$$\mathcal{R}(z; J_j) = \operatorname{diag}\{\mathcal{R}(z; J_{j,i_1}), \dots, \mathcal{R}(z; J_{j,i_{n_j}})\}$$

  for all $1 \le j \le p$.
- For each $1 \le k \le p$ and every Cauchy contour γ_k satisfying

$$\gamma_k \cap \sigma(A) = \varnothing \quad \text{and} \quad \sigma(A) \cap D_{\gamma_k} = \{\lambda_k\},$$

  the following identities are satisfied:

$$\frac{1}{2\pi i}\int_{\gamma_k} \mathcal{R}(z; A)\, dz = P\frac{1}{2\pi i}\int_{\gamma_k} \mathcal{R}(z; J)\, dz\, P^{-1},$$

$$\frac{1}{2\pi i}\int_{\gamma_k} \mathcal{R}(z; J)\, dz = \operatorname{diag}\left\{\frac{1}{2\pi i}\int_{\gamma_k} \mathcal{R}(z; J_1)\, dz, \dots, \frac{1}{2\pi i}\int_{\gamma_k} \mathcal{R}(z; J_p)\, dz\right\},$$

  and, for each $1 \le j \le p$,

$$\frac{1}{2\pi i}\int_{\gamma_k} \mathcal{R}(z; J_j)\, dz = \operatorname{diag}\left\{\frac{1}{2\pi i}\int_{\gamma_k} \mathcal{R}(z; J_{j,i_1})\, dz, \dots, \frac{1}{2\pi i}\int_{\gamma_k} \mathcal{R}(z; J_{j,i_{n_j}})\, dz\right\}.$$

- For each $1 \le j \le p$ and $1 \le h \le n_j$,

$$\mathcal{R}(z; J_{j,h}) = \operatorname{trng}\left\{(z-\lambda_j)^{-1}, \dots, (z-\lambda_j)^{-h}\right\}.$$

- For each $1 \le j, k \le p$, the matrix

$$\frac{1}{2\pi i}\int_{\gamma_k} \mathcal{R}(z; J_j)\, dz$$

  is the identity matrix if $j = k$, and is the zero matrix if $j \ne k$. Moreover,

$$\sum_{k=1}^{p} \frac{1}{2\pi i}\int_{\gamma_k} \mathcal{R}(z; J)\, dz = I.$$

- For each $1 \leq k \leq p$ define

$$\hat{\mathcal{P}}_k = \frac{1}{2\pi i} \int_{\gamma_k} \mathcal{R}(z; J)\, dz.$$

Then, $\hat{\mathcal{P}}_1, \dots, \hat{\mathcal{P}}_p$ are mutually disjoint projections whose sum is the identity. Moreover, for each $1 \leq k \leq p$ the ascent $\nu(\lambda_k)$ of the eigenvalue λ_k equals its order as a pole of $\mathcal{R}(\cdot; J)$, and

$$R[\hat{\mathcal{P}}_k] = N[(J - \lambda_k I)^{\nu(\lambda_k)}].$$

- For each $1 \leq k \leq p$ define

$$\mathcal{P}_k = \frac{1}{2\pi i} \int_{\gamma_k} \mathcal{R}(z; A)\, dz.$$

Then, $\mathcal{P}_1, \dots, \mathcal{P}_p$ are mutually disjoint projections whose sum is the identity. Moreover, for each $1 \leq k \leq p$ the ascent $\nu(\lambda_k)$ of the eigenvalue λ_k equals its order as a pole of $\mathcal{R}(\cdot; A)$, and

$$R[\mathcal{P}_k] = N[(A - \lambda_k I)^{\nu(\lambda_k)}].$$

3.6 Comments on Chapter 3

Although this chapter contains standard material, its approach differs substantially from the one adopted in classic textbooks, such as Gohberg & Kreĭn [45], Taylor [122], Dunford & Schwartz [27], Yosida [130], Kato [65], and Gohberg, Goldberg & Kaashoek [43], where the fact that the eigenvalues are poles of the resolvent is proved once the spectral projections are constructed. Actually, in most of the available texts the Jordan theorem is derived from the properties of the spectral projections, and at a later step it is shown that any eigenvalue must be a pole of the resolvent. This chapter has adopted the converse methodology. The use of the elementary Proposition 3.1.1, attributable to Kato [65], to prove from the very beginning that any eigenvalue is a pole of the resolvent, simplifies tremendously the mathematical apparatus required to cover these topics. As a by-product, it facilitates the teaching of these topics at an undergraduate level. Nevertheless, this is not the unique advantage of the approach chosen in this chapter. Indeed, the fact that all spectral properties can be easily obtained from the fact that the eigenvalues are poles of the resolvent suggests that for general one-parameter families of operators $\mathfrak{L}$, not necessarily of the special form

$$\mathfrak{L}(z) = zI - A, \qquad z \in \mathbb{K},$$

most of their local spectral properties at any singular value $z_0 \in \mathbb{K}$ should follow, by some method, by simply imposing the growth condition

$$\|\mathfrak{L}(z)^{-1}\| \leq \frac{C}{|z - z_0|^{\nu}}, \qquad z \sim z_0, \;\; z \neq z_0,$$

even if $\mathfrak{L}$ is not real analytic. This crucial observation is on the foundations of the general abstract theory that is going to be developed in the next chapters. It actually triggered it from the very beginning.

In most textbooks it is usual to introduce the spectral projections in the infinite-dimensional setting for the special case when A is compact and, possibly, selfadjoint. Although this introductory chapter has dealt exclusively with the finite-dimensional setting, in Chapter 9 we will generalize these results to cover the general case when $\mathfrak{L}$ is a one-parameter family of Fredholm operators of index zero.

As in Chapters 1 and 2, the approach to this chapter has been based on the books by López-Gómez [83, 84].

Exercise 2 is based on the finite-dimensional approach to operator calculus given in the book by Gohberg, Lancaster & Rodman [50].

The theory of Laurent series also holds for vector-valued holomorphic functions. In the book by Gohberg, Goldberg & Kaashoek [43] a general way of reasoning to prove results on vector-valued functions from the corresponding ones for scalar-valued functions is explained.

Applications of the Laurent series to linear ordinary differential equations, from the point of view of matrix theory, can be found in Gantmacher [39].

Part II

Algebraic Multiplicities

Part II: Summary

Banach spaces and linear bounded operators

Throughout Part II, we consider the field $\mathbb{K} \in \{\mathbb{R}, \mathbb{C}\}$, and $U, V, W, X, Y \dots$ will denote *non-zero* $\mathbb{K}$*-Banach spaces.* Now we introduce the main classes of linear bounded operators used all through the book. Given two $\mathbb{K}$-Banach spaces U and V, the set $\mathcal{L}(U,V)$ denotes the $\mathbb{K}$-Banach space of the linear bounded operators from U to V, and $\mathcal{L}(U) := \mathcal{L}(U,U)$. As U and V are Banach spaces, they are equipped with norms. Both norms are always denoted by $\|\cdot\|$, as well as the operator norm subordinated to those norms in $\mathcal{L}(U,V)$.

For any $T \in \mathcal{L}(U,V)$, the sets $N[T]$ and $R[T]$ stand for the *null space* (kernel) and the *range* (image) of T, respectively, and the *rank* of T is defined by

$$\operatorname{rank} T := \dim R[T].$$

An operator $T \in \mathcal{L}(U,V)$ is said to be a *finite rank* operator if

$$\operatorname{rank} T < \infty.$$

An operator $P \in \mathcal{L}(U)$ is said to be a *projection* if $P^2 = P$. In such a case, P is said to project U onto $R[P]$.

An operator $T \in \mathcal{L}(U,V)$ is said to be a *Fredholm operator* when

$$\dim N[T] < \infty \quad \text{and} \quad \operatorname{codim} R[T] < \infty.$$

In such a case, $R[T]$ must be closed. Throughout Part II, we denote by $\operatorname{Fred}(U,V)$ the set of all Fredholm operators in $\mathcal{L}(U,V)$, and $\operatorname{Fred}(U) := \operatorname{Fred}(U,U)$. Moreover, for every $T \in \operatorname{Fred}(U,V)$, the *index* of T is defined through

$$\operatorname{ind} T := \dim N[T] - \operatorname{codim} R[T].$$

The set of *Fredholm operators of index zero* of $\mathcal{L}(U,V)$ is denoted by $\operatorname{Fred}_0(U,V)$. Consequently, $T \in \operatorname{Fred}_0(U,V)$ if and only if

$$T \in \mathcal{L}(U,V) \quad \text{and} \quad \dim N[T] = \operatorname{codim} R[T] < \infty.$$

We also denote $\operatorname{Fred}_0(U) := \operatorname{Fred}_0(U,U)$. The identity operator of U, always denoted by I_U, lies in $\operatorname{Fred}_0(U)$, of course. But $0 \notin \operatorname{Fred}(U,V)$ if U or V are infinite dimensional. In most of Part II we will assume that $\operatorname{Fred}_0(U,V) \neq \varnothing$; then U and V must be isomorphic. Therefore, assuming $U = V$ is far from being a serious restriction. In such a case, the most paradigmatic subset of $\operatorname{Fred}_0(U)$ consists of the set of compact perturbations of the identity I_U. An operator $T \in \mathcal{L}(U,V)$ is said to be *compact* if the closure $\overline{T(B)}$ is a compact subset of V for all bounded sets $B \subset U$. We denote by $\mathcal{K}(U,V)$ the subset of $\mathcal{L}(U,V)$ formed by all compact operators. It is well known that $I_U \in \mathcal{K}(U)$ if and only if $\dim U < \infty$. Another important subset of $\operatorname{Fred}_0(U,V)$ is the set $\operatorname{Iso}(U,V)$ consisting of all the isomorphisms from U to V. As usual,

$$\mathcal{K}(U) := \mathcal{K}(U,U) \quad \text{and} \quad \operatorname{Iso}(U) := \operatorname{Iso}(U,U).$$

Continuous and smooth maps

Throughout Part II, given $\mathbb{K} \in \{\mathbb{R}, \mathbb{C}\}$, an open subset $\Omega \subset \mathbb{K}$, and a $\mathbb{K}$-Banach space W, we denote by $\mathcal{C}(\Omega, W)$ the set of continuous maps from Ω to W, and, more generally, for each integer $r \geq 1$, the set $\mathcal{C}^r(\Omega, W)$ stands for the subset of $\mathcal{C}(\Omega, W)$ of those maps having a continuous rth differential. We also consider

$$\mathcal{C}^0(\Omega, W) := \mathcal{C}(\Omega, W), \qquad \mathcal{C}^\infty(\Omega, W) := \bigcap_{r \in \mathbb{N}} \mathcal{C}^r(\Omega, W),$$

and denote by $\mathcal{C}^\omega(\Omega, W)$ the set of analytic maps from Ω to W. It should be noted that if $\mathbb{K} = \mathbb{C}$, then

$$\mathcal{C}^1(\Omega, W) = \mathcal{C}^\omega(\Omega, W) = \mathcal{H}(\Omega, W),$$

all these sets being the space of holomorphic maps from Ω to W. Finally, $\mathcal{C}^r(\Omega)$ will denote $\mathcal{C}^r(\Omega, \mathbb{K})$, for each $r \in \mathbb{N} \cup \{\infty, \omega\}$.

Given $\lambda_0 \in \Omega$, a map $f : \Omega \setminus \{\lambda_0\} \to W$, and $k \in \mathbb{Z}$, it is said that

$$f(\lambda) = o((\lambda - \lambda_0)^k) \quad \text{as } \lambda \to \lambda_0$$

if

$$\lim_{\lambda \to \lambda_0} \frac{f(\lambda)}{(\lambda - \lambda_0)^k} = 0.$$

When λ_0 is clear from the context, it is simply said that

$$f(\lambda) = o((\lambda - \lambda_0)^k).$$

Given $\lambda_0 \in \Omega$, a map $f \in \mathcal{C}^r(\Omega, W)$ for some $r \in \mathbb{N} \cup \{\infty\}$, and an integer $0 \leq k \leq r$, we will write

$$\operatorname{ord}_{\lambda_0} f := k$$

when $f^{j)}(\lambda_0) = 0$ for all $0 \leq j < k$ and $f^{k)}(\lambda_0) \neq 0$, while

$$\operatorname{ord}_{\lambda_0} f := \infty$$

if $r = \infty$ and $f^{j)}(\lambda_0) = 0$ for all $j \geq 0$. The expression $\operatorname{ord}_{\lambda_0} f$ will be read: *the order of f at λ_0*. We leave $\operatorname{ord}_{\lambda_0} f$ undefined when, for some integer $k \geq 0$, the function f is not of class $\mathcal{C}^{k+1}$ in any neighborhood of λ_0, and $f^{j)}(\lambda_0)$ exists and equals zero for all $0 \leq j \leq k$.

For the sake of notation in Taylor developments, throughout this part, from its beginning to its end, given $\lambda_0 \in \Omega$, a natural number k and $f \in \mathcal{C}^k(\Omega, W)$, we use the notation

$$f_k := \frac{1}{k!} f^{k)}(\lambda_0).$$

It should not cause any confusion, because λ_0 will always be fixed beforehand. In particular, we will subsequently set

$$f_0 := f(\lambda_0),$$

if there is no ambiguity. *It is important to keep this notation in mind from now on, as it will be used throughout the remainder of the book.*

Operator families and generalized spectrum

Throughout Part II, given two $\mathbb{K}$-Banach spaces U, V, and $r \in \mathbb{N} \cup \{\infty, \omega\}$, any map

$$\mathfrak{L} \in \mathcal{C}^r(\Omega, \mathcal{L}(U, V))$$

will be referred to as an *operator family* of class $\mathcal{C}^r$ in Ω from U to V. They will be denoted hereafter by $\mathfrak{L}, \mathfrak{M}, \hat{\mathfrak{L}} \ldots$ Among them, polynomial operator families will play a pivotal role. A map $\Phi \in \mathcal{C}^\omega(\mathbb{K}, \mathcal{L}(U, V))$ is said to be a *polynomial operator family* if there exist $d \in \mathbb{N}$ and $F_0, \ldots F_d \in \mathcal{L}(U, V)$ such that

$$\Phi(\lambda) = \sum_{j=0}^{d} \lambda^j F_j, \qquad \lambda \in \mathbb{K}.$$

The set of all polynomial operator families from U to V will be denoted by $\mathbb{P}[\mathcal{L}(U, V)]$, and $\mathbb{P}[\mathcal{L}(U)] := \mathbb{P}[\mathcal{L}(U, U)]$. Usually, polynomial operator families will be named by Greek capital letters $\Phi, \Psi, \Upsilon \ldots$

The product of two operator families is defined in the natural way by means of their composition. Precisely, if Ω and Ω' are two open subsets of $\mathbb{K}$ and

$$\mathfrak{L} : \Omega \to \mathcal{L}(U, V), \qquad \mathfrak{M} : \Omega' \to \mathcal{L}(W, U)$$

are two arbitrary maps, then the product map

$$\mathfrak{L}\mathfrak{M} : \Omega \cap \Omega' \to \mathcal{L}(W, V)$$

is defined by

$$\mathfrak{L}\mathfrak{M}(\lambda) := \mathfrak{L}(\lambda) \circ \mathfrak{M}(\lambda), \qquad \lambda \in \Omega \cap \Omega',$$

though, in practice, the composition symbol $\circ$ will not be written.

Given an operator family $\mathfrak{L} \in \mathcal{C}(\Omega, \mathcal{L}(U, V))$, the point $\lambda_0 \in \Omega$ is said to be a *singular value* of $\mathfrak{L}$ if

$$\mathfrak{L}_0 := \mathfrak{L}(\lambda_0) \notin \operatorname{Iso}(U, V),$$

while it is said to be an *eigenvalue* of $\mathfrak{L}$ if

$$\dim N[\mathfrak{L}_0] \geq 1.$$

The *spectrum* of $\mathfrak{L}$, denoted by $\Sigma(\mathfrak{L})$, is defined as the set of singular values of $\mathfrak{L}$, while $\operatorname{Eig}(\mathfrak{L})$ stands for the set of eigenvalues of $\mathfrak{L}$. Obviously,

$$\operatorname{Eig}(\mathfrak{L}) \subset \Sigma(\mathfrak{L}).$$

Thanks to the open mapping theorem, in the special case when $\mathfrak{L}_0 \in \operatorname{Fred}_0(U, V)$, one has that $\lambda_0 \in \Sigma(\mathfrak{L})$ if and only if $\lambda_0 \in \operatorname{Eig}(\mathfrak{L})$. Consequently, if $\mathfrak{L}(\Omega) \subset \operatorname{Fred}_0(U, V)$, then $\Sigma(\mathfrak{L}) = \operatorname{Eig}(\mathfrak{L})$. Such concepts should not be confused with the classic concept of *spectrum* of the operator $\mathfrak{L}(\lambda)$, denoted by $\sigma(\mathfrak{L}(\lambda))$, for a particular value $\lambda \in \Omega$. By definition, for every $T \in \mathcal{L}(U)$,

$$\sigma(T) = \Sigma(\mathfrak{L}^T),$$

where

$$\mathfrak{L}^T(\lambda) := \lambda I_U - T, \quad \lambda \in \mathbb{K}.$$

Operations and order in $\mathbb{N} \cup \{\infty, \omega\}$

We will sometimes use the sets $\mathbb{N} \cup \{\infty\}$ and $\mathbb{N} \cup \{\infty, \omega\}$; for example, to consider maps of class $\mathcal{C}^r(\Omega, W)$ for some r belonging to $\mathbb{N} \cup \{\infty\}$ or to $\mathbb{N} \cup \{\infty, \omega\}$. Consequently, it is appropriate to fix the following conventions. The sum, difference and total order in $\mathbb{N} \cup \{\infty\}$ are the usual ones. The sum, difference and partial order in $\mathbb{N} \cup \{\infty, \omega\}$ are defined so that $\mathbb{N} \cup \{\infty\}$ preserves them, and, in addition, for all $k \in \mathbb{N}$,

$$k + \omega = \omega, \quad k < \omega, \quad \omega - k = \omega.$$

Also, ω becomes a synonym for ∞ when it is used for labeling sequences, or it indicates cardinality, so that the following expressions have the same meaning:

$$\{x_n\}_{n \in \mathbb{N}}, \qquad \{x_n\}_{n=0}^{\infty}, \qquad \{x_n\}_{n=0}^{\omega}.$$

Main general goal

Essentially, classic spectral theory studies holomorphic operator families

$$\mathfrak{L} \in \mathcal{H}(\mathbb{C}, \mathcal{M}_N(\mathbb{C}))$$

of the special type

$$\mathfrak{L}(\lambda) = \lambda I_{\mathbb{C}^N} - A, \qquad \lambda \in \mathbb{C},$$

for a given $A \in \mathcal{M}_N(\mathbb{C})$. These are the families already studied in Part I. More generally, the Riesz–Fredholm theory deals with holomorphic families $\mathfrak{L} \in \mathcal{H}(\mathbb{C}, \mathcal{L}(U))$ of the form

$$\mathfrak{L}(\lambda) = \lambda I_U - T, \qquad \lambda \in \mathbb{C},$$

for a given $T \in \mathcal{L}(U)$, where U is a complex Banach space. In particular, it analyzes the concepts of algebraic multiplicity, ascent... at any eigenvalue $\lambda_0 \in \sigma(T)$, and studies their main properties.

Part II introduces a generalized concept of algebraic multiplicity at every $\lambda_0 \in \mathrm{Eig}(\mathfrak{L})$ for a general operator family $\mathfrak{L} \in \mathcal{C}^r(\Omega, \mathcal{L}(U, V))$, for some $r \in \mathbb{N}$ such that $\mathfrak{L}_0 \in \mathrm{Fred}_0(U, V)$, and studies its most relevant properties; among them, the *product formula* (see Chapter 5). Actually, the product formula and a certain normalization condition determine the algebraic multiplicity uniquely (see Chapter 6). As a by-product, we will get an extremely simple axiomatic for all existing concepts of algebraic multiplicity available in the specialized literature. According to that axiomatic, it will become apparent that the generalized concept of algebraic multiplicity introduced in Part II indeed extends all classic concepts of algebraic multiplicity. The most significant consequences of the theory developed in this part will be discussed throughout each of the chapters forming it.

Prerequisites

The reading of Part II requires more mathematical background than the one necessary for Part I. But not that much, as only the most basic concepts and results of Functional Analysis, such as the Hahn–Banach theorem, are needed. In addition, the knowledge of the most basic features of the theory of (bounded) Fredholm operators is required. The following list collects the main results that will be invoked throughout Part II:

- If $L \in \mathrm{Fred}_0(U,V)$ and $T \in \mathcal{K}(U,V)$, then $L+T \in \mathrm{Fred}_0(U,V)$.
- If $L \in \mathrm{Fred}_0(U,V)$ and $M \in \mathrm{Fred}_0(W,U)$, then $LM \in \mathrm{Fred}_0(W,V)$.
- $\mathrm{Fred}_0(U,V)$ is an open subset of $\mathcal{L}(U,V)$.
- For each $L \in \mathrm{Fred}_0(U,V)$ there exists a finite-rank operator $F \in \mathcal{L}(U,V)$ such that $L+F \in \mathrm{Iso}(U,V)$.
- The Fredholm index defines a continuous map
$$T \mapsto \operatorname{ind} T$$
from $\mathrm{Fred}(U,V)$ to $\mathbb{Z}$.

Two good sources for the theory of Fredholm operators are the review paper by Gohberg & Kreĭn [45] and the comprehensive treatise of Kato [65]. The book of Gohberg, Goldberg & Kaashoek [43] also provides an elementary introduction to that theory.

Although the Riesz–Fredholm theory of compact operators and their algebraic multiplicities is useful to understand why the generalized concept of algebraic multiplicity introduced in Part II is indeed a generalization of the classic concept of multiplicity for a compact operator, some of its most significant results will follow in a series of corollaries of the abstract theory developed herein. Consequently, no previous knowledge of it is required.

Except in Section 11.3, where certain features about non-bounded operators will be discussed, throughout Part II, by an *operator* is always meant a linear and bounded operator.

Structure

Part II is divided into eight chapters. Chapters 4 and 5 construct two different but equivalent algebraic multiplicities. Chapter 6 gives an axiomatic for the algebraic multiplicity. Chapter 7 characterizes the existence of the local Smith form of a family at a singular value. Chapter 8 focuses attention on holomorphic and classical operator families, obtaining the most important results of the classic theory of Riesz–Fredholm. Chapter 9 constructs finite Laurent developments for the inverses of wide classes of non-analytic families, and ascertains some important properties concerning the logarithmic residues. Chapter 10 gives a proof of the spectral theorem for matrix polynomials. Finally, in Chapter 11 we describe some further developments of the algebraic multiplicity to enhance further research in some related areas of great interest.

Chapter 4

Algebraic Multiplicity Through Transversalization

Throughout this chapter we will consider $\mathbb{K} \in \{\mathbb{R}, \mathbb{C}\}$, two $\mathbb{K}$-Banach spaces U and V, an open subset $\Omega \subset \mathbb{K}$, a point $\lambda_0 \in \Omega$, and a family

$$\mathfrak{L} \in \mathcal{C}^r(\Omega, \mathcal{L}(U, V)),$$

for some $r \in \mathbb{N} \cup \{\infty\}$, such that

$$\mathfrak{L}_0 := \mathfrak{L}(\lambda_0) \in \operatorname{Fred}_0(U, V).$$

When $\lambda_0 \in \operatorname{Eig}(\mathfrak{L})$, the point λ_0 is said to be an *algebraic eigenvalue* of $\mathfrak{L}$ if there exist $\delta, C > 0$ and $m \geq 1$ such that, for each $0 < |\lambda - \lambda_0| < \delta$, the operator $\mathfrak{L}(\lambda)$ is an isomorphism and

$$\|\mathfrak{L}(\lambda)^{-1}\| \leq \frac{C}{|\lambda - \lambda_0|^m}.$$

The main goal of this chapter is to introduce the concept of algebraic multiplicity of $\mathfrak{L}$ at any algebraic eigenvalue λ_0. This algebraic multiplicity will be denoted by $\chi[\mathfrak{L}; \lambda_0]$, and will be defined through the auxiliary concept of *transversal eigenvalue.* Such concept will be motivated in Section 4.1 and will be formally defined in Section 4.2. Essentially, λ_0 is a transversal eigenvalue of $\mathfrak{L}$ when it is an algebraic eigenvalue for which the perturbed eigenvalues $a(\lambda) \in \sigma(\mathfrak{L}(\lambda))$ from $0 \in \sigma(\mathfrak{L}_0)$, as λ moves from λ_0, can be determined through standard perturbation techniques; these perturbed eigenvalues $a(\lambda)$ are those satisfying $a(\lambda_0) = 0$. This feature will be clarified in Sections 4.1 and 4.4, where we study the behavior of the eigenvalue $a(\lambda)$ and its associated eigenvector in the special case when 0 is a simple eigenvalue of $\mathfrak{L}_0$. In such a case, the multiplicity of $\mathfrak{L}$ at λ_0 equals the order of the function a at λ_0. Section 4.3 shows that every operator family $\mathfrak{L}$ satisfying the general requirements of this chapter can be *tranversalized* through a polynomial family $\Phi \in \mathbb{P}[\mathcal{L}(U,V)]$ with $\Phi_0 \in \mathrm{Iso}(U,V)$, provided λ_0 is an algebraic eigenvalue of $\mathfrak{L}$; that means that the product family $\mathfrak{L}\Phi$ is transversal at λ_0. The algebraic multiplicity of $\mathfrak{L}$ at λ_0 is then defined through $\chi[\mathfrak{L}\Phi;\lambda_0]$. The consistency of this concept of multiplicity relies upon the fact that it does not depend on the local transversalizing family Φ.

4.1 Motivating the concept of transversality

Originally, the interest of the concept of transversality goes back to the fact that *transversality makes the resolution of a special class of triangular systems trivial.* Although the precise meaning of this claim will not become apparent until Chapters 7 and 9, we consider now a simple example to illustrate it. The deepest significance of the concept of transversal eigenvalue will be revealed later; through, e.g., Proposition 7.1.3 and the results of Section 7.2, where it will be made apparent that, at transversal eigenvalues, the structure of the *canonical sets of Jordan chains* are themselves *canonical.* According to Theorem 4.3.2, which establishes the existence of transversalizing isomorphisms at any eigenvalue λ_0, it seems rather superfluous to give non-trivial examples for transversality (see Theorem 5.3.1 as well). Indeed, by Theorem 4.3.2, through the appropriate local change of coordinates at λ_0, any operator family can be transversalized.

Suppose we are given a natural number $k \geq 0$ and $k+1$ operators

$$\mathfrak{L}_0, \dots, \mathfrak{L}_k \in \mathcal{L}(U,V),$$

and consider the homogeneous linear system

$$\sum_{i=0}^{j} \mathfrak{L}_i u_{j-i} = 0, \qquad 0 \leq j \leq k, \tag{4.1}$$

in the unknowns $u_0, \dots, u_k \in U$. The next lemma reveals that under a special condition, closely related to the concept of transversality, system (4.1) becomes the uncoupled system

$$\mathfrak{L}_i u_{j-i} = 0, \qquad 0 \leq i \leq j \leq k.$$

Lemma 4.1.1. *Let $k \in \mathbb{N}$ and $\mathfrak{L}_0, \dots, \mathfrak{L}_k \in \mathcal{L}(U,V)$ be such that the spaces*

$$R[\mathfrak{L}_0],\ \mathfrak{L}_1(N[\mathfrak{L}_0]),\ \dots,\ \mathfrak{L}_k\Big(\bigcap_{j=0}^{k-1} N[\mathfrak{L}_j]\Big) \tag{4.2}$$

are in direct sum. Then, (4.1) *is satisfied if and only if*

$$u_j \in N[\mathfrak{L}_0] \cap \dots \cap N[\mathfrak{L}_{k-j}], \qquad 0 \leq j \leq k. \tag{4.3}$$

Proof. Obviously, (4.3) implies (4.1). To prove the converse implication, we will proceed by induction on k. Suppose (4.1). Then, $\mathfrak{L}_0 u_0 = 0$ and, hence, (4.3) holds true for $k = 0$. Assume that the facts

$$(y_0, \dots, y_{k-1}) \in U^k \quad \text{and} \quad \sum_{i=0}^{j} \mathfrak{L}_i y_{j-i} = 0, \qquad 0 \le j \le k-1,$$

imply

$$y_j \in \bigcap_{i=0}^{k-1-j} N[\mathfrak{L}_i], \qquad 0 \le j \le k-1.$$

Then, applying this induction hypothesis to $(u_0, \dots, u_{k-1})$ shows that

$$u_j \in \bigcap_{i=0}^{k-1-j} N[\mathfrak{L}_i], \qquad 0 \le j \le k-1. \tag{4.4}$$

On the other hand, particularizing (4.1) at $j = k$ gives

$$\mathfrak{L}_0 u_k + \sum_{i=1}^{k} \mathfrak{L}_i u_{k-i} = 0. \tag{4.5}$$

Moreover, by (4.4), we have that

$$\mathfrak{L}_0 u_k \in R[\mathfrak{L}_0], \qquad \mathfrak{L}_i u_{k-i} \in \mathfrak{L}_i \left(N[\mathfrak{L}_0] \cap \cdots \cap N[\mathfrak{L}_{i-1}] \right), \quad 1 \le i \le k.$$

Thus, since the spaces (4.2) are in direct sum, (4.5) implies

$$\mathfrak{L}_0 u_k = 0 \quad \text{and} \quad \mathfrak{L}_i u_{k-i} = 0 \quad \text{for each} \quad 1 \le i \le k.$$

Equivalently,

$$\mathfrak{L}_{k-j} u_j = 0 \quad \text{for each} \quad 0 \le j \le k.$$

Combining these relations with (4.4) gives

$$u_j \in \bigcap_{i=0}^{k-j} N[\mathfrak{L}_i] \quad \text{for each} \quad 0 \le j \le k,$$

and, therefore, (4.1) is satisfied. □

There are many situations where systems of the form (4.1) appear; for instance, in setting the concept of generalized Jordan chain. But those applications are postponed until Chapters 7 and 9.

Now we will analyze another important example closely related to the problem of ascertaining the perturbed eigenvalues from 0 for the family $\mathfrak{L}(\lambda)$ as λ moves from λ_0. As already announced, this problem will be dealt with in Section 4.4. We use the notation introduced in the general presentation of Part II.

Lemma 4.1.2. *Suppose $\mathfrak{L} \in \mathcal{C}^r(\Omega, \mathcal{L}(U,V))$ and $\varphi \in \mathcal{C}^r(\Omega, U)$ for some $r \in \mathbb{N} \cup \{\infty\}$. Consider $\lambda_0 \in \Omega$ and assume that the spaces (4.2) are in direct sum for some* *integer* $1 \leq k \leq r$. *Then the following properties are equivalent:*

- $\mathfrak{L}(\lambda)\varphi(\lambda) = o((\lambda - \lambda_0)^k)$.
- $\sum_{i=0}^{j} \mathfrak{L}_i \varphi_{j-i} = 0$ *for all* $0 \leq j \leq k$.
- $\mathfrak{L}_i \varphi_{j-i} = 0$ *for all* $0 \leq i \leq j \leq k$.

Proof. As $\mathfrak{L}\varphi \in \mathcal{C}^r(\Omega, V)$, the condition

$$\mathfrak{L}(\lambda)\varphi(\lambda) = o((\lambda - \lambda_0)^k)$$

is equivalent to

$$(\mathfrak{L}\varphi)^{j)}(\lambda_0) = 0, \qquad 0 \leq j \leq k.$$

By the Leibniz rule, this is equivalent to

$$\sum_{i=0}^{j} \mathfrak{L}_i \varphi_{j-i} = 0, \qquad 0 \leq j \leq k. \tag{4.6}$$

Finally, according to Lemma 4.1.1, (4.6) is satisfied if and only if

$$\mathfrak{L}_i \varphi_{j-i} = 0, \qquad 0 \leq i \leq j \leq k,$$

which concludes the proof. □

Based on Lemma 4.1.2, we can easily obtain the following perturbation result. It clearly reveals how the transversality of the spaces (4.2) tremendously facilitates the calculations in ascertaining the perturbations of eigenvalues and associated eigenvectors for linear operators. Its full significance will become apparent in Section 4.4.

Lemma 4.1.3. *Suppose that for some* $r \in \mathbb{N} \cup \{\infty\}$ *the functions*

$$\mathfrak{L} \in \mathcal{C}^r(\Omega, \mathcal{L}(U)), \quad \varphi \in \mathcal{C}^r(\Omega, U), \quad a \in \mathcal{C}^r(\Omega)$$

satisfy

$$\mathfrak{L}(\lambda)\varphi(\lambda) = a(\lambda)\varphi(\lambda), \qquad \lambda \in \Omega,$$

and, for some point $\lambda_0 \in \Omega$ *and some integer* $1 \leq k \leq r$, *the spaces* (4.2) *are in direct sum, and*

$$a_i = 0, \qquad 0 \leq i \leq k.$$

Then

$$\mathfrak{L}_i \varphi_{j-i} = 0, \qquad 0 \leq i \leq j \leq k.$$

Proof. Define the auxiliary map $\mathfrak{M} \in \mathcal{C}^r(\Omega, \mathcal{L}(U))$ through

$$\mathfrak{M}(\lambda) := \mathfrak{L}(\lambda) - a(\lambda)I, \qquad \lambda \in \Omega.$$

Then, $\mathfrak{M}\varphi = 0$ in Ω, and

$$\mathfrak{M}_i = \mathfrak{L}_i, \qquad 0 \leq i \leq k.$$

Consequently, Lemma 4.1.2 concludes the proof. □

4.2 The concept of transversal eigenvalue

The next definition introduces the concept of transversal eigenvalue. It should not be forgotten that

$$j!\mathfrak{L}_j = \mathfrak{L}^{j)}(\lambda_0), \qquad 0 \le j < r+1.$$

Definition 4.2.1. Let $\mathfrak{L} \in \mathcal{C}^r(\Omega, \mathcal{L}(U,V))$ satisfy $\mathfrak{L}_0 \in \mathrm{Fred}_0(U,V)$ for some $r \ge 1$; consider $\lambda_0 \in \mathrm{Eig}(\mathfrak{L})$, and $k \in \mathbb{N} \cap [1,r]$. Then, λ_0 is said to be a k-transversal eigenvalue of $\mathfrak{L}$ if the following topological direct sum decomposition is satisfied:

$$\bigoplus_{j=1}^{k} \mathfrak{L}_j(N[\mathfrak{L}_0] \cap \cdots \cap N[\mathfrak{L}_{j-1}]) \oplus R[\mathfrak{L}_0] = V, \tag{4.7}$$

with

$$\mathfrak{L}_k(N[\mathfrak{L}_0] \cap \cdots \cap N[\mathfrak{L}_{k-1}]) \neq \{0\}. \tag{4.8}$$

In such a case, the integer k is called the order of transversality of $\mathfrak{L}$ at λ_0.

More generally, given $k \in \mathbb{N} \cap [0,r]$, the point λ_0 is said to be a k-transversal value of $\mathfrak{L}$ when either $k=0$ and $\mathfrak{L}_0 \in \mathrm{Iso}(U,V)$, or $k \ge 1$ and λ_0 is a k-transversal eigenvalue of $\mathfrak{L}$.

Condition (4.8) makes the integer k uniquely determined. The next result provides us with some useful relationships between the dimensions of the vector spaces arising in (4.7), as well as with a characterization of transversal eigenvalues.

Proposition 4.2.2. *Let $r \in \mathbb{N}^* \cup \{\infty\}$ and $\mathfrak{L} \in \mathcal{C}^r(\Omega, \mathcal{L}(U,V))$ be such that $\mathfrak{L}_0 \in \mathrm{Fred}_0(U,V)$, with $\lambda_0 \in \mathrm{Eig}(\mathfrak{L})$, and the spaces (4.2) are in direct sum for some integer $1 \le k \le r$. Set*

$$n_i := \dim \bigcap_{j=0}^{i} N[\mathfrak{L}_j], \qquad 0 \le i \le k,$$

$$\ell_i := \dim \mathfrak{L}_i\Big(\bigcap_{j=0}^{i-1} N[\mathfrak{L}_j]\Big), \qquad 1 \le i \le k.$$

Then

$$n_{i-1} = n_i + \ell_i, \qquad 1 \le i \le k, \tag{4.9}$$

and, consequently,

$$n_i = n_k + \sum_{j=i+1}^{k} \ell_j, \qquad 0 \le i \le k-1. \tag{4.10}$$

Moreover, λ_0 is a k-transversal eigenvalue of $\mathfrak{L}$ if and only if

$$n_k = 0, \qquad \ell_k \ge 1, \tag{4.11}$$

and, in such a case,

$$n_i = \sum_{j=i+1}^{k} \ell_j, \qquad 0 \le i \le k-1. \tag{4.12}$$

Proof. Since

$$\bigcap_{j=0}^{i} N[\mathfrak{L}_j] \subset \bigcap_{j=0}^{i-1} N[\mathfrak{L}_j] \subset N[\mathfrak{L}_0], \qquad 1 \le i \le k,$$

and $\dim N[\mathfrak{L}_0] < \infty$, for each $1 \le i \le k$ there exists a finite-dimensional subspace $U_i \subset U$ such that

$$N[\mathfrak{L}_0] \cap \cdots \cap N[\mathfrak{L}_{i-1}] = U_i \oplus (N[\mathfrak{L}_0] \cap \cdots \cap N[\mathfrak{L}_i]) \tag{4.13}$$

and, hence,

$$N[\mathfrak{L}_0] = U_1 \oplus \cdots \oplus U_k \oplus (N[\mathfrak{L}_0] \cap \cdots \cap N[\mathfrak{L}_k])\,. \tag{4.14}$$

By construction,

$$\mathfrak{L}_i(N[\mathfrak{L}_0] \cap \cdots \cap N[\mathfrak{L}_{i-1}]) = \mathfrak{L}_i(U_i), \qquad 1 \le i \le k, \tag{4.15}$$

and, in addition,

$$\dim U_i = \dim \mathfrak{L}_i(U_i), \qquad 1 \le i \le k, \tag{4.16}$$

because the restriction of $\mathfrak{L}_i$ from U_i to $\mathfrak{L}_i(U_i)$ is an isomorphism. Thanks to (4.13), (4.15) and (4.16), we find that

$$n_{i-1} = \dim U_i + n_i = \dim \mathfrak{L}_i(U_i) + n_i = \ell_i + n_i, \qquad 1 \le i \le k,$$

which provides us with (4.9). Identity (4.10) is a direct consequence of the identities of (4.9).

Now, assume that λ_0 is a k-transversal eigenvalue of $\mathfrak{L}$. Then,

$$\mathfrak{L}_1(U_1) \oplus \cdots \oplus \mathfrak{L}_k(U_k) \oplus R[\mathfrak{L}_0] = V \tag{4.17}$$

and, hence,

$$\dim N[\mathfrak{L}_0] = \operatorname{codim} R[\mathfrak{L}_0] = \sum_{i=1}^{k} \dim \mathfrak{L}_i(U_i) = \sum_{i=1}^{k} \dim U_i,$$

because $\mathfrak{L}_0 \in \text{Fred}_0(U,V)$. Thus, we find from (4.14) that

$$N[\mathfrak{L}_0] = U_1 \oplus \cdots \oplus U_k \quad \text{and} \quad N[\mathfrak{L}_0] \cap \cdots \cap N[\mathfrak{L}_k] = \{0\}.$$

Therefore, $n_k = 0$. Note that $\ell_k \ge 1$ holds from (4.8), and that (4.12) is just (4.10) with $n_k = 0$.

To show the converse implication, suppose (4.11). Then, (4.10) implies (4.12). In particular,

$$\dim N[\mathfrak{L}_0] = n_0 = \sum_{j=1}^{k} \ell_j$$

and, hence,

$$n_0 = \operatorname{codim} R[\mathfrak{L}_0] \geq \sum_{i=1}^{k} \dim \mathfrak{L}_i(U_i) = n_0.$$

Consequently,

$$\operatorname{codim} R[\mathfrak{L}_0] = \sum_{i=1}^{k} \dim \mathfrak{L}_i(U_i)$$

and (4.17) holds. As $\ell_k \geq 1$, the point λ_0 must be a k-transversal eigenvalue of $\mathfrak{L}$. This concludes the proof. □

4.3 Algebraic eigenvalues and transversalization

The following definition fixes the pivotal concept of algebraic eigenvalue.

Definition 4.3.1. Suppose $r \in \mathbb{N} \cup \{\infty\}$, and $\mathfrak{L} \in \mathcal{C}^r(\Omega, \mathcal{L}(U,V))$ satisfy

$$\mathfrak{L}_0 \in \mathrm{Fred}_0(U,V),$$

with $\lambda_0 \in \mathrm{Eig}(\mathfrak{L})$. Then, λ_0 is said to be an algebraic eigenvalue of $\mathfrak{L}$ if there are an integer $\nu \geq 1$ and two positive constants $\delta > 0$ and $C > 0$ such that, for each $0 < |\lambda - \lambda_0| < \delta$, the operator $\mathfrak{L}(\lambda)$ is an isomorphism and

$$\|\mathfrak{L}(\lambda)^{-1}\| \leq \frac{C}{|\lambda - \lambda_0|^{\nu}}. \tag{4.18}$$

The least integer $\nu \geq 1$ satisfying these requirements will be called the order of λ_0 as an algebraic eigenvalue of $\mathfrak{L}$.

More generally, for a given $\nu \in \mathbb{N}$, the point λ_0 is said to be a ν-algebraic value of $\mathfrak{L}$ when either $\nu = 0$ and $\mathfrak{L}_0 \in \mathrm{Iso}(U,V)$, or $\nu \geq 1$ and λ_0 is an algebraic eigenvalue of order ν of $\mathfrak{L}$.

For every $\nu \in \mathbb{N}$, we denote by $\mathrm{Alg}_\nu(\mathfrak{L})$ the set of all ν-algebraic values of $\mathfrak{L}$ in Ω, and

$$\mathrm{Alg}(\mathfrak{L}) := \bigcup_{\nu \in \mathbb{N}} \mathrm{Alg}_\nu(\mathfrak{L}).$$

It must be emphasized that 0-algebraic values of $\mathfrak{L}$ are not eigenvalues of $\mathfrak{L}$; however, they will be useful to extend the concept of algebraic multiplicity to cover all values of $\lambda \in \Omega$ for which $\mathfrak{L}(\lambda) \in \mathrm{Iso}(U,V)$. We also note

$$\mathrm{Alg}_0(\mathfrak{L}) = \Omega \setminus \Sigma(\mathfrak{L}) \quad \text{and} \quad \mathrm{Alg}(\mathfrak{L}) \setminus \mathrm{Alg}_0(\mathfrak{L}) \subset \mathrm{Eig}(\mathfrak{L}).$$

The order of λ_0 as an algebraic eigenvalue of $\mathfrak{L}$ equals the order of λ_0 as a pole of $\mathfrak{L}^{-1}$ if $\mathfrak{L}$ is analytic. Rather strikingly, although the concept of pole is not available for non-analytic families, the theory developed in Chapter 9 will reveal that the concept of order of λ_0 as an algebraic eigenvalue of $\mathfrak{L}$ in the non-analytic case shares most of the local properties with the concept of order of λ_0 as a pole of $\mathfrak{L}^{-1}$ in the analytic case. Among them, the existence of Laurent expansions of $\mathfrak{L}^{-1}$, necessarily finite if $\mathfrak{L}$ is not analytic.

The main result of this section establishes that if $\lambda_0 \in \Omega$ is a ν-algebraic eigenvalue of $\mathfrak{L} \in \mathcal{C}^r(\Omega, \mathcal{L}(U,V))$, for some $r \geq 1$ and some integer $1 \leq \nu \leq r$, then there exist a natural number $1 \leq k \leq \nu$ and a polynomial family $\Phi \in \mathbb{P}[\mathcal{L}(U,V)]$ with $\Phi_0 \in \mathrm{Iso}(U,V)$ such that λ_0 is a k-transversal eigenvalue of the transformed family $\mathfrak{L}^\Phi$ defined as

$$\mathfrak{L}^\Phi(\lambda) := \mathfrak{L}(\lambda)\Phi(\lambda), \qquad \lambda \in \Omega. \tag{4.19}$$

Moreover, the dimensions

$$\dim \mathfrak{L}_j^\Phi(N[\mathfrak{L}_0^\Phi] \cap \cdots \cap N[\mathfrak{L}_{j-1}^\Phi]), \qquad 0 \leq j < r+1,$$

do not depend on the family of isomorphisms Φ and, consequently, they may be regarded as algebraic invariants of $\mathfrak{L}$. This result can be stated briefly by simply saying that $\mathfrak{L}$ can be *transversalized* at λ_0 if λ_0 is an algebraic eigenvalue of order $1 \leq \nu \leq r$. In such a case, any local family of isomorphisms Φ for which λ_0 is a transversal eigenvalue of (4.19) will be referred to as a *transversalizing* family for $\mathfrak{L}$ at λ_0.

It turns out that if $r = \infty$, then λ_0 is an algebraic eigenvalue of $\mathfrak{L}$ if and only if $\mathfrak{L}$ can be transversalized at λ_0, but the complete proof of this characterization is postponed until Chapter 5, as it uses another concept of multiplicity to be introduced there.

The main result of this chapter can be stated as follows.

Theorem 4.3.2 (of transversalization). *Suppose* $r \in \mathbb{N} \cup \{\infty\}$. *Let* $\mathfrak{L} \in \mathcal{C}^r(\Omega, \mathcal{L}(U,V))$ *satisfy* $\mathfrak{L}_0 \in \mathrm{Fred}_0(U,V)$ *and* $\lambda_0 \in \mathrm{Eig}(\mathfrak{L})$. *Then, for each integer* $1 \leq k \leq r$ *there exists* $\Phi^k \in \mathbb{P}[\mathcal{L}(U)]$ *with* $\Phi_0^k = I_U$ *such that the product family* $\mathfrak{L}^{\Phi^k}$ *defined by*

$$\mathfrak{L}^{\Phi^k}(\lambda) := \mathfrak{L}(\lambda)\Phi^k(\lambda), \qquad \lambda \in \Omega,$$

satisfies

$$V_k := \bigoplus_{j=1}^{k} \mathfrak{L}_j^{\Phi^k}(N[\mathfrak{L}_0^{\Phi^k}] \cap \cdots \cap N[\mathfrak{L}_{j-1}^{\Phi^k}]) \oplus R[\mathfrak{L}_0^{\Phi^k}] \subset V.$$

If, in addition, $\lambda_0 \in \mathrm{Alg}_\nu(\mathfrak{L})$ *for some integer* $1 \leq \nu \leq r$, *then there exists a least* $1 \leq k \leq \nu$ *for which* $V_k = V$ *and, therefore,* λ_0 *is a* k*-transversal eigenvalue of* $\mathfrak{L}^{\Phi^k}$.

Moreover, if $\Phi, \Psi \in \mathcal{C}^r(\Omega, \mathcal{L}(U))$ *are two families for which* $\Phi_0, \Psi_0 \in \mathrm{Iso}(U)$ *and, for some* $1 \le k, \ell \le \nu$, *the transformed families*

$$\mathfrak{L}^\Phi := \mathfrak{L}\Phi, \qquad \mathfrak{L}^\Psi := \mathfrak{L}\Psi,$$

satisfy

$$\bigoplus_{j=1}^{k} \mathfrak{L}_j^\Phi\Big(\bigcap_{i=0}^{j-1} N[\mathfrak{L}_i^\Phi]\Big) \oplus R[\mathfrak{L}_0^\Phi] = V, \qquad \mathfrak{L}_k^\Phi\Big(\bigcap_{i=0}^{k-1} N[\mathfrak{L}_i^\Phi]\Big) \neq \{0\},$$

$$\bigoplus_{j=1}^{\ell} \mathfrak{L}_j^\Psi\Big(\bigcap_{i=0}^{j-1} N[\mathfrak{L}_i^\Psi]\Big) \oplus R[\mathfrak{L}_0^\Psi] = V, \qquad \mathfrak{L}_\ell^\Psi\Big(\bigcap_{i=0}^{\ell-1} N[\mathfrak{L}_i^\Psi]\Big) \neq \{0\},$$

then $k = \ell$ *and, for each* $1 \le j \le k$,

$$\dim \mathfrak{L}_j^\Phi\Big(\bigcap_{i=0}^{j-1} N[\mathfrak{L}_i^\Phi]\Big) = \dim \mathfrak{L}_j^\Psi\Big(\bigcap_{i=0}^{j-1} N[\mathfrak{L}_i^\Psi]\Big).$$

Before giving the proof of Theorem 4.3.2, we discuss some of its main consequences. According to Theorem 4.3.2, the following concept of algebraic multiplicity is consistent.

Definition 4.3.3. Suppose $r \in \mathbb{N} \cup \{\infty\}$, the family $\mathfrak{L} \in \mathcal{C}^r(\Omega, \mathcal{L}(U, V))$ satisfies $\mathfrak{L}_0 \in \mathrm{Fred}_0(U, V)$, and either $r = \infty$, or $\lambda_0 \in \mathrm{Alg}_\nu(\mathfrak{L})$ for some integer $0 \le \nu < r + 1$. Then, the algebraic multiplicity of $\mathfrak{L}$ at λ_0, denoted by $\chi[\mathfrak{L}; \lambda_0]$, is defined as follows:

- $\chi[\mathfrak{L}; \lambda_0] := \sum_{j=0}^{k} j \cdot \dim \mathfrak{L}_j^\Phi\Big(\bigcap_{i=0}^{j-1} N[\mathfrak{L}_i^\Phi]\Big)$, if $\lambda_0 \in \mathrm{Alg}_\nu(\mathfrak{L})$ for some $1 \le \nu < r+1$, where $\Phi \in \mathcal{C}^r(\Omega, \mathcal{L}(U))$ is any family for which λ_0 is a k-transversal eigenvalue of $\mathfrak{L}^\Phi := \mathfrak{L}\Phi$, for a certain $1 \le k \le \nu$.
- $\chi[\mathfrak{L}; \lambda_0] := 0$, if $\lambda_0 \in \mathrm{Alg}_0(\mathfrak{L})$; equivalently, if $\mathfrak{L}_0 \in \mathrm{Iso}(U, V)$.
- $\chi[\mathfrak{L}; \lambda_0] := \infty$, if $r = \infty$ and $\lambda_0 \notin \mathrm{Alg}(\mathfrak{L})$.

The multiplicity $\chi[\mathfrak{L}; \lambda_0]$ is left undefined in all remaining cases, that is, when

$$r \subset \mathbb{N} \quad \text{and} \quad \lambda_0 \notin \bigcup_{0 \le \nu \le r} \mathrm{Alg}_\nu(\mathfrak{L}).$$

As commented earlier, if $r = \infty$ and $\lambda_0 \in \mathrm{Eig}(\mathfrak{L})$, then $\mathfrak{L}$ can be transversalized at λ_0 if and only if $\lambda_0 \in \mathrm{Alg}(\mathfrak{L})$. Actually, in such a case, as a consequence of the theory developed in Chapter 5, for any local transversalizing isomorphism $\Phi \in \mathcal{C}^\infty(\Omega, \mathcal{L}(U))$, the order of λ_0 as an algebraic eigenvalue of $\mathfrak{L}$ equals the order of transversality of λ_0 as a transversal eigenvalue of the transversalized family

$$\mathfrak{L}^\Phi := \mathfrak{L}\Phi.$$

Consequently, the hypothesis $\lambda_0 \in \mathrm{Alg}_\nu(\mathfrak{L})$, for some $1 \leq \nu \leq r$, in the second assertion of Theorem 4.3.2 is not only sufficient for its validity, but also necessary. This is an extremely sharp result, because it entails that V_k must be a proper subspace of V for all $k \geq 1$ if $\lambda_0 \notin \mathrm{Alg}(\mathfrak{L})$ (see the statement of Theorem 4.3.2).

Simple examples show that the assumption that λ_0 is an algebraic eigenvalue of $\mathfrak{L}$ cannot be removed from the statement of Theorem 4.3.2 without seriously affecting its validity. Indeed, the family $\mathfrak{L} \in \mathcal{C}^\infty(\mathbb{R})$ defined by

$$\mathfrak{L}(\lambda) := \begin{cases} e^{-\lambda^{-2}} & \text{if } \lambda \in \mathbb{R} \setminus \{0\}, \\ 0 & \text{if } \lambda = 0 \end{cases} \tag{4.20}$$

is real analytic in $\mathbb{R} \setminus \{0\}$, but $0 \notin \mathrm{Alg}(\mathfrak{L})$. Moreover, $\mathfrak{L}_j = 0$ for all $j \geq 0$, and, therefore, 0 cannot be a k-transversal eigenvalue of $\mathfrak{L}\Phi$ for any integer $k \geq 0$ and $\Phi \in \mathcal{C}^\infty(\mathbb{R})$.

The proof of Theorem 4.3.2 is based on the following lemma.

Lemma 4.3.4. *Let $m \in \mathbb{N}^*$ and $\mathfrak{L} \in \mathcal{C}^{m+1}(\Omega, \mathcal{L}(U,V))$ be such that $\mathfrak{L}_0 \in \mathrm{Fred}_0(U,V)$ and*

$$V_m := \bigoplus_{j=1}^{m} \mathfrak{L}_j(N[\mathfrak{L}_0] \cap \cdots \cap N[\mathfrak{L}_{j-1}]) \oplus R[\mathfrak{L}_0] \subset V. \tag{4.21}$$

Then there exists $\Phi \in \mathbb{P}[\mathcal{L}(U)]$ with $\Phi_0 = I_U$ for which the new family

$$\mathfrak{M} := \mathfrak{L}\Phi$$

satisfies

$$V_{m+1} := \bigoplus_{j=1}^{m+1} \mathfrak{M}_j(N[\mathfrak{M}_0] \cap \cdots \cap N[\mathfrak{M}_{j-1}]) \oplus R[\mathfrak{M}_0] \subset V \tag{4.22}$$

and

$$\mathfrak{M}_h = \mathfrak{L}_h, \qquad 0 \leq h \leq m. \tag{4.23}$$

Proof. Let Y be a closed subspace of U such that

$$U = N[\mathfrak{L}_0] \oplus Y.$$

Arguing as in the proof of Proposition 4.2.2, for each $1 \leq j \leq m$ there exists a subspace U_j of $N[\mathfrak{L}_0]$ such that

$$N[\mathfrak{L}_0] \cap \cdots \cap N[\mathfrak{L}_{j-1}] = U_j \oplus (N[\mathfrak{L}_0] \cap \cdots \cap N[\mathfrak{L}_j]). \tag{4.24}$$

Moreover, these subspaces satisfy

$$\mathfrak{L}_j(U_j) = \mathfrak{L}_j(N[\mathfrak{L}_0] \cap \cdots \cap N[\mathfrak{L}_{j-1}]), \qquad 1 \leq j \leq m,$$

and

$$\dim U_j = \dim \mathfrak{L}_j(U_j), \qquad 1 \leq j \leq m.$$

Thus, (4.21) can be expressed as

$$V_m = \bigoplus_{j=1}^{m} \mathfrak{L}_j(U_j) \oplus R[\mathfrak{L}_0].$$

From now on, we set

$$X_m := \bigoplus_{j=1}^{m} U_j \oplus Y.$$

Clearly,

$$\operatorname{codim} X_m = \operatorname{codim} V_m,$$

since $\mathfrak{L}_0 \in \mathrm{Fred}_0(U,V)$. Actually, the operators

$$\mathfrak{L}_0|_Y : Y \to R[\mathfrak{L}_0] \quad \text{and} \quad \mathfrak{L}_j|_{U_j} : U_j \to \mathfrak{L}_j(U_j), \quad 1 \le j \le m,$$

are isomorphisms, and

$$U_j \subset N[\mathfrak{L}_h] \quad \text{whenever} \quad 0 \le h < j \le m.$$

Now, consider the projections

$$\mathcal{P}_0, \dots, \mathcal{P}_m \in \mathcal{L}(U) \quad \text{and} \quad \mathcal{Q}_0, \dots, \mathcal{Q}_m \in \mathcal{L}(V)$$

such that

$$\begin{aligned}
&R[\mathcal{P}_0] = Y, && N[\mathcal{P}_0] = N[\mathfrak{L}_0], && \\
&R[\mathcal{P}_h] = U_h, && N[\mathcal{P}_h] = \bigoplus_{\substack{j=1 \\ j\neq h}}^{m} U_j \oplus Y, && 1 \le h \le m, \\
&R[\mathcal{Q}_0] = R[\mathfrak{L}_0], && N[\mathcal{Q}_0] = \bigoplus_{j=1}^{m} \mathfrak{L}_j(U_j), && \\
&R[\mathcal{Q}_h] = \mathfrak{L}_h(U_h), && N[\mathcal{Q}_h] = \bigoplus_{\substack{j=1 \\ j\neq h}}^{m} \mathfrak{L}_j(U_j) \oplus R[\mathfrak{L}_0], && 1 \le h \le m.
\end{aligned}$$

Then, by definition of direct sum, in order to obtain (4.22) it suffices to construct $\Phi \in \mathbb{P}[\mathcal{L}(U)]$ such that $\mathfrak{M} := \mathfrak{L}\Phi$ satisfies (4.23) and

$$\mathcal{Q}_h \mathfrak{M}_{m+1} = 0, \qquad 0 \le h \le m. \tag{4.25}$$

From the definition of $\mathfrak{M}$ it is apparent that

$$\mathfrak{M}_h = \sum_{j=0}^{h} \mathfrak{L}_{h-j}\Phi_j, \qquad 0 \le h \le m+1. \tag{4.26}$$

Thus, to obtain (4.23) it is sufficient to choose Φ such that $\Phi_0 = I_U$ and

$$\mathfrak{L}_{h-j}\Phi_j = 0, \qquad 1 \le j \le h \le m.$$

These relations can be equivalently written in the form

$$R[\Phi_j] \subset N[\mathfrak{L}_0] \cap \cdots \cap N[\mathfrak{L}_{m-j}], \qquad 1 \le j \le m. \tag{4.27}$$

By construction, for each $1 \le j \le m$,

$$N[\mathfrak{L}_0] \cap \cdots \cap N[\mathfrak{L}_{m-j}] = U_{m-j+1} \oplus \cdots \oplus U_m \oplus (N[\mathfrak{L}_0] \cap \cdots \cap N[\mathfrak{L}_m])$$

and, hence, (4.27) is accomplished for every family Φ satisfying

$$\mathcal{P}_{h-j}\Phi_j = 0, \qquad 1 \le j \le h \le m. \tag{4.28}$$

Indeed, (4.28) implies

$$\mathcal{P}_0\Phi_j = 0, \qquad 1 \le j \le m,$$

and, hence,

$$R[\Phi_j] \subset N[\mathfrak{L}_0] = U_1 \oplus \cdots \oplus U_m \oplus (N[\mathfrak{L}_0] \cap \cdots \cap N[\mathfrak{L}_m]).$$

Moreover,

$$\mathcal{P}_1\Phi_j = \cdots = \mathcal{P}_{m-j}\Phi_j = 0.$$

Therefore, (4.28) implies (4.27). Assume $\Phi_0 = I_U$ and (4.27). Using (4.26), we find that (4.25) is satisfied if and only if

$$\sum_{j=0}^{m+1} \mathcal{Q}_h \mathfrak{L}_{m-j+1}\Phi_j = 0, \qquad 0 \le h \le m. \tag{4.29}$$

Moreover, it follows from (4.27) that, for each $1 \le j \le m$,

$$R[\mathfrak{L}_{m-j+1}\Phi_j] \subset \mathfrak{L}_{m-j+1}(U_{m-j+1}),$$

since

$$R[\Phi_j] \subset N[\mathfrak{L}_0] \cap \cdots \cap N[\mathfrak{L}_{m-j}] = U_{m-j+1} \oplus (N[\mathfrak{L}_0] \cap \cdots \cap N[\mathfrak{L}_{m-j+1}]).$$

Thus, for each $1 \le j \le m$ and $0 \le h \le m$ with $h \ne m-j+1$,

$$\mathcal{Q}_h \mathfrak{L}_{m-j+1}\Phi_j = 0$$

and, hence, (4.29) can be written in the form

$$\mathcal{Q}_h \mathfrak{L}_{m+1} = -\mathcal{Q}_h \mathfrak{L}_h \Phi_{m+1-h}, \qquad 0 \le h \le m. \tag{4.30}$$

Further, since the operators

$$\mathcal{Q}_0\mathfrak{L}_0|_Y : Y \to R[\mathfrak{L}_0],$$
$$\mathcal{Q}_h\mathfrak{L}_h|_{U_h} : U_h \to \mathfrak{L}_h(U_h), \quad 1 \le h \le m,$$

are isomorphisms, (4.30) is satisfied provided

$$\Phi_{m+1} = -\mathcal{P}_0(\mathcal{Q}_0\mathfrak{L}_0|_Y)^{-1}\mathcal{Q}_0\mathfrak{L}_{m+1} \tag{4.31}$$

and

$$\Phi_{m+1-h} = -\mathcal{P}_h(\mathcal{Q}_h\mathfrak{L}_h|_{U_h})^{-1}\mathcal{Q}_h\mathfrak{L}_{m+1}, \qquad 1 \le h \le m. \tag{4.32}$$

As (4.28) does not entail any restriction on Φ_{m+1}, it is clear that (4.31) is compatible with (4.28). Moreover, choosing (4.32) yields

$$R[\Phi_{m+1-h}] \subset U_h \subset N[\mathfrak{L}_0] \cap \cdots \cap N[\mathfrak{L}_{h-1}], \qquad 1 \le h \le m.$$

Equivalently,

$$R[\Phi_j] \subset N[\mathfrak{L}_0] \cap \cdots \cap N[\mathfrak{L}_{m-j}], \qquad 1 \le j \le m.$$

Therefore, (4.27) holds true. This concludes the proof. □

Throughout the rest of this book, for any integer $n \ge 1$, vector spaces $X_1, \dots, X_n$ and vectors $x_j \in X_j$, for $1 \le j \le n$, we will denote

$$\operatorname{col}[x_1, \dots, x_n] := \begin{pmatrix} x_1 \\ \vdots \\ x_n \end{pmatrix} \in X_1 \times \cdots \times X_n. \tag{4.33}$$

Also, we use the notation trng, which has been partially introduced in (2.24). Given an integer $n \ge 1$, two Banach spaces U, V, and n operators

$$B_1, \dots, B_n \in \mathcal{L}(U, V),$$

the operator

$$\operatorname{trng}\{B_1, B_2, \dots, B_n\} := \begin{pmatrix} B_1 & 0 & \cdots & 0 \\ B_2 & B_1 & \cdots & 0 \\ \vdots & \vdots & \ddots & \vdots \\ B_n & B_{n-1} & \cdots & B_1 \end{pmatrix} \tag{4.34}$$

will denote the element of $\mathcal{L}(U^n, V^n)$ defined through

$$\begin{aligned} &\operatorname{trng}\{B_1, B_2, \dots, B_n\} \operatorname{col}[u_1, \dots, u_n] \\ &\quad := \operatorname{col}[B_1u_1, B_2u_1 + B_1u_2, \dots, B_nu_1 + B_{n-1}u_2 + \cdots + B_1u_n] \end{aligned}$$

for each $u_1, \dots, u_n \in U$.

The following result, of technical nature, will be useful in the proof of Theorem 4.3.2.

Lemma 4.3.5. *Let $r \in \mathbb{N}$ and $\mathfrak{L}, \mathfrak{M} \in \mathcal{C}^r(\Omega, \mathcal{L}(U,V))$ be such that, for some integer $1 \leq k \leq r$, the spaces*

$$R[\mathfrak{L}_0],\ \mathfrak{L}_1(N[\mathfrak{L}_0]),\ \ldots,\ \mathfrak{L}_k(\bigcap_{j=0}^{k-1} N[\mathfrak{L}_j]),$$

are in direct sum, and so are

$$R[\mathfrak{M}_0],\ \mathfrak{M}_1(N[\mathfrak{M}_0]),\ \ldots,\ \mathfrak{M}_k(\bigcap_{j=0}^{k-1} N[\mathfrak{M}_j]).$$

Suppose, in addition, that there exists $\Upsilon \in \mathcal{C}^r(\Omega, \mathcal{L}(U))$ such that

$$\Upsilon_0 \in \operatorname{Iso}(U) \quad \text{and} \quad \mathfrak{L} = \mathfrak{M}\Upsilon. \tag{4.35}$$

Then, for every $0 \leq j \leq k$,

$$\Upsilon_0 \left(N[\mathfrak{L}_0] \cap \cdots \cap N[\mathfrak{L}_j]\right) = N[\mathfrak{M}_0] \cap \cdots \cap N[\mathfrak{M}_j]$$

and, consequently,

$$\dim N[\mathfrak{L}_0] \cap \cdots \cap N[\mathfrak{L}_j] = \dim N[\mathfrak{M}_0] \cap \cdots \cap N[\mathfrak{M}_j].$$

Proof. Applying the Leibniz rule to (4.35) yields

$$\operatorname{trng}\{\mathfrak{L}_0, \ldots, \mathfrak{L}_s\} = \operatorname{trng}\{\mathfrak{M}_0, \ldots, \mathfrak{M}_s\} \cdot \operatorname{trng}\{\Upsilon_0, \ldots, \Upsilon_s\} \tag{4.36}$$

for all integers $0 \leq s \leq r$. Now pick $0 \leq j \leq k$ and

$$x \in N[\mathfrak{L}_0] \cap \cdots \cap N[\mathfrak{L}_j].$$

Then

$$\operatorname{col}[x, \ldots, x] \in N\left[\operatorname{trng}\{\mathfrak{L}_0, \ldots, \mathfrak{L}_j\}\right],$$

and, thanks to (4.36), we find that

$$\operatorname{trng}\{\Upsilon_0, \ldots, \Upsilon_j\} \operatorname{col}[x, \ldots, x] \in N\left[\operatorname{trng}\{\mathfrak{M}_0, \ldots, \mathfrak{M}_j\}\right].$$

Consequently, Lemma 4.1.1 implies that

$$\Upsilon_0 x \in \dim N[\mathfrak{M}_0] \cap \cdots \cap N[\mathfrak{M}_j]$$

and, therefore, for each $0 \leq j \leq k$,

$$\Upsilon_0(N[\mathfrak{L}_0] \cap \cdots \cap N[\mathfrak{L}_j]) \subset N[\mathfrak{M}_0] \cap \cdots \cap N[\mathfrak{M}_j].$$

As Υ_0 is an isomorphism, there exists an open neighborhood Ω' of λ_0 such that the function Υ^{-1} defined by

$$\Upsilon^{-1}(\lambda) := \Upsilon(\lambda)^{-1} \quad \text{for} \quad \lambda \in \Omega'$$

belongs to $\mathcal{C}^r(\Omega', \mathcal{L}(U))$. In Ω' we have $\mathfrak{M} = \mathfrak{L}\Upsilon^{-1}$, and, therefore, according to the result that we have just proved, for each $0 \leq j \leq k$,

$$\Upsilon_0^{-1}(N[\mathfrak{M}_0] \cap \cdots \cap N[\mathfrak{M}_j]) \subset N[\mathfrak{L}_0] \cap \cdots \cap N[\mathfrak{L}_j]$$

and, consequently,

$$N[\mathfrak{M}_0] \cap \cdots \cap N[\mathfrak{M}_j] \subset \Upsilon_0(N[\mathfrak{L}_0] \cap \cdots \cap N[\mathfrak{L}_j]),$$

which concludes the proof. □

Now, we proceed to prove Theorem 4.3.2.

Proof of Theorem 4.3.2. Let Y be a closed subspace of U, and Z a finite-dimensional subspace of V such that

$$U = N[\mathfrak{L}_0] \oplus Y \quad \text{and} \quad V = Z \oplus R[\mathfrak{L}_0].$$

Then, since $\mathfrak{L}_0 \in \operatorname{Fred}_0(U, V)$, we have

$$\dim Z = \dim N[\mathfrak{L}_0].$$

Consider continuous projections $\mathcal{P}_0 \in \mathcal{L}(U)$ and $\mathcal{Q}_0 \in \mathcal{L}(V)$ such that

$$R[\mathcal{P}_0] = Y, \quad N[\mathcal{P}_0] = N[\mathfrak{L}_0], \quad R[\mathcal{Q}_0] = R[\mathfrak{L}_0], \quad N[\mathcal{Q}_0] = Z.$$

The construction of the transversalizing local isomorphisms of U will be accomplished iteratively. We begin by constructing a polynomial $\Phi^1 \in \mathbb{P}[\mathcal{L}(U)]$ of the form

$$\Phi^1(\lambda) := I_U + (\lambda - \lambda_0)\Phi_1^1, \qquad \lambda \in \mathbb{K},$$

so that the family

$$\mathfrak{L}^1 := \mathfrak{L}\Phi^1 \tag{4.37}$$

satisfies

$$V_1 := \mathfrak{L}_1^1(N[\mathfrak{L}_0^1]) \oplus R[\mathfrak{L}_0^1] \subset V.$$

Necessarily, by (4.37),

$$\mathfrak{L}_0^1 = \mathfrak{L}_0 \quad \text{and} \quad \mathfrak{L}_1^1 = \mathfrak{L}_1 + \mathfrak{L}_0\Phi_1^1.$$

Clearly, it suffices to construct $\Phi_1^1 \in \mathcal{L}(U)$ such that

$$\mathcal{Q}_0\mathfrak{L}_1 + \mathcal{Q}_0\mathfrak{L}_0\Phi_1^1 = 0,$$

which can be accomplished by making the choice

$$\Phi_1^1 := -\mathcal{P}_0(\mathcal{Q}_0\mathfrak{L}_0|_Y)^{-1}\mathcal{Q}_0\mathfrak{L}_1.$$

Now consider an integer $1 \leq k \leq r$. Then, based on the previous result and applying Lemma 4.3.4 recursively, there exist $\Psi^1, \dots, \Psi^k \in \mathbb{P}[\mathcal{L}(U)]$ such that

$$\Psi_0^m := \Psi^m(\lambda_0) = I_U, \qquad 1 \leq m \leq k,$$

for which the family $\mathfrak{L}^m$ defined by

$$\mathfrak{L}^m(\lambda) := \mathfrak{L}(\lambda)\Psi^1(\lambda) \cdots \Psi^m(\lambda), \qquad \lambda \in \Omega,$$

satisfies

$$V_m := \bigoplus_{j=1}^{m} \mathfrak{L}_j^m(N[\mathfrak{L}_0^m] \cap \cdots \cap N[\mathfrak{L}_{j-1}^m]) \oplus R[\mathfrak{L}_0^m] \subset V.$$

This ends the proof of the first part of the theorem.

Now, we shall show that $V_\nu = V$ if $\lambda_0 \in \mathrm{Alg}_\nu(\mathfrak{L})$ for some $1 \leq \nu \leq r$. As in the proof of Lemma 4.3.4, for each $1 \leq j \leq \nu$ let U_j be a subspace of $N[\mathfrak{L}_0]$ satisfying (4.24), and set

$$X_m := \bigoplus_{j=1}^{m} U_j \oplus Y, \qquad 1 \leq m \leq \nu.$$

Arguing as in the proof of Lemma 4.3.4 shows that

$$\operatorname{codim} X_m = \operatorname{codim} V_m, \qquad 1 \leq m \leq \nu. \tag{4.38}$$

Recall that

$$\mathfrak{L}_0 \in \mathrm{Fred}_0(U, V) \quad \text{and} \quad \mathfrak{L}_0^m = \mathfrak{L}_0, \qquad 1 \leq m \leq \nu.$$

Looking for a contradiction, assume that $\lambda_0 \in \mathrm{Alg}_\nu(\mathfrak{L})$ with $1 \leq \nu \leq r$ and $V_\nu \subsetneq V$. Then, it follows from (4.38) that X_m is a proper subspace of U for each $1 \leq m \leq \nu$. Moreover, by construction, we already know that

$$N[\mathfrak{L}_0] = N[\mathfrak{L}_0^\nu] = U_1 \oplus \cdots \oplus U_\nu \oplus (N[\mathfrak{L}_0^\nu] \cap \cdots \cap N[\mathfrak{L}_\nu^\nu]).$$

Thus, since

$$U = N[\mathfrak{L}_0] \oplus Y \quad \text{and} \quad X_\nu \subsetneq U,$$

we find that

$$N[\mathfrak{L}_0^\nu] \cap \cdots \cap N[\mathfrak{L}_\nu^\nu] \neq \{0\}$$

and, therefore, we can pick a vector

$$u_0 \in N[\mathfrak{L}_0^\nu] \cap \cdots \cap N[\mathfrak{L}_\nu^\nu] \setminus \{0\}.$$

Necessarily,

$$\mathfrak{L}^\nu(\lambda)u_0 = o((\lambda - \lambda_0)^\nu) \quad \text{as } \lambda \to \lambda_0. \tag{4.39}$$

Consequently, since $\lambda_0 \in \mathrm{Alg}_\nu(\mathfrak{L})$ and, by construction,

$$\Psi^1(\lambda)\cdots\Psi^\nu(\lambda)u_0 = \mathfrak{L}(\lambda)^{-1}\mathfrak{L}^\nu(\lambda)u_0,$$

we find from (4.39) that

$$0 \neq u_0 = \lim_{\lambda\to\lambda_0} \Psi^1(\lambda)\cdots\Psi^\nu(\lambda)u_0 = \lim_{\lambda\to\lambda_0} \mathfrak{L}(\lambda)^{-1}\mathfrak{L}^\nu(\lambda)u_0 = 0,$$

which is impossible. This contradiction shows that $V_\nu = V$ and concludes the proof of the second assertion of the theorem.

Finally, suppose that $\lambda_0 \in \mathrm{Alg}_\nu(\mathfrak{L})$, with $1 \leq \nu \leq r$, and let $\Phi, \Psi \in \mathcal{C}^r(\Omega, \mathcal{L}(U))$ be such that $\Phi_0, \Psi_0 \in \mathrm{Iso}(U)$ and, for some $1 \leq k, \ell \leq \nu$, the point λ_0 is a k-transversal eigenvalue of $\mathfrak{L}^\Phi := \mathfrak{L}\Phi$, while λ_0 is an ℓ-transversal eigenvalue of $\mathfrak{L}^\Psi := \mathfrak{L}\Psi$. Then,

$$\mathfrak{L}^\Phi = \mathfrak{L}^\Psi\Psi^{-1}\Phi$$

and, according to Lemma 4.3.5, it is apparent that

$$\dim N[\mathfrak{L}_0^\Phi]\cap\cdots\cap N[\mathfrak{L}_j^\Phi] = \dim N[\mathfrak{L}_0^\Psi]\cap\cdots\cap N[\mathfrak{L}_j^\Psi] \tag{4.40}$$

for all $0 \leq j \leq \min\{k,\ell\}$.

On the other hand, according to Proposition 4.2.2, we have that, for each $1 \leq j \leq \min\{k,\ell\}$,

$$\dim \bigcap_{i=0}^{j-1} N[\mathfrak{L}_i^\Phi] = \dim \mathfrak{L}_j^\Phi\Big(\bigcap_{i=0}^{j-1} N[\mathfrak{L}_i^\Phi]\Big) + \dim \bigcap_{i=0}^{j} N[\mathfrak{L}_i^\Phi],$$

$$\dim \bigcap_{i=0}^{j-1} N[\mathfrak{L}_i^\Psi] = \dim \mathfrak{L}_j^\Psi\Big(\bigcap_{i=0}^{j-1} N[\mathfrak{L}_i^\Psi]\Big) + \dim \bigcap_{i=0}^{j} N[\mathfrak{L}_i^\Psi].$$

Consequently, putting together the above equalities with (4.40) shows

$$\dim \mathfrak{L}_j^\Phi(N[\mathfrak{L}_0^\Phi]\cap\cdots\cap N[\mathfrak{L}_{j-1}^\Phi]) = \dim \mathfrak{L}_j^\Psi(N[\mathfrak{L}_0^\Psi]\cap\cdots\cap N[\mathfrak{L}_{j-1}^\Psi])$$

for all $0 \leq j \leq \min\{k,\ell\}$. As, by Definition 4.2.1, we have that

$$\mathfrak{L}_k^\Phi\Big(\bigcap_{i=0}^{k-1} N[\mathfrak{L}_i^\Phi]\Big) \neq \{0\} \quad \text{and} \quad \mathfrak{L}_\ell^\Psi\Big(\bigcap_{i=0}^{\ell-1} N[\mathfrak{L}_i^\Psi]\Big) \neq \{0\},$$

necessarily $k = \ell$. This concludes the proof. □

4.4 Perturbation from simple eigenvalues

Throughout this section we assume that U is a subspace of V with continuous inclusion, that $\mathfrak{L} \in \mathcal{C}^r(\Omega, \mathcal{L}(U,V))$ for some $r \geq 1$, and that 0 is a simple eigenvalue of $\mathfrak{L}_0$, which means that

$$\dim N[\mathfrak{L}_0] = 1 \quad \text{and} \quad N[\mathfrak{L}_0] \oplus R[\mathfrak{L}_0] = V. \tag{4.41}$$

Exercises 2 and 3 will provide a list of necessary and sufficient conditions for the transversality condition of (4.41) to hold. In particular, it will be shown that $\mathfrak{L}_0 \in \mathrm{Fred}_0(U,V)$ if

$$\dim N[\mathfrak{L}_0] < \infty \quad \text{and} \quad N[\mathfrak{L}_0] \oplus R[\mathfrak{L}_0] = V.$$

Consequently, in this case, it is not necessary to impose the Fredholm condition $\mathfrak{L}_0 \in \mathrm{Fred}_0(U,V)$, as we have done in this chapter.

Under these general assumptions, the zero eigenvalue of $\mathfrak{L}_0$ perturbs into a unique eigenvalue $a(\lambda)$ of $\mathfrak{L}(\lambda)$. Precisely, the following perturbation result is satisfied.

Lemma 4.4.1. *Suppose U is a subspace of V with continuous inclusion, $r \in \mathbb{N}^* \cup \{\infty, \omega\}$, the family $\mathfrak{L} \in \mathcal{C}^r(\Omega, \mathcal{L}(U,V))$ satisfies (4.41), and $\lambda_0 \in \Omega$. Let Y be a closed subspace of U and $\varphi_0 \in N[\mathfrak{L}_0] \setminus \{0\}$ such that*

$$N[\mathfrak{L}_0] = \mathrm{span}[\varphi_0] \quad \textit{and} \quad N[\mathfrak{L}_0] \oplus Y = U.$$

Then there exist an open neighborhood Ω' of λ_0 and two unique functions $a \in \mathcal{C}^r(\Omega', \mathbb{K})$ and $\varphi \in \mathcal{C}^r(\Omega', U)$ such that, for each $\lambda \in \Omega'$,

$$\mathfrak{L}(\lambda)\varphi(\lambda) = a(\lambda)\varphi(\lambda) \tag{4.42}$$

and

$$a(\lambda_0) = 0, \quad \varphi(\lambda_0) = \varphi_0, \qquad \varphi(\lambda) - \varphi_0 \in Y.$$

Proof. Let $\mathcal{U}$ be a neighborhood of $(\lambda_0, 0, 0)$ in $\mathbb{K} \times \mathbb{K} \times Y$ such that the operator $\mathfrak{F} : \mathcal{U} \to V$ defined by

$$\mathfrak{F}(\lambda, a, z) := \mathfrak{L}(\lambda)(\varphi_0 + z) - a(\varphi_0 + z), \qquad (\lambda, a, z) \in \mathcal{U},$$

is well defined. Then, $\mathfrak{F} \in \mathcal{C}^r(\mathcal{U}, V)$ and $\mathfrak{F}(\lambda_0, 0, 0) = 0$. Moreover, the derivative of $\mathfrak{F}(\lambda_0, \cdot, \cdot)$ at $(0,0)$ equals the operator $T \in \mathcal{L}(\mathbb{K} \times Y, V)$ defined through

$$T(a, z) := \mathfrak{L}_0 z - a\varphi_0, \qquad (a, z) \in \mathbb{K} \times Y.$$

Condition (4.41) entails T being an isomorphism. The implicit function theorem easily completes the proof. $\square$

One of the main goals of perturbation theory is to ascertain the first non-vanishing term of the Taylor expansion of the perturbed eigenvalue function a, as the knowledge of those terms is imperative in most applications of the underlying theory. The next result shows that $a_1 := a'(\lambda_0) \neq 0$ if and only if λ_0 is a 1-transversal eigenvalue of $\mathfrak{L}$.

Proposition 4.4.2. *Under the same hypotheses and notation of Lemma 4.4.1, $a_1 \neq 0$ if and only if λ_0 is a 1-transversal eigenvalue of $\mathfrak{L}$.*

Proof. Differentiating (4.42) and particularizing at λ_0 gives

$$\mathfrak{L}_1\varphi_0 + \mathfrak{L}_0\varphi_1 = a_1\varphi_0. \tag{4.43}$$

Thus, $\mathfrak{L}_1\varphi_0 = -\mathfrak{L}_0\varphi_1 \in R[\mathfrak{L}_0]$ if $a_1 = 0$.

Conversely, $\mathfrak{L}_1\varphi_0 \in R[\mathfrak{L}_0]$ implies that

$$\mathfrak{L}_1\varphi_0 + \mathfrak{L}_0\varphi_1 \in R[\mathfrak{L}_0]$$

and, hence, due to (4.43), $a_1\varphi_0 \in R[\mathfrak{L}_0]$. Therefore, $a_1 = 0$, because we are assuming that $\varphi_0 \notin R[\mathfrak{L}_0]$, by (4.41). □

The following result shows that the algebraic multiplicity $\chi[\mathfrak{L};\lambda_0]$ of $\mathfrak{L}$ at λ_0 equals $\mathrm{ord}_{\lambda_0} a$, where a is the perturbed eigenvalue function from the zero eigenvalue of $\mathfrak{L}_0$, whose existence was established by Lemma 4.4.1.

Theorem 4.4.3. *Under the same hypotheses and notation of Lemma* 4.4.1*, suppose, in addition, that* $\lambda_0 \in \mathrm{Alg}_\nu(\mathfrak{L})$ *for some integer* $1 \leq \nu \leq r$*. Then, for every choice of* Y*,*

$$\chi[\mathfrak{L};\lambda_0] = \mathrm{ord}_{\lambda_0} a.$$

Moreover, $\chi[\mathfrak{L};\lambda_0]$ *equals the order of transversality of all transversalized families from* $\mathfrak{L}$ *at* λ_0*.*

Proof. According to Theorem 4.3.2, there exist $k \in \{1,\dots,\nu\}$ and $\Phi \in \mathbb{P}[\mathcal{L}(U)]$ with $\Phi_0 = I_U$ for which λ_0 is a k-transversal eigenvalue of

$$\mathfrak{L}^\Phi := \mathfrak{L}\Phi.$$

According to Definitions 4.2.1 and 4.3.3, we have that $\chi[\mathfrak{L};\lambda_0] = k$, because

$$\mathrm{codim}\, R[\mathfrak{L}_0] = 1,$$

by (4.41). Consequently, to prove the theorem it suffices to show the identity

$$\mathrm{ord}_{\lambda_0} a = k.$$

Note that, by Definition 4.2.1,

$$\mathfrak{L}_k^\Phi(N[\mathfrak{L}_0^\Phi] \cap \cdots \cap N[\mathfrak{L}_{k-1}^\Phi]) \neq \{0\}$$

and, hence, $N[\mathfrak{L}_0^\Phi] \cap \cdots \cap N[\mathfrak{L}_{k-1}^\Phi]$ is a non-zero space contained in

$$N[\mathfrak{L}_j^\Phi] = \mathrm{span}[\varphi_0],$$

so

$$\bigcap_{j=0}^{k-1} N[\mathfrak{L}_j^\Phi] = \mathrm{span}[\varphi_0].$$

Set

$$\psi(\lambda) := \Phi(\lambda)^{-1}\varphi(\lambda) \quad \text{for} \quad \lambda \sim \lambda_0.$$

Then, identity (4.42) can be expressed in the form

$$\mathfrak{L}^{\Phi}(\lambda)\psi(\lambda) = a(\lambda)\varphi(\lambda) \quad \text{for} \quad \lambda \sim \lambda_0. \tag{4.44}$$

We now show that $a_m \neq 0$ for some $1 \leq m \leq k$. Suppose the opposite to be true, so that $a_j = 0$ for each $0 \leq j \leq k$. Then, by (4.44),

$$\mathfrak{L}^{\Phi}(\lambda)\psi(\lambda) = o((\lambda - \lambda_0)^k) \quad \text{for} \quad \lambda \sim \lambda_0$$

and, hence, according to Lemma 4.1.2 applied to $\mathfrak{L}^{\Phi}$,

$$\psi_0 \in \bigcap_{j=0}^{k} N[\mathfrak{L}_j^{\Phi}].$$

On the other hand, by Proposition 4.2.2 applied to $\mathfrak{L}^{\Phi}$, we have

$$\bigcap_{j=0}^{k} N[\mathfrak{L}_j^{\Phi}] = \{0\}.$$

Therefore, $\psi_0 = 0$, which is a contradiction, since

$$\psi_0 = \Phi_0^{-1}\varphi_0 = \varphi_0 \neq 0.$$

Consequently,

$$1 \leq m := \operatorname{ord}_{\lambda_0} a \leq k$$

and, hence, it follows from (4.44) that

$$\mathfrak{L}^{\Phi}(\lambda)\psi(\lambda) = o((\lambda - \lambda_0)^{m-1}) \quad \text{for} \quad \lambda \sim \lambda_0.$$

Thus, Lemma 4.1.2 applied to $\mathfrak{L}^{\Phi}$ shows that

$$\psi_0, \dots, \psi_{m-1} \in N[\mathfrak{L}_0^{\Phi}] = \bigcap_{j=0}^{m-1} N[\mathfrak{L}_j^{\Phi}] = \operatorname{span}[\varphi_0].$$

Combining this with (4.44) and using the transversality of $\mathfrak{L}^{\Phi}$ yields

$$\mathfrak{L}_0^{\Phi}\psi_m + \mathfrak{L}_m^{\Phi}\psi_0 = a_m\varphi_0 \neq 0$$

and, hence, $\mathfrak{L}_m^{\Phi}\varphi_0 \neq 0$, because of $\mathfrak{L}_0^{\Phi} = \mathfrak{L}_0$ and (4.41). Therefore,

$$\dim \mathfrak{L}_m^{\Phi}(N[\mathfrak{L}_0^{\Phi}] \cap \cdots \cap N[\mathfrak{L}_{m-1}^{\Phi}]) = 1.$$

This implies $k = m$ and ends the proof. □

Theorem 5.7.1 will provide us with an improvement of Theorem 4.4.3 by simply assuming that $\chi[\mathfrak{L}; \lambda_0]$ is well defined, even if $\lambda_0 \notin \operatorname{Alg}(\mathfrak{L})$ (see also Exercise 5).

4.5 Exercises

1. Identify $\mathcal{L}(\mathbb{K}^2)$ with $\mathcal{M}_2(\mathbb{K})$, and let $\mathfrak{L}, \mathfrak{L}^* \in \mathcal{L}(\mathbb{K}, \mathcal{L}(\mathbb{K}^2))$ be the operator families defined by

$$\mathfrak{L}(\lambda) := \begin{pmatrix} \lambda & \lambda^2 \\ 1 & 0 \end{pmatrix}, \qquad \mathfrak{L}^*(\lambda) := \mathfrak{L}(\lambda)^*, \qquad \lambda \in \mathbb{K}.$$

Prove that 0 is a 2-transversal eigenvalue of $\mathfrak{L}$, but it is not a transversal eigenvalue of $\mathfrak{L}^*$.

2. Suppose U is a subspace of V with continuous inclusion and let $B \in \mathcal{L}(U, V)$ satisfy

$$\dim N[B] < \infty \quad \text{and} \quad N[B] \oplus R[B] = V.$$

Prove that $B \in \mathrm{Fred}_0(U, V)$.

3. Prove that the following conditions are equivalent for every $A \in \mathrm{Fred}_0(U, V)$ and $\alpha \in \mathbb{N}$:

- $N[A^\alpha] \oplus R[A^\alpha] = V$.
- $N[A^\alpha] \cap R[A^\alpha] = \{0\}$.
- $N[A^\alpha] + R[A^\alpha] = V$.
- $N[A^\alpha] \oplus (U \cap R[A^\alpha]) = U$.
- The restriction of A^α from $U \cap R[A^\alpha]$ to V is injective.
- The restriction of A^α from $U \cap R[A^\alpha]$ to $R[A^\alpha]$ is an isomorphism.
- $N[A^\alpha] = N[A^{\alpha+1}]$.
- $N[A^\alpha] = N[A^\beta]$ for all $\beta \geq \alpha$.
- $R[A^\alpha] = R[A^{\alpha+1}]$.
- $R[A^\alpha] = R[A^\beta]$ for all $\beta \geq \alpha$.

4. Given $T \in \mathcal{L}(U)$, consider the family $\mathfrak{L} \in \mathcal{C}^\omega(\mathbb{K}, \mathcal{L}(U))$ defined by

$$\mathfrak{L}(\lambda) := \lambda I - T, \qquad \lambda \in \mathbb{K},$$

and let $\lambda_0 \in \mathrm{Eig}(\mathfrak{L})$ be such that $\mathfrak{L}_0 \in \mathrm{Fred}_0(U)$. Prove the equivalence of the following assertions:

- λ_0 is a transversal eigenvalue of $\mathfrak{L}$.
- λ_0 is a 1-transversal eigenvalue of $\mathfrak{L}$.
- $N[\mathfrak{L}_0] \oplus R[\mathfrak{L}_0] = U$.

In such a case, $\chi[\mathfrak{L}; \lambda_0] - \dim N[\mathfrak{L}_0]$.

5. Identify $\mathcal{L}(\mathbb{R})$ with $\mathbb{R}$, and suppose $\mathfrak{L} \in \mathcal{C}^r(\Omega, \mathcal{L}(\mathbb{R}))$ for some $r \in \mathbb{N} \cup \{\infty\}$. Prove that $\mathrm{ord}_{\lambda_0} \mathfrak{L}$ is defined if and only if so is $\chi[\mathfrak{L}; \lambda_0]$, and that, in such a case,

$$\chi[\mathfrak{L}; \lambda_0] = \mathrm{ord}_{\lambda_0} \mathfrak{L}.$$

6. Suppose $\mathfrak{L} \in \mathcal{C}^r(\Omega, \mathcal{L}(U, V))$ for some $r \in \mathbb{N} \cup \{\infty\}$, and for a certain $\lambda_0 \in \Omega$ one has $\mathfrak{L}_0 \in \mathrm{Fred}_0(U, V)$, and $\chi[\mathfrak{L}; \lambda_0]$ is defined. Prove that

$$\chi[\mathfrak{L}; \lambda_0] \geq \dim N[\mathfrak{L}_0].$$

7. Let $\mathfrak{L} \in \mathcal{C}^r(\Omega, \mathcal{L}(U, V))$ be a family for which $\chi[\mathfrak{L}; \lambda_0]$ is defined, and consider the new family $\mathfrak{M}$ defined by

$$\mathfrak{M}(\lambda) := \mathfrak{L}(\lambda + \lambda_0), \qquad \text{for } \lambda + \lambda_0 \in \Omega.$$

Prove that

$$\chi[\mathfrak{M}; 0] = \chi[\mathfrak{L}; \lambda_0].$$

8. Let U, V, W be Banach spaces, $r \in \mathbb{N} \cup \{\infty\}$, and

$$\mathfrak{L} \in \mathcal{C}^r(\Omega, \mathcal{L}(U, V)), \qquad \Phi \in \mathcal{C}^r(\Omega, \mathcal{L}(V, W))$$

such that $\mathfrak{L}_0 \in \text{Fred}_0(U, V)$ and $\Phi_0 \in \text{Iso}(V, W)$ for some $\lambda_0 \in \Omega$. Given an integer $0 \leq k \leq r$, prove the following:

- λ_0 is a k-transversal eigenvalue of $\mathfrak{L}$ if and only if it is a k-transversal eigenvalue of $\Phi\mathfrak{L}$.
- $\chi[\mathfrak{L}; \lambda_0]$ is defined if and only if $\chi[\Phi\mathfrak{L}; \lambda_0]$ is defined, and, in such a case,

$$\chi[\mathfrak{L}; \lambda_0] = \chi[\Phi\mathfrak{L}; \lambda_0].$$

9. Let U_1, U_2, V_1, V_2 be Banach spaces and $\lambda_0 \in \Omega \subset \mathbb{K}$. Consider

$$\mathfrak{L} \in \mathcal{C}^r(\Omega, \mathcal{L}(U_1, V_1)) \quad \text{and} \quad \mathfrak{M} \in \mathcal{C}^r(\Omega, \mathcal{L}(U_2, V_2))$$

such that $\mathfrak{L}_0 \in \text{Fred}_0(U_1, V_1)$ and $\mathfrak{M}_0 \in \text{Fred}_0(U_2, V_2)$. Define the product family $\mathfrak{L} \times \mathfrak{M}$ as

$$(\mathfrak{L} \times \mathfrak{M})(\lambda) = \mathfrak{L}(\lambda) \times \mathfrak{M}(\lambda) \in \mathcal{L}(U_1 \times U_2, V_1 \times V_2), \qquad \lambda \in \Omega.$$

Prove, step by step, the following:

- For all integers $0 \leq s \leq r$,

$$(\mathfrak{L} \times \mathfrak{M})_s\Big(\bigcap_{j=0}^{s-1} N[(\mathfrak{L} \times \mathfrak{M})_j]\Big) = \mathfrak{L}_s\Big(\bigcap_{j=0}^{s-1} N[\mathfrak{L}_j]\Big) \times \mathfrak{M}_s\Big(\bigcap_{j=0}^{s-1} N[\mathfrak{M}_j]\Big).$$

- If λ_0 is a k-transversal eigenvalue of $\mathfrak{L}$ and λ_0 is an ℓ-transversal eigenvalue of $\mathfrak{M}$, then λ_0 is a $\max\{k, \ell\}$-transversal eigenvalue of $\mathfrak{L} \times \mathfrak{M}$.
- If λ_0 is a k-algebraic eigenvalue of $\mathfrak{L}$ and λ_0 is an ℓ-algebraic eigenvalue of $\mathfrak{M}$, then λ_0 is a $\max\{k, \ell\}$-algebraic eigenvalue of $\mathfrak{L} \times \mathfrak{M}$.
- Suppose $r = \infty$. If $\lambda_0 \notin \text{Alg}(\mathfrak{L})$ or $\lambda_0 \notin \text{Alg}(\mathfrak{M})$, then $\lambda_0 \notin \text{Alg}(\mathfrak{L} \times \mathfrak{M})$.
- $\chi[\mathfrak{L} \times \mathfrak{M}; \lambda_0]$ is defined if and only if both $\chi[\mathfrak{L}; \lambda_0]$ and $\chi[\mathfrak{M}; \lambda_0]$ are defined, and, in such a case,

$$\chi[\mathfrak{L} \times \mathfrak{M}; \lambda_0] = \chi[\mathfrak{L}; \lambda_0] + \chi[\mathfrak{M}; \lambda_0].$$

4.6 Comments on Chapter 4

The construction of the algebraic multiplicity presented in this chapter goes back to Esquinas & López-Gómez [29, 30, 31, 82]. Originally, it was introduced in the context of bifurcation theory to characterize the nonlinear eigenvalues of $\mathfrak{L}$, which further enhanced a huge industry in topological global analysis. A value $\lambda_0 \in \Omega$ is said to be a *nonlinear eigenvalue* of $\mathfrak{L} \in \mathcal{C}^r(\Omega, \mathcal{L}(U,V))$, for some $r \geq 2$, if for every map $\mathfrak{N} \in \mathcal{C}^2(\Omega \times U, V)$ such that

$$\mathfrak{N}(\lambda, 0) = 0, \qquad \lim_{u \to 0} \frac{\mathfrak{N}(\lambda, u)}{\|u\|} = 0, \qquad \lambda \sim \lambda_0,$$

there exists a sequence $\{(\xi_n, u_n)\}_{n \geq 1}$ in $\Omega \times (U \setminus \{0\})$ such that

$$\lim_{n \to \infty} (\xi_n, u_n) = (\lambda_0, 0) \quad \text{and} \quad \mathfrak{L}(\xi_n)u_n + \mathfrak{N}(\xi_n, u_n) = 0, \qquad n \geq 1.$$

The theory of Esquinas & López-Gómez established that, when $\chi[\mathfrak{L}; \lambda_0]$ is defined, a $\lambda_0 \in \mathrm{Eig}(\mathfrak{L})$ is a nonlinear eigenvalue of $\mathfrak{L}$ if and only if

$$\chi[\mathfrak{L}; \lambda_0] \in 2\mathbb{N} + 1.$$

These topics will be treated in Chapter 12, where we shall give a proof of the theorem of characterization of nonlinear eigenvalues through the local Smith form studied in Chapter 7 and the topological degree, which will be introduced there. So, we stop this discussion here.

The concept of algebraic eigenvalue goes back to López-Gómez [82]. It was introduced to overcome a serious gap in the proof of the main theorem of Esquinas [29], which is true as stated there in if $\mathfrak{L} \in \mathcal{C}^\omega(\Omega, \mathcal{L}(U,V))$, because in such a case all isolated eigenvalues must be algebraic, as will become apparent in Chapter 8.

Since at least Poincaré, the study of eigenvalue perturbations has been a milestone in the generation of new problems and ideas in spectral theory. Possibly, the paradigm of the perturbation theory of linear operators is the celebrated book of Kato [65]. A briefer text in the theory of perturbation theory of Fredholm operators is the review paper by Gohberg & Kreĭn [45]. In fact, Lemma 4.4.1 is attributable to Kato [65], where, in addition, the problem of constructing generalized algebraic multiplicities for families of the form

$$\mathfrak{L}(\lambda) = \mathfrak{L}_0 + (\lambda - \lambda_0)\mathfrak{L}_1, \qquad \lambda \in \mathbb{K},$$

in the context of perturbation theory, was originally addressed. Theorem 4.4.3 gives a rather satisfactory answer to the problem of analyzing spectral perturbations from simple eigenvalues; it goes back to López-Gómez [82], where, actually, a completely general answer to the perturbation problem was proposed by means of the topological degree. Essentially, when $\mathbb{K} = \mathbb{R}$, the algebraic multiplicity $\chi[\mathfrak{L}; \lambda_0]$ measures the parity of the Poincaré *crossing number* of $\mathfrak{L}(\lambda)$ as λ crosses λ_0. By crossing number is meant the total number of *classic eigenvalues* of $\mathfrak{L}(\lambda)$ that crosses the imaginary axis as λ crosses λ_0. This feature is crucial in the context of the theory of dynamical systems. According to Theorem 4.4.3, $a(\lambda)$ changes sign as λ crosses λ_0 if and only if $\chi[\mathfrak{L}; \lambda_0]$ is odd, and, consequently, $\chi[\mathfrak{L}; \lambda_0]$ provides us with an algebraic invariant characterizing whether or not $a(\lambda)$ crosses the imaginary axis along the real axis. Such characterizations are imperative to study the physical plausibility of $(\lambda, 0)$, as a mathematical solution, in a number of nonlinear equations describing real world models of the form

$$\mathfrak{L}(\lambda)u + \mathfrak{N}(\lambda, u) = 0,$$

where $\mathfrak{N}(\lambda, u) = o(\|u\|)$ is some nonlinear operator. Further details on these and some other closely related nonlinear topics, directed to an interdisciplinary audience, can be found in López-Gómez [85, 82].

Proposition 4.4.2 can be stated by saying that if zero is a simple eigenvalue of $\mathfrak{L}_0$, then $a_1 := a'(\lambda_0) \neq 0$ (a transversality condition of dynamical nature) if and only if

$$\mathfrak{L}_1(N[\mathfrak{L}_0]) \oplus R[\mathfrak{L}_0] = V \tag{4.45}$$

(a transversality condition of algebraic nature). This crucial feature goes back to Crandall & Rabinowitz [22, 23], and is on the roots of their extremely celebrated bifurcation theorem and exchange stability principle. Actually, (4.45) is referred to in the specialized literature on nonlinear analysis as the *transversality condition of Crandall and Rabinowitz.* From the very beginning, (4.45) triggered the abstract theory developed in this chapter. By its huge number of applications in the context of nonlinear partial differential equations, there is no doubt that the results of Crandall & Rabinowitz [22, 23] are among the most important mathematical works of the last decades.

Although the general exposition of this chapter has been based on [82], the notation and materials included here have been considerably polished and refined.

Chapter 5

Algebraic Multiplicity Through Polynomial Factorization

This chapter describes an equivalent approach to the concept of multiplicity $\chi[\mathfrak{L};\lambda_0]$ introduced in Chapter 4; in this occasion by means of an appropriate polynomial factorization of $\mathfrak{L}$ at λ_0. However, at first glance these approaches are seemingly completely different.

The approach adopted in this chapter provides two extremely relevant properties of $\chi[\mathfrak{L};\lambda_0]$, not discussed in Chapter 4, and most of the uniqueness results of Chapter 6. Precisely, it might be viewed as a technical device to prove the following:

- Given $\lambda_0 \in \Omega$ and $\mathfrak{L} \in \mathcal{C}^\infty(\Omega, \mathcal{L}(U,V))$, the family $\mathfrak{L}$ can be transversalized at $\lambda_0 \in \Omega$ if and only if $\lambda_0 \in \mathrm{Alg}(\mathfrak{L})$, equivalently, if and only if $\chi[\mathfrak{L};\lambda_0] < \infty$.
- The algebraic multiplicity of a product equals the sum of the algebraic multiplicities (product formula).

Chapter 5 is organized as follows. Section 5.1 constructs the new concept of multiplicity, denoted by $\mu[\mathfrak{L};\lambda_0]$, and proves its most elementary properties. Section 5.2 establishes some hidden connections between the multiplicities μ and χ. Section 5.3 shows that $\mu = \chi$ and uses this equivalence to prove that a $\mathcal{C}^\infty$ family $\mathfrak{L}$ can be transversalized at λ_0 if and only if $\lambda_0 \in \mathrm{Alg}(\mathfrak{L})$, which is an improvement of the second assertion of Theorem 4.3.2. Section 5.4 gives a formula for the *partial multiplicities* introduced in Section 5.1. Section 5.5 consists of the proof of the following real non-analytic counterpart of the celebrated theorem of Riemann on removable singularities.

Theorem 5.0.1 (theorem of removable singularities). *Let $r \in \mathbb{N} \cup \{\infty\}$, consider $\lambda_0 \in \Omega$ and $\mathfrak{L} \in \mathcal{C}^r(\Omega, \mathcal{L}(U,V))$ with $\mathfrak{L}(\lambda_0) \in \mathrm{Fred}_0(U,V)$. Suppose $(\lambda - \lambda_0)^\nu \mathfrak{L}(\lambda)^{-1}$ exists and is bounded for λ in $\Omega \setminus \{\lambda_0\}$, for some $\nu \in [0,r] \cap \mathbb{Z}$. Then* *there exists an open subset $\Omega' \subset \Omega$ with $\lambda_0 \in \Omega'$ for which the function*

$$\lambda \mapsto (\lambda - \lambda_0)^\nu \mathfrak{L}(\lambda)^{-1}, \qquad \lambda \in \Omega' \setminus \{\lambda_0\},$$

admits an extension of class $\mathcal{C}^{r-\nu}$ to Ω'.

Theorem 5.0.1 is a deep result whose proof is based on the main theorem of this chapter (Theorem 5.3.1). The condition $\mathfrak{L}_0 \in \mathrm{Fred}_0(U,V)$ cannot be removed from its statement (see Exercise 9). The assumption

> $(\lambda - \lambda_0)^\nu \mathfrak{L}(\lambda)^{-1}$ exists and is bounded for λ in $\Omega \setminus \{\lambda_0\}$, for some $\nu \in [0,r] \cap \mathbb{Z}$

holds if and only if $\lambda_0 \in \mathrm{Alg}_k(\mathfrak{L})$ for some $k \in [0,r] \cap \mathbb{Z}$. In such a case, the theorem applies for all $\nu \in [k,r] \cap \mathbb{Z}$. Chapter 8 shows that this is actually the case when $\mathfrak{L}$ is analytic and λ_0 is an isolated singularity of $\mathfrak{L}^{-1}$.

Section 5.6 is devoted to the proof of the product formula. Finally, Section 5.7 gives the sharp version of Theorem 4.4.3 announced at the end of Section 4.4.

As in Chapter 4, throughout this chapter we consider $\mathbb{K} \in \{\mathbb{R}, \mathbb{C}\}$, two $\mathbb{K}$-Banach spaces U and V, an open subset $\Omega \subset \mathbb{K}$, a point $\lambda_0 \in \Omega$, and a family $\mathfrak{L} \in \mathcal{C}^r(\Omega, \mathcal{L}(U,V))$ for some $r \in \mathbb{N} \cup \{\infty\}$, such that

$$\mathfrak{L}_0 := \mathfrak{L}(\lambda_0) \in \mathrm{Fred}_0(U,V),$$

even if not mentioned. As $\mathrm{Fred}_0(U,V)$ is an open set of $\mathcal{L}(U,V)$, the previous conditions actually imply

$$\mathfrak{L}(\lambda) \in \mathrm{Fred}_0(U,V) \qquad \text{for} \quad \lambda \sim \lambda_0.$$

5.1 Derived families and factorization

Given the family $\mathfrak{L}$ we consider the $r+1$ auxiliary families

$$\mathfrak{M}_{\{j\}} : \Omega \to \mathcal{L}(U,V), \qquad 0 \leq j < r+1,$$

defined inductively as follows:

1. $\mathfrak{M}_{\{0\}} := \mathfrak{L}$.
2. Let $0 \leq i < r$ and assume that a family

$$\mathfrak{M}_{\{i\}} \in \mathcal{C}^{r-i}(\Omega, \mathcal{L}(U,V))$$

has been defined. Consider a projection $\Pi_i \in \mathcal{L}(U)$ with range $N[\mathfrak{M}_{\{i\}}(\lambda_0)]$. Then, we define

$$\mathfrak{M}_{\{i+1\}}(\lambda) := \begin{cases} \mathfrak{M}_{\{i\}}(\lambda)\left[(\lambda-\lambda_0)^{-1}\Pi_i + I_U - \Pi_i\right] & \text{if } \lambda \neq \lambda_0, \\ \mathfrak{M}'_{\{i\}}(\lambda_0)\Pi_i + \mathfrak{M}_{\{i\}}(\lambda_0) & \text{if } \lambda = \lambda_0. \end{cases}$$

Note that $\mathfrak{M}_{\{i\}}(\lambda_0)\Pi_i = 0$ and $\mathfrak{M}_{\{i+1\}} \in \mathcal{C}^{r-i-1}(\Omega, \mathcal{L}(U,V))$.

Definition 5.1.1. The (finite or infinite) sequence $\{\mathfrak{M}_{\{j\}}\}_{j=0}^{r}$ is called a derived sequence from $\mathfrak{L}$ at λ_0. For each $0 \leq j < r+1$, the family $\mathfrak{M}_{\{j\}}$ is said to be a jth derived family from $\mathfrak{L}$ at λ_0. The sequence $\{\Pi_j\}_{j=0}^{r}$ is said to be the sequence of projections corresponding to $\{\mathfrak{M}_{\{j\}}\}_{j=0}^{r}$.

From now on, given a finite-rank projection $P \in \mathcal{L}(U)$, we will consider the polynomial family $\mathfrak{R}_P \in \mathbb{P}[\mathcal{L}(U)]$ defined by

$$\mathfrak{R}_P(t) := P + t(I_U - P), \qquad t \in \mathbb{K}. \tag{5.1}$$

Note that $\mathfrak{R}_P(t)$ is invertible for all $t \neq 0$ and that

$$\mathfrak{R}_P(t)^{-1} = \mathfrak{R}_P(t^{-1}), \qquad t \in \mathbb{K} \setminus \{0\}. \tag{5.2}$$

Using this notation, one may express

$$\mathfrak{M}_{\{i+1\}}(\lambda) = (\lambda - \lambda_0)^{-1}\mathfrak{M}_{\{i\}}(\lambda)\mathfrak{R}_{\Pi_i}(\lambda - \lambda_0), \qquad \lambda \in \Omega \setminus \{\lambda_0\}.$$

In the construction of $\mathfrak{M}_{\{j+1\}}$ from $\mathfrak{M}_{\{j\}}$, the projection Π_j can be chosen in many different ways. Consequently, there are, in fact, infinitely many derived sequences from $\mathfrak{L}$ at λ_0. The following result collects their main properties. In particular, it justifies the way they have been constructed. Note that, in Step 2 above, the projection Π_j exists if $\dim N[\mathfrak{M}_{\{j\}}(\lambda_0)] < \infty$, since every finite-dimensional subspace of U possesses a topological supplement.

Lemma 5.1.2. *Let $\{\mathfrak{M}_{\{j\}}\}_{j=0}^{r}$ be a derived sequence from $\mathfrak{L}$ at λ_0 with corresponding sequence of projections $\{\Pi_j\}_{j=0}^{r}$. Take $j \in \mathbb{N} \cap [0, r]$. Then*

$$\mathfrak{M}_{\{j\}}(\lambda) \in \mathrm{Fred}_0(U, V)$$

for all $\lambda \in \Omega$ satisfying $\mathfrak{L}(\lambda) \in \mathrm{Fred}_0(U, V)$. Moreover,

$$\mathfrak{M}_{\{j\}}(\lambda) = (\lambda - \lambda_0)^{-j}\mathfrak{L}(\lambda)\mathfrak{R}_{\Pi_0}(\lambda - \lambda_0)\cdots\mathfrak{R}_{\Pi_{j-1}}(\lambda - \lambda_0) \tag{5.3}$$

for all $\lambda \in \Omega \setminus \{\lambda_0\}$. Furthermore, for every $j \in \mathbb{N} \cap [0, r-1]$,

$$N[\Pi_j] \cap N[\mathfrak{M}_{\{j+1\}}(\lambda_0)] = \{0\} \tag{5.4}$$

and

$$\dim N[\mathfrak{M}_{\{j+1\}}(\lambda_0)] \leq \dim N[\mathfrak{M}_{\{j\}}(\lambda_0)]. \tag{5.5}$$

Proof. First, we show that $\mathfrak{M}_{\{j\}}(\lambda) \in \mathrm{Fred}_0(U, V)$ if $\mathfrak{L}(\lambda) \in \mathrm{Fred}_0(U, V)$. The proof is a simple induction on j. As $\mathfrak{M}_{\{0\}} = \mathfrak{L}$, there is nothing to prove when $j = 0$. Suppose $\mathfrak{M}_{\{j\}}(\lambda) \in \Phi_0(U, V)$ for some $0 \leq j < r$ and all $\lambda \in \Omega$ satisfying $\mathfrak{L}(\lambda) \in \Phi_0(U, V)$. Then, $N[\mathfrak{M}_{\{j\}}(\lambda_0)]$ is finite dimensional and, hence, there exists a projection $\Pi_j \in \mathcal{L}(U)$ with

$$R[\Pi_j] = N[\mathfrak{M}_{\{j\}}(\lambda_0)].$$

Thus, by definition,

$$\mathfrak{M}_{\{j+1\}}(\lambda_0) = \mathfrak{M}'_{\{i\}}(\lambda_0)\Pi_i + \mathfrak{M}_{\{i\}}(\lambda_0)$$

is a sum of a finite-rank operator plus an element of $\mathrm{Fred}_0(U,V)$, and, hence, it belongs to $\mathrm{Fred}_0(U,V)$. If $\lambda \neq \lambda_0$ and $\mathfrak{L}(\lambda) \in \mathrm{Fred}_0(U,V)$, then, by definition,

$$\mathfrak{M}_{\{j+1\}}(\lambda) := (\lambda-\lambda_0)^{-1}\mathfrak{M}_{\{j\}}(\lambda)\mathfrak{R}_{\Pi_j}(\lambda-\lambda_0)$$

and, hence, due to the induction hypothesis, $\mathfrak{M}_{\{j+1\}}(\lambda)$ is given by a product of a Fredholm operator of index zero with an invertible operator. Therefore,

$$\mathfrak{M}_{\{j+1\}}(\lambda) \in \mathrm{Fred}_0(U,V)$$

(see the properties of Fredholm operators listed in the presentation of Part II).

We now prove (5.3). By definition, for each $\lambda \in \Omega \setminus \{\lambda_0\}$,

$$\mathfrak{M}_{\{1\}}(\lambda) = (\lambda-\lambda_0)^{-1}\mathfrak{L}(\lambda)\mathfrak{R}_{\Pi_0}(\lambda-\lambda_0).$$

Suppose (5.3) is satisfied by a particular j. Then,

$$\begin{aligned}\mathfrak{M}_{\{j+1\}}(\lambda) =&(\lambda-\lambda_0)^{-1}\mathfrak{M}_{\{j\}}(\lambda)\mathfrak{R}_{\Pi_j}(\lambda-\lambda_0)\\ =&(\lambda-\lambda_0)^{-(j+1)}\mathfrak{L}(\lambda)\mathfrak{R}_{\Pi_0}(\lambda-\lambda_0)\cdots\mathfrak{R}_{\Pi_j}(\lambda-\lambda_0),\end{aligned}$$

which proves (5.3).

To prove (5.4), let

$$x \in N[\Pi_j] \cap N[\mathfrak{M}_{\{j+1\}}(\lambda_0)].$$

Then,

$$0 = \mathfrak{M}_{\{j+1\}}(\lambda_0)x = \mathfrak{M}'_{\{i\}}(\lambda_0)\Pi_i x + \mathfrak{M}_{\{i\}}(\lambda_0)x = \mathfrak{M}_{\{j\}}(\lambda_0)x$$

and, hence, $x \in N[\mathfrak{M}_{\{j\}}(\lambda_0)]$. Therefore,

$$0 = \Pi_j x = x.$$

This proves (5.4), and, hence, Π_j establishes a linear injection from $N[\mathfrak{M}_{\{j+1\}}(\lambda_0)]$ to $N[\mathfrak{M}_{\{j\}}(\lambda_0)]$ and, consequently, (5.5) holds. □

Suppose $\mathfrak{M}_{\{j\}}(\lambda_0)$ is invertible for some $j \geq 0$. Then, it follows from (5.5) that for each $j \leq k < r+1$,

$$N[\mathfrak{M}_{\{k\}}(\lambda_0)] = N[\mathfrak{M}_{\{j\}}(\lambda_0)] = \{0\}$$

and, hence, $\mathfrak{M}_{\{k\}} = \mathfrak{M}_{\{j\}}$, by definition. Bearing this in mind, we can introduce the following concepts of ascent and multiplicity.

Definition 5.1.3. Let $\{\mathfrak{M}_{\{j\}}\}_{j=0}^{r}$ be a derived sequence from $\mathfrak{L}$ at λ_0. Then, the ascent of $\mathfrak{L}$ at λ_0, denoted by $\alpha[\mathfrak{L};\lambda_0]$, is defined as follows:

- $\alpha[\mathfrak{L};\lambda_0]$ equals the least $j \in \mathbb{N} \cap [0,r]$ for which $\mathfrak{M}_{\{j\}}(\lambda_0) \in \mathrm{Iso}(U,V)$, if such a j exists.
- $\alpha[\mathfrak{L};\lambda_0] = \infty$, if $r = \infty$ and $\dim N[\mathfrak{M}_{\{j\}}(\lambda_0)] \geq 1$ for all $j \in \mathbb{N}$.

For every integer $0 \leq j \leq r$, the jth partial μ-multiplicity of $\mathfrak{L}$ at λ_0 is defined through

$$\mu_j[\mathfrak{L};\lambda_0] := \sum_{i=0}^{j} \dim N[\mathfrak{M}_{\{i\}}(\lambda_0)],$$

and the μ-multiplicity of the operator family $\mathfrak{L}$ at λ_0, denoted by $\mu[\mathfrak{L};\lambda_0]$, is defined by:

- $\mu[\mathfrak{L};\lambda_0] := \mu_{\alpha[\mathfrak{L};\lambda_0]}[\mathfrak{L};\lambda_0]$, if $\alpha[\mathfrak{L};\lambda_0] < \infty$.
- $\mu[\mathfrak{L};\lambda_0] := \infty$, if $\alpha[\mathfrak{L};\lambda_0] = \infty$.

The ascent $\alpha[\mathfrak{L};\lambda_0]$ and the μ-multiplicity $\mu[\mathfrak{L};\lambda_0]$ are left undefined when $r \in \mathbb{N}$, the family $\mathfrak{L}$ is of class $\mathcal{C}^r$, but $\mathfrak{L}$ is not of class $\mathcal{C}^{r+1}$ in any neighborhood of λ_0, and $\mathfrak{M}_{\{r\}}(\lambda_0)$ is not invertible. Moreover, we have actually defined

$$\alpha[\mathfrak{L};\lambda_0] = \mu[\mathfrak{L};\lambda_0] = 0$$

if $\mathfrak{L}_0 \in \mathrm{Iso}(U,V)$. Conversely, $\mathfrak{L}_0 \in \mathrm{Iso}(U,V)$ if $\alpha[\mathfrak{L};\lambda_0] = 0$ or $\mu[\mathfrak{L};\lambda_0] = 0$.

The concepts introduced by Definition 5.1.3 are consistent, in the sense that they do not vary when the sequence of derived families changes. Indeed, let $\{\mathfrak{M}_{\{j\}}\}_{j=0}^{r}$ and $\{\hat{\mathfrak{M}}_{\{j\}}\}_{j=0}^{r}$ be two arbitrary derived sequences from $\mathfrak{L}$ at λ_0 with corresponding projections $\{\Pi_j\}_{j=0}^{r}$ and $\{\hat{\Pi}_j\}_{j=0}^{r}$, respectively. We claim that for each $0 \leq j < r+1$ there exists a polynomial $\Theta_j \in \mathbb{P}[\mathcal{L}(U)]$ such that $\Theta_j(\lambda_0) \in \mathrm{Iso}(U)$ and

$$\hat{\mathfrak{M}}_{\{j\}}(\lambda) = \mathfrak{M}_{\{j\}}(\lambda)\Theta_j(\lambda), \quad \lambda \in \Omega. \tag{5.6}$$

This actually shows the consistency of Definition 5.1.3, because it implies

$$\dim N[\hat{\mathfrak{M}}_{\{j\}}(\lambda_0)] = \dim N[\mathfrak{M}_{\{j\}}(\lambda_0)], \qquad j \in \mathbb{N} \cap [0,r].$$

To prove the previous claim we proceed by induction on j. Assume $j = 0$. Then, (5.6) is satisfied with

$$\Theta_0(\lambda) = I_U, \qquad \lambda \in \mathbb{K},$$

since

$$\hat{\mathfrak{M}}_{\{0\}} = \mathfrak{M}_{\{0\}} = \mathfrak{L}.$$

Now, assume that (5.6) is satisfied for some particular $j \geq 0$ and that Θ_j and Θ_j^{-1} are polynomials. Then,

$$\Theta_j(\lambda_0)(N[\hat{\mathfrak{M}}_{\{j\}}(\lambda_0)]) = N[\mathfrak{M}_{\{j\}}(\lambda_0)]$$

and, hence,

$$\Pi_j \Theta_j(\lambda_0)\hat{\Pi}_j = \Theta_j(\lambda_0)\hat{\Pi}_j, \quad \hat{\Pi}_j \Theta_j(\lambda_0)^{-1}\Pi_j = \Theta_j(\lambda_0)^{-1}\Pi_j. \tag{5.7}$$

Thus, using the induction hypothesis and (5.2), one obtains that, for each $\lambda \in \Omega \setminus \{\lambda_0\}$,

$$\begin{aligned} \hat{\mathfrak{M}}_{\{j+1\}}(\lambda) &= (\lambda - \lambda_0)^{-1} \hat{\mathfrak{M}}_j(\lambda) \mathfrak{R}_{\hat{\Pi}_j}(\lambda - \lambda_0) \\ &= (\lambda - \lambda_0)^{-1} \mathfrak{M}_{\{j\}}(\lambda) \Theta_j(\lambda) \mathfrak{R}_{\hat{\Pi}_j}(\lambda - \lambda_0) \\ &= (\lambda - \lambda_0)^{-1} \mathfrak{M}_{\{j\}}(\lambda) \mathfrak{R}_{\Pi_j}(\lambda - \lambda_0) \Theta_{j+1}(\lambda) \\ &= \mathfrak{M}_{\{j+1\}}(\lambda) \Theta_{j+1}(\lambda), \end{aligned} \tag{5.8}$$

where Θ_{j+1} is defined through

$$\Theta_{j+1}(\lambda) := \mathfrak{R}_{\Pi_j}((\lambda - \lambda_0)^{-1}) \Theta_j(\lambda) \mathfrak{R}_{\hat{\Pi}_j}(\lambda - \lambda_0), \quad \lambda \in \mathbb{K} \setminus \{\lambda_0\},$$

whose inverse is given by

$$\Theta_{j+1}(\lambda)^{-1} = \mathfrak{R}_{\hat{\Pi}_j}((\lambda - \lambda_0)^{-1}) \Theta_j(\lambda)^{-1} \mathfrak{R}_{\Pi_j}(\lambda - \lambda_0), \quad \lambda \in \mathbb{K} \setminus \{\lambda_0\}.$$

It remains to show that the families Θ_{j+1} and Θ_{j+1}^{-1} are indeed polynomials when extended to $\mathbb{K}$ by passing to the limit as $\lambda \to \lambda_0$. The unique doubtful terms in their developments are

$$(\lambda - \lambda_0)^{-1}(I_U - \Pi_j)\Theta_j(\lambda)\hat{\Pi}_j \quad \text{and} \quad (\lambda - \lambda_0)^{-1}(I_U - \hat{\Pi}_j)\Theta_j(\lambda)^{-1}\Pi_j,$$

respectively, but, actually, owing to (5.7), we have that

$$(I_U - \Pi_j)\Theta_j(\lambda_0)\hat{\Pi}_j = 0 \quad \text{and} \quad (I_U - \hat{\Pi}_j)\Theta_j(\lambda_0)^{-1}\Pi_j = 0.$$

As, according to the induction hypothesis Θ_j and Θ_j^{-1} are polynomials, necessarily so are Θ_{j+1} and Θ_{j+1}^{-1}. Finally, passing to the limit as $\lambda \to \lambda_0$ in (5.8) shows also that

$$\hat{\mathfrak{M}}_{\{j+1\}}(\lambda_0) = \mathfrak{M}_{\{j+1\}}(\lambda_0)\Theta_{j+1}(\lambda_0).$$

This concludes the induction and proves the consistency of Definition 5.1.3.

The next result shows that the ascent $\alpha[\mathfrak{L}; \lambda_0]$ and the multiplicity $\mu[\mathfrak{L}; \lambda_0]$ are invariant by pre- and post-multiplication of local isomorphism families.

Proposition 5.1.4. *Suppose $r \geq 1$ and $\mathfrak{L}, \hat{\mathfrak{L}} \in \mathcal{C}^r(\Omega, \mathcal{L}(U, V))$ with $\mathfrak{L}_0 \in \mathrm{Fred}_0(U, V)$. If there exist $\Phi \in \mathcal{C}^r(\Omega, \mathcal{L}(U))$ such that $\Phi_0 \in \mathrm{Iso}(U)$ and $\hat{\mathfrak{L}} = \mathfrak{L}\Phi$, then*

$$\mu_j[\hat{\mathfrak{L}}; \lambda_0] = \mu_j[\mathfrak{L}; \lambda_0], \qquad 0 \leq j < r + 1. \tag{5.9}$$

Consequently,

$$\alpha[\hat{\mathfrak{L}}; \lambda_0] = \alpha[\mathfrak{L}; \lambda_0] \quad \text{and} \quad \mu[\hat{\mathfrak{L}}; \lambda_0] = \mu[\mathfrak{L}; \lambda_0] \tag{5.10}$$

if $\mu[\mathfrak{L}; \lambda_0]$ is defined.

Similarly, (5.9) and (5.10) are satisfied if there exists $\Psi \in \mathcal{C}^r(\Omega, \mathcal{L}(V))$ with $\Psi_0 \in \mathrm{Iso}(V)$ such that $\hat{\mathfrak{L}} = \Psi\mathfrak{L}$.

Proof. Note that

$$\hat{\mathfrak{L}}_0 = \mathfrak{L}_0 \Phi_0 \in \mathrm{Fred}_0(U,V).$$

By making Ω smaller, if necessary, we can suppose, without loss of generality, that $\Phi(\lambda) \in \mathrm{Iso}(U)$ for all $\lambda \in \Omega$.

Let $\{\mathfrak{M}_{\{j\}}\}_{j=0}^r$ and $\{\hat{\mathfrak{M}}_{\{j\}}\}_{j=0}^r$ be two derived sequences from $\mathfrak{L}$ and $\hat{\mathfrak{L}}$, respectively, at λ_0 with corresponding projections $\{\Pi_j\}_{j=0}^r$ and $\{\hat{\Pi}_j\}_{j=0}^r$. As in the proof of the consistency of Definition 5.1.3, it suffices to show that, for each $0 \le j < r+1$, there exists $\Theta_j \in \mathcal{C}^{r-j}(\Omega, \mathcal{L}(U))$ such that $\Theta_j(\lambda_0) \in \mathrm{Iso}(U)$ and

$$\hat{\mathfrak{M}}_{\{j\}} = \mathfrak{M}_{\{j\}} \Theta_j. \tag{5.11}$$

Similarly, the proof of (5.11) involves an induction argument. Clearly, when $j = 0$, equality (5.11) holds true with

$$\Theta_0 := \Phi \in \mathcal{C}^r(\Omega, \mathcal{L}(U)).$$

Suppose (5.11) holds for a particular $0 \le j < r$ and some $\Theta_j \in \mathcal{C}^{r-j}(\Omega, \mathcal{L}(U))$ with $\Theta(\lambda_0) \in \mathrm{Iso}(U)$. Then, for each $\lambda \in \Omega \setminus \{\lambda_0\}$, it follows from (5.11) and the definition of derived family that

$$\hat{\mathfrak{M}}_{\{j+1\}}(\lambda) = \mathfrak{M}_{\{j+1\}}(\lambda)\Theta_{j+1}(\lambda), \tag{5.12}$$

where

$$\Theta_{j+1}(\lambda) := \mathfrak{R}_{\Pi_j}((\lambda-\lambda_0)^{-1})\Theta_j(\lambda)\mathfrak{R}_{\hat{\Pi}_j}(\lambda-\lambda_0).$$

Note that, for each $\lambda \in \Omega \setminus \{\lambda_0\}$,

$$\Theta_{j+1}(\lambda)^{-1} = \mathfrak{R}_{\hat{\Pi}_j}((\lambda-\lambda_0)^{-1})\Theta_j(\lambda)^{-1}\mathfrak{R}_{\Pi_j}(\lambda-\lambda_0),$$

and that $\Theta_j^{-1} \in \mathcal{C}^{r-j}(\Omega, \mathcal{L}(U))$ as well. As in the proof of the consistency, it is apparent from the induction hypothesis that

$$\Pi_j\Theta_j(\lambda_0)\hat{\Pi}_j = \Theta_j(\lambda_0)\hat{\Pi}_j, \qquad \hat{\Pi}_j\Theta_j(\lambda_0)^{-1}\Pi_j = \Theta_j(\lambda_0)^{-1}\Pi_j,$$

and, therefore,

$$\lim_{\lambda\to\lambda_0}(\lambda-\lambda_0)^{-1}(I_U-\Pi_j)\Theta_j(\lambda)\hat{\Pi}_j = (I_U-\Pi_j)\Theta_j'(\lambda_0)\hat{\Pi}_j,$$

$$\lim_{\lambda\to\lambda_0}(\lambda-\lambda_0)^{-1}(I_U-\hat{\Pi}_j)\Theta_j(\lambda)^{-1}\Pi_j = (I_U-\hat{\Pi}_j)[\Theta_j^{-1}]'(\lambda_0)\Pi_j.$$

In particular, the operators

$$\Theta_{j+1}(\lambda_0) := \lim_{\lambda\to\lambda_0}\Theta_{j+1}(\lambda) \quad \text{and} \quad \lim_{\lambda\to\lambda_0}\Theta_{j+1}(\lambda)^{-1}$$

are well defined and mutually inverse. Consequently, the families Θ_{j+1} and Θ_{j+1}^{-1} are of class $\mathcal{C}^{r-j-1}$ in Ω. Passing to the limit as $\lambda \to \lambda_0$ in (5.12) completes the proof of the first claim.

The proof of the last assertion of the statement is proposed in Exercise 1. $\square$

Chapter 7 will provide a general version of Proposition 5.1.4, where the regularity assumptions imposed on the isomorphisms families are shown to be superfluous.

5.2 Connections between χ and μ

This section reveals the rather hidden connections between the multiplicity $\mu[\mathfrak{L};\lambda_0]$ introduced by Definition 5.1.3 and the multiplicity $\chi[\mathfrak{L};\lambda_0]$ introduced by Definition 4.3.3. As a by-product, it will become apparent that $\mu[\mathfrak{L};\lambda_0] < \infty$ if $\lambda_0 \in \mathrm{Alg}_\nu(\mathfrak{L})$ for some integer $0 \leq \nu < r+1$. In such a case, according to Definition 4.3.3, the multiplicity $\chi[\mathfrak{L};\lambda_0]$ is well defined and finite. It turns out that

$$\mu[\mathfrak{L};\lambda_0] = \chi[\mathfrak{L};\lambda_0] \quad \text{if} \quad \lambda_0 \in \bigcup_{0\leq\nu<r+1} \mathrm{Alg}_\nu(\mathfrak{L}). \tag{5.13}$$

The proof of (5.13) is based on the next result.

Theorem 5.2.1. *Suppose $r \geq 0$, the family $\mathfrak{L} \in \mathcal{C}^r(\Omega, \mathcal{L}(U,V))$ satisfies*

$$\mathfrak{L}_0 \in \mathrm{Fred}_0(U,V),$$

and, for some $k \in \mathbb{N} \cap [0,r]$, the linear spaces

$$R[\mathfrak{L}_0],\ \mathfrak{L}_1(N[\mathfrak{L}_0]),\ \ldots,\ \mathfrak{L}_k\Big(\bigcap_{j=0}^{k-1} N[\mathfrak{L}_j]\Big) \tag{5.14}$$

are in direct sum. Then, for every sequence $\{\mathfrak{M}_{\{j\}}\}_{j=0}^r$ of derived families from $\mathfrak{L}$ at λ_0,

$$N[\mathfrak{M}_{\{i\}}(\lambda_0)] = N[\mathfrak{L}_0] \cap \cdots \cap N[\mathfrak{L}_i], \tag{5.15}$$

for all $i \in \{0,\ldots,k\}$, and

$$\mu_{k-1}[\mathfrak{L};\lambda_0] = \sum_{j=1}^{k} j \cdot \dim \mathfrak{L}_j\Big(\bigcap_{i=0}^{j-1} N[\mathfrak{L}_i]\Big) + k \cdot \dim N[\mathfrak{M}_{\{k\}}(\lambda_0)]. \tag{5.16}$$

Proof. Let $\{\mathfrak{M}_{\{j\}}\}_{j=0}^r$ be a sequence of derived families from $\mathfrak{L}$ at λ_0 with corresponding sequence of projections $\{\Pi_j\}_{j=0}^r$. Proving (5.15) requires induction. By construction,

$$N[\mathfrak{M}_{\{0\}}(\lambda_0)] = N[\mathfrak{L}_0].$$

Now, take $1 \leq j \leq k$ and assume that (5.15) is satisfied for each $0 \leq i \leq j-1$. Then, for each $0 \leq \ell \leq i \leq j-1$,

$$N[\mathfrak{M}_{\{i\}}(\lambda_0)] \subset N[\mathfrak{M}_{\{\ell\}}(\lambda_0)]$$

and, hence,

$$\begin{aligned} &\Pi_\ell \Pi_i = \Pi_i, && (I_U - \Pi_\ell)\Pi_i = 0, \\ &\Pi_\ell (I_U - \Pi_i) = \Pi_\ell - \Pi_i, && (I_U - \Pi_\ell)(I_U - \Pi_i) = I_U - \Pi_\ell. \end{aligned} \tag{5.17}$$

Now, it follows from (5.1) and (5.17) that, for every $\lambda \in \Omega$,

$$\begin{aligned} &\mathfrak{R}_{\Pi_0}(\lambda - \lambda_0)\mathfrak{R}_{\Pi_1}(\lambda - \lambda_0) \cdots \mathfrak{R}_{\Pi_{j-1}}(\lambda - \lambda_0) \\ &= \Pi_{j-1} + \sum_{i=1}^{j-1} (\lambda - \lambda_0)^i \Pi_{j-i-1}(I_U - \Pi_{j-i}) + (\lambda - \lambda_0)^j (I_U - \Pi_0). \end{aligned} \tag{5.18}$$

On the other hand, according to (5.3),

$$\begin{aligned} \mathfrak{M}_{\{j\}}(\lambda) = (\lambda - \lambda_0)^{-j} \Big[\sum_{i=0}^{j} (\lambda - \lambda_0)^i \mathfrak{L}_i + o((\lambda - \lambda_0)^j) \Big] \\ \cdot \mathfrak{R}_{\Pi_0}(\lambda - \lambda_0)\mathfrak{R}_{\Pi_1}(\lambda - \lambda_0) \cdots \mathfrak{R}_{\Pi_{j-1}}(\lambda - \lambda_0). \end{aligned} \tag{5.19}$$

Therefore, substituting (5.18) into (5.19) and using the induction hypothesis demonstrate that

$$\mathfrak{M}_{\{j\}}(\lambda_0) = \mathfrak{L}_0 (I_U - \Pi_0) + \sum_{i=1}^{j-1} \mathfrak{L}_i \Pi_{i-1}(I_U - \Pi_i) + \mathfrak{L}_j \Pi_{j-1}. \tag{5.20}$$

In particular, by (5.20),

$$R[\mathfrak{M}_{\{j\}}(\lambda_0)] \subset \bigoplus_{i=1}^{j} \mathfrak{L}_i (N[\mathfrak{L}_0] \cap \cdots \cap N[\mathfrak{L}_{i-1}]) \oplus R[\mathfrak{L}_0]. \tag{5.21}$$

Now we will prove that

$$N[\mathfrak{M}_{\{j\}}(\lambda_0)] \subset N[\mathfrak{L}_0] \cap \cdots \cap N[\mathfrak{L}_j]. \tag{5.22}$$

Take $u_0 \in N[\mathfrak{M}_{\{j\}}(\lambda_0)]$. Then, using (5.20) and (5.21), we observe that, for each $1 \leq i \leq j-1$,

$$0 = \mathfrak{L}_0 (I - \Pi_0) u_0 = \mathfrak{L}_i \Pi_{i-1} (I - \Pi_i) u_0 = \mathfrak{L}_j \Pi_{j-1} u_0. \tag{5.23}$$

On the other hand, it follows from the definition of Π_0 that

$$\Pi_0 u_0 \in N[\mathfrak{M}_{\{0\}}(\lambda_0)] = N[\mathfrak{L}_0]$$

and so $\mathfrak{L}_0 \Pi_0 u_0 = 0$. Thus, the first equality of (5.23) implies $\mathfrak{L}_0 u_0 = 0$. In particular, $\Pi_0 u_0 = u_0$.

Now, fix $0 \leq \ell \leq j-2$ and assume

$$u_0 \in N[\mathfrak{L}_0] \cap \cdots \cap N[\mathfrak{L}_\ell]. \tag{5.24}$$

By (5.17), equation (5.23) implies that

$$0 = \mathfrak{L}_{\ell+1}\Pi_\ell(I_U - \Pi_{\ell+1})u_0 = \mathfrak{L}_{\ell+1}(\Pi_\ell u_0 - \Pi_{\ell+1}u_0) = \mathfrak{L}_{\ell+1}\Pi_\ell u_0,$$

since

$$\Pi_{\ell+1}u_0 \in N[\mathfrak{M}_{\{\ell+1\}}(\lambda_0)] = N[\mathfrak{L}_0] \cap \cdots \cap N[\mathfrak{L}_{\ell+1}] \subset N[\mathfrak{L}_{\ell+1}],$$

because of the induction hypothesis. Thus,

$$\Pi_\ell u_0 \in N[\mathfrak{L}_{\ell+1}].$$

On the other hand, using the induction hypothesis, we find from (5.24) that

$$u_0 \in N[\mathfrak{M}_{\{\ell\}}(\lambda_0)].$$

Hence, $\Pi_\ell u_0 = u_0$ and, therefore,

$$u_0 \in N[\mathfrak{L}_0] \cap \cdots \cap N[\mathfrak{L}_{\ell+1}].$$

We have just proved that

$$u_0 \in N[\mathfrak{L}_0] \cap \cdots \cap N[\mathfrak{L}_{j-1}]. \tag{5.25}$$

Moreover, it follows from the last identity of (5.23) that

$$\Pi_{j-1}u_0 \in N[\mathfrak{L}_j],$$

and, using the induction assumption, (5.25) implies

$$u_0 \in N[\mathfrak{M}_{\{j-1\}}(\lambda_0)].$$

Therefore, $\Pi_{j-1}u_0 = u_0$ and

$$u_0 \in N[\mathfrak{L}_0] \cap \cdots \cap N[\mathfrak{L}_j], \tag{5.26}$$

which concludes the proof of (5.22).

Now, we prove the converse inclusion of (5.22). Suppose (5.26). By the induction hypothesis,

$$u_0 \in N[\mathfrak{M}_{\{0\}}(\lambda_0)] \cap \cdots \cap N[\mathfrak{M}_{\{j-1\}}(\lambda_0)],$$

and, therefore,

$$\Pi_0 u_0 = \cdots = \Pi_{j-1}u_0 = u_0.$$

By (5.20), these equalities imply

$$\mathfrak{M}_{\{j\}}(\lambda_0)u_0 = \mathfrak{L}_j u_0 = 0,$$

which concludes the proof of (5.15).

Finally, by (5.15), Proposition 4.2.2 shows that, for each $0 \le i \le k-1$,

$$\dim N[\mathfrak{M}_{\{i\}}(\lambda_0)] = \dim N[\mathfrak{M}_{\{i+1\}}(\lambda_0)] + \dim \mathfrak{L}_{i+1}(N[\mathfrak{M}_{\{i\}}(\lambda_0)]).$$

Thus, for each $0 \le \ell \le k-1$,

$$\begin{aligned}\sum_{i=\ell}^{k-1} \dim N[\mathfrak{M}_{\{i\}}(\lambda_0)] &= \sum_{i=\ell}^{k-1} \left[\dim \mathfrak{L}_{i+1}(N[\mathfrak{M}_{\{i\}}(\lambda_0)]) + \dim N[\mathfrak{M}_{\{i+1\}}(\lambda_0)]\right] \\ &= \sum_{i=\ell+1}^{k} \dim \mathfrak{L}_i(N[\mathfrak{M}_{\{i-1\}}(\lambda_0)]) \\ &\quad + \sum_{i=\ell+1}^{k-1} \dim N[\mathfrak{M}_{\{i\}}(\lambda_0)] + \dim N[\mathfrak{M}_{\{k\}}(\lambda_0)].\end{aligned}$$

Therefore, iterating the previous identity gives

$$\sum_{i=0}^{k-1} \dim N[\mathfrak{M}_{\{i\}}(\lambda_0)] = \sum_{j=1}^{k} j \cdot \dim \mathfrak{L}_j(N[\mathfrak{M}_{\{j-1\}}(\lambda_0)]) + k \cdot \dim N[\mathfrak{M}_{\{k\}}(\lambda_0)],$$

since

$$\dim N[\mathfrak{M}_{\{k-1\}}(\lambda_0)] = \dim \mathfrak{L}_k(N[\mathfrak{M}_{\{k-1\}}(\lambda_0)]) + \dim N[\mathfrak{M}_{\{k\}}(\lambda_0)].$$

This proves (5.16) and concludes the proof. □

Based on Theorem 5.2.1, Proposition 5.1.4 and Theorem 4.3.2, we can obtain the main result of this section.

Theorem 5.2.2. *Suppose* $r \in \mathbb{N} \cup \{\infty\}$, *the family* $\mathfrak{L} \in \mathcal{C}^r(\Omega, \mathcal{L}(U,V))$ *satisfies* $\mathfrak{L}_0 \in \mathrm{Fred}_0(U,V)$, *and* $\lambda_0 \in \mathrm{Alg}_\nu(\mathfrak{L})$ *for some integer* $1 \le \nu \le r$. *Let* $\Phi \in \mathcal{C}^r(\Omega, \mathcal{L}(U))$ *satisfy* $\Phi_0 \in \mathrm{Iso}(U)$ *and* λ_0 *is a* k*-transversal eigenvalue of* $\mathfrak{L}^\Phi := \mathfrak{L}\Phi$ *for some* $k \subset \{1, \dots, \nu\}$. *Then*

$$\alpha[\mathfrak{L}; \lambda_0] = k$$

and

$$\mu_j[\mathfrak{L}; \lambda_0] = \mu[\mathfrak{L}; \lambda_0] = \chi[\mathfrak{L}; \lambda_0], \qquad k-1 \le j < r+1,$$

Proof. Let $\{\mathfrak{M}_{\{j\}}\}_{j=0}^r$ be a sequence of derived families from $\mathfrak{L}^\Phi$ at λ_0. By Definition 4.3.3,

$$\chi[\mathfrak{L}; \lambda_0] = \sum_{h=1}^{k} h \dim \mathfrak{L}_h^\Phi(N[\mathfrak{L}_0^\Phi] \cap \cdots \cap N[\mathfrak{L}_{h-1}^\Phi])$$

and, according to Proposition 5.1.4,

$$\alpha[\mathfrak{L};\lambda_0] = \alpha[\mathfrak{L}^\Phi;\lambda_0]$$

and

$$\mu_j[\mathfrak{L};\lambda_0] = \mu_j[\mathfrak{L}^\Phi;\lambda_0] \qquad 0 \le j < r+1.$$

Thus, to complete the proof, it suffices to show $\alpha[\mathfrak{L}^\Phi;\lambda_0] = k$ and

$$\mu_j[\mathfrak{L}^\Phi;\lambda_0] = \sum_{h=1}^{k} h \dim \mathfrak{L}_h^\Phi (N[\mathfrak{L}_0^\Phi] \cap \cdots \cap N[\mathfrak{L}_{h-1}^\Phi]) \tag{5.27}$$

for each integer $k-1 \le j \le r$. Indeed, as λ_0 is a k-transversal eigenvalue of $\mathfrak{L}^\Phi$, by Proposition 4.2.2,

$$N[\mathfrak{L}_0^\Phi] \cap \cdots \cap N[\mathfrak{L}_k^\Phi] = \{0\}$$

and, hence, due to Theorem 5.2.1,

$$N[\mathfrak{M}_{\{j\}}(\lambda_0)] = \{0\}, \qquad k \le j < r+1. \tag{5.28}$$

On the other hand, since λ_0 is a k-transversal eigenvalue of $\mathfrak{L}^\Phi$,

$$N[\mathfrak{L}_0^\Phi] \cap \cdots \cap N[\mathfrak{L}_{k-1}^\Phi] \ne \{0\}$$

and, hence, thanks again to Theorem 5.2.1, we find that

$$N[\mathfrak{M}_{\{k-1\}}(\lambda_0)] \ne \{0\}.$$

Therefore, k is the least natural number $j \ge 0$ for which $\mathfrak{M}_{\{j\}}(\lambda_0)$ is invertible. Consequently, $\alpha[\mathfrak{L}^\Phi;\lambda_0] = k$, by Definition 5.1.3. Finally, (5.27) follows straight from (5.16) and (5.28). □

5.3 Coincidence of the multiplicities χ and μ

This section shows that both multiplicities χ and μ are defined if and only if either $r = \infty$, or $r \in \mathbb{N}$ and $\lambda_0 \in \mathrm{Alg}_\nu(\mathfrak{L})$ for some integer $\nu \in \{0, \dots, r\}$; in any of these cases, $\mu = \chi$. Moreover, it reveals that the multiplicity of $\mathfrak{L}$ at λ_0 exists and is finite if and only if $\lambda_0 \in \mathrm{Alg}_\nu(\mathfrak{L})$ for some $0 \le \nu < r+1$. In particular, in the second assertion of Theorem 4.3.2, the assumption that λ_0 is an algebraic eigenvalue of $\mathfrak{L}$ is not only sufficient for its validity, but also necessary. Precisely, the following result is satisfied.

Theorem 5.3.1 (of coincidence and finiteness of χ and μ). *Let $r \in \mathbb{N} \cup \{\infty\}$ and $\mathfrak{L} \in \mathcal{C}^r(\Omega, \mathcal{L}(U,V))$ be with $\mathfrak{L}_0 \in \mathrm{Fred}_0(U,V)$. Then the following assertions are equivalent:*

i) $\lambda_0 \in \mathrm{Alg}_\nu(\mathfrak{L})$ *for some integer* $0 \le \nu \le r$.

ii) $\mu[\mathfrak{L};\lambda_0]$ *is defined and finite.*

iii) *There exist an integer* $0 \leq k \leq r$ *and a family* $\Phi \in \mathcal{C}^r(\Omega, \mathcal{L}(U))$ *with* $\Phi_0 \in \mathrm{Iso}(U)$ *for which* λ_0 *is a* k*-transversal value of* $\mathfrak{L}^\Phi := \mathfrak{L}\Phi$.

iv) *There exist an integer* $0 \leq k \leq r$ *and a polynomial* $\Phi \in \mathbb{P}[\mathcal{L}(U)]$ *with* $\Phi_0 = I_U$ *for which* λ_0 *is a* k*-transversal value of* $\mathfrak{L}^\Phi := \mathfrak{L}\Phi$.

Moreover, in such a case,

$$\alpha[\mathfrak{L}; \lambda_0] = k = \nu \quad \textit{and} \quad \mu[\mathfrak{L}; \lambda_0] = \chi[\mathfrak{L}; \lambda_0]. \tag{5.29}$$

Furthermore, $\mu[\mathfrak{L}; \lambda_0] = \infty$ *if and only if* $\chi[\mathfrak{L}; \lambda_0] = \infty$*; and* $\mu[\mathfrak{L}; \lambda_0]$ *is undefined if and only if so is* $\chi[\mathfrak{L}; \lambda_0]$*. In any of these cases, for every integer* $1 \leq k \leq r$*, the subspace* V_k *constructed in Theorem* 4.3.2 *is a proper subspace of* V.

Proof. The conclusions follow straight away from the definitions if $\mathfrak{L}_0$ is an isomorphism. So, throughout the proof, we will assume that

$$\lambda_0 \in \mathrm{Eig}(\mathfrak{L}).$$

Suppose $\mu[\mathfrak{L}; \lambda_0] < \infty$ and set

$$\ell := \alpha[\mathfrak{L}; \lambda_0] \geq 1.$$

Let $\{\mathfrak{M}_{\{j\}}\}_{j=0}^r$ be any sequence of derived families from $\mathfrak{L}$ at λ_0, with corresponding sequence of projections $\{\Pi_j\}_{j=0}^r$. By Definition 5.1.3, the operator $\mathfrak{M}_{\{\ell\}}(\lambda_0)$ is an isomorphism. Moreover, according to (5.3), for each $\lambda \in \Omega \setminus \{\lambda_0\}$,

$$\mathfrak{M}_{\{\ell\}}(\lambda) = (\lambda - \lambda_0)^{-\ell} \mathfrak{L}(\lambda) \mathfrak{R}_{\Pi_0}(\lambda - \lambda_0) \cdots \mathfrak{R}_{\Pi_{\ell-1}}(\lambda - \lambda_0).$$

Since $\mathfrak{M}_{\{\ell\}}(\lambda_0)$ is invertible, then $\mathfrak{M}_{\{\ell\}}(\lambda) \in \mathrm{Iso}(U, V)$ for each $\lambda \sim \lambda_0$, and, therefore, $\mathfrak{L}(\lambda) \in \mathrm{Iso}(U, V)$ if $\lambda \sim \lambda_0$ with $\lambda \neq \lambda_0$. Thus, for each $\lambda \sim \lambda_0$ with $\lambda \neq \lambda_0$,

$$\mathfrak{L}(\lambda)^{-1} = (\lambda - \lambda_0)^{-\ell} \mathfrak{R}_{\Pi_0}(\lambda - \lambda_0) \cdots \mathfrak{R}_{\Pi_{\ell-1}}(\lambda - \lambda_0) \mathfrak{M}_{\{\ell\}}(\lambda)^{-1}. \tag{5.30}$$

Consequently, by (5.30), there exists a constant $C > 0$ such that

$$\|\mathfrak{L}(\lambda)^{-1}\| \leq C |\lambda - \lambda_0|^{-\ell}, \qquad \lambda \sim \lambda_0, \ \lambda \neq \lambda_0.$$

This shows that $\lambda_0 \in \mathrm{Alg}_\nu(\mathfrak{L})$ for some $\nu \in \{1, \ldots, \ell\}$. Moreover, according to Theorem 4.3.2, there exist $k \in \{1, \ldots, \nu\}$ and a polynomial $\Phi \in \mathbb{P}[\mathcal{L}(U)]$ with $\Phi_0 = I_U$ for which λ_0 is a k-transversal eigenvalue of $\mathfrak{L}\Phi$. In particular,

$$k \leq \nu \leq \ell = \alpha[\mathfrak{L}; \lambda_0].$$

According to Theorem 5.2.2, we actually have $k = \alpha[\mathfrak{L}; \lambda_0]$ and

$$\mu[\mathfrak{L}; \lambda_0] = \mu_{k-1}[\mathfrak{L}; \lambda_0] = \chi[\mathfrak{L}; \lambda_0].$$

Therefore, (5.29) holds.

Reciprocally, when $\lambda_0 \in \mathrm{Alg}_\nu(\mathfrak{L})$ for some integer $1 \leq \nu < r+1$, by Theorems 4.3.2 and 5.2.2, $\mathfrak{L}$ admits a transversalizing polynomial Φ and

$$k = \alpha[\mathfrak{L}; \lambda_0] \leq \nu \quad \text{and} \quad \mu[\mathfrak{L}; \lambda_0] = \chi[\mathfrak{L}; \lambda_0] < \infty.$$

This shows (5.29).

Finally, assume that either $\alpha[\mathfrak{L}; \lambda_0]$ equals infinity or it is not defined. Then

$$N[\mathfrak{M}_{\{j\}}(\lambda_0)] \neq \{0\} \qquad \text{for each} \quad 0 \leq j < r+1. \tag{5.31}$$

For each integer $1 \leq k \leq r$, let V_k be any of the subspaces of V constructed in Theorem 4.3.2. If $V_k = V$ for some $1 \leq k < r+1$, then, owing to Propositions 4.2.2, 5.1.4 and Theorem 5.2.1, the operator $\mathfrak{M}_{\{k\}}(\lambda_0)$ must be an isomorphism, which contradicts (5.31). Therefore,

$$V_k \subsetneq V \qquad \text{for each integer} \quad 1 \leq k \leq r.$$

Consequently, in the case $r = \infty$,

$$\alpha[\mathfrak{L}; \lambda_0] = \infty \quad \text{and} \quad \mu[\mathfrak{L}; \lambda_0] = \chi[\mathfrak{L}; \lambda_0] = \infty,$$

by definition, whereas neither α, μ nor χ are defined if $r \in \mathbb{N}$. □

According to Theorem 5.3.1, throughout the rest of this book we may use the notation $\chi[\mathfrak{L}; \lambda_0]$ for the μ-multiplicity introduced by Definition 5.1.3. We will sometimes set

$$\chi_j[\mathfrak{L}; \lambda_0] := \mu_j[\mathfrak{L}; \lambda_0], \qquad 0 \leq j < r+1. \tag{5.32}$$

For each integer $0 \leq j \leq r$, the number $\chi_j[\mathfrak{L}; \lambda_0]$ will be referred to as the *jth partial algebraic μ-multiplicity* of $\mathfrak{L}$ at λ_0, or as the *jth partial algebraic χ-multiplicity* of $\mathfrak{L}$ at λ_0.

From now on, given a finite-rank projection $P \in \mathcal{L}(U)$, we will consider the polynomial family $\mathfrak{C}_P \in \mathbb{P}[\mathcal{L}(U)]$ defined by

$$\mathfrak{C}_P(t) := tP + I_U - P, \qquad t \in \mathbb{K}. \tag{5.33}$$

According to (5.1) and (5.2),

$$\mathfrak{R}_P(t)^{-1} = \mathfrak{R}_P(t^{-1}) = t^{-1}\mathfrak{C}_P(t), \qquad t \in \mathbb{K} \setminus \{0\}. \tag{5.34}$$

Consequently, $\mathfrak{C}_P(t)$ is invertible for all $t \in \mathbb{K} \setminus \{0\}$, and (5.3) can be equivalently written in the form

$$\mathfrak{L}(\lambda) = \mathfrak{M}_{\{j\}}(\lambda)\mathfrak{C}_{\Pi_{j-1}}(\lambda - \lambda_0) \cdots \mathfrak{C}_{\Pi_0}(\lambda - \lambda_0). \tag{5.35}$$

As a by-product of Theorem 5.3.1, the next corollary is found.

Corollary 5.3.2. *Let $r \in \mathbb{N} \cup \{\infty\}$ and $\mathfrak{L} \in \mathcal{C}^r(\Omega, \mathcal{L}(U,V))$ be with $\mathfrak{L}_0 \in \mathrm{Fred}_0(U,V)$. Then, the following characterizations, describing mutually exclusive options, hold true:*

(a) $\mathfrak{L}_0 \in \mathrm{Iso}(U,V)$ *if and only if* $\chi[\mathfrak{L};\lambda_0] = 0$.

(b) $\lambda_0 \in \mathrm{Alg}_\nu(\mathfrak{L})$ *for some integer* $1 \leq \nu \leq r$ *if and only if there exist* $\mathfrak{M} \in \mathcal{C}^{r-\nu}(\Omega, \mathcal{L}(U,V))$ *with* $\mathfrak{M}_0 \in \mathrm{Iso}(U,V)$, *and* ν *finite-rank projections*

$$P_0, \dots, P_{\nu-1} \in \mathcal{L}(U) \setminus \{0\}$$

such that

$$\mathfrak{L}(\lambda) = \mathfrak{M}(\lambda)\mathfrak{C}_{P_{\nu-1}}(\lambda-\lambda_0)\cdots\mathfrak{C}_{P_0}(\lambda-\lambda_0), \qquad \lambda \in \Omega. \tag{5.36}$$

Moreover, in such a case,

$$\chi[\mathfrak{L};\lambda_0] = \sum_{i=0}^{\nu-1} \mathrm{rank}\, P_i.$$

(c) $\lambda_0 \notin \bigcup_{0 \leq \nu < r+1} \mathrm{Alg}_\nu(\mathfrak{L})$ *if and only if, for each integer* $1 \leq \nu \leq r$, *there exist* $\mathfrak{M} \in \mathcal{C}^{r-\nu}(\Omega, \mathcal{L}(U,V))$, *with* $\dim N[\mathfrak{M}_0] \geq 1$, *and* ν *finite-rank projections* $P_0, \dots, P_{\nu-1} \in \mathcal{L}(U)\setminus\{0\}$ *satisfying* (5.36). *Moreover, in such a case,* $\chi[\mathfrak{L};\lambda_0]$ *equals infinity if* $r = \infty$, *whereas it is undefined if* $r \in \mathbb{N}$.

5.4 A formula for the partial μ-multiplicities

The following result provides us with a finite algorithm to calculate the jth partial χ-multiplicities of a family $\mathfrak{L}$ of class $\mathcal{C}^r$ at any $\lambda_0 \in \mathrm{Alg}_\nu(\mathfrak{L})$ with $\nu < r+1$.

Theorem 5.4.1. *Let $r \in \mathbb{N} \cup \{\infty\}$ and $\mathfrak{L} \in \mathcal{C}^r(\Omega, \mathcal{L}(U,V))$ be with $\mathfrak{L}_0 \in \mathrm{Fred}_0(U,V)$. Suppose $\lambda_0 \in \mathrm{Alg}_\nu(\mathfrak{L})$ for some integer $1 \leq \nu \leq r$. Let $\Phi \in \mathcal{C}^r(\Omega, \mathcal{L}(U))$ be such that $\Phi_0 \in \mathrm{Iso}(U)$, for which λ_0 is a ν-transversal eigenvalue of $\mathfrak{L}^\Phi := \mathfrak{L}\Phi$, and set*

$$\ell_j := \dim \mathfrak{L}_j^\Phi \Big(\bigcap_{i=0}^{j-1} N[\mathfrak{L}_i^\Phi] \Big), \qquad 1 \leq j \leq \nu,$$

$$A := (a_{ij})_{1 \leq i,j \leq \nu} \in \mathcal{M}_\nu(\mathbb{R}), \qquad a_{ij} := \min\{i,j\}, \qquad 1 \leq i,j \leq \nu.$$

Then, the matrix A is invertible and

$$\mathrm{col}[\chi_0[\mathfrak{L};\lambda_0], \dots, \chi_{\nu-1}[\mathfrak{L};\lambda_0]] = A\, \mathrm{col}[\ell_1, \dots, \ell_\nu].$$

In particular,

$$\chi[\mathfrak{L};\lambda_0] = \chi_{\nu-1}[\mathfrak{L};\lambda_0] = \sum_{j=1}^{\nu} j\, \ell_j.$$

Proof. The existence of the transversalizing family Φ is guaranteed by Theorem 5.3.1. Moreover, by (5.32) and Proposition 5.1.4,

$$\chi_j[\mathfrak{L};\lambda_0] = \chi_j[\mathfrak{L}^\Phi;\lambda_0], \qquad 0 \le j \le \nu - 1.$$

Let $\{\mathfrak{M}_{\{j\}}\}_{j=0}^r$ be a sequence of derived families from $\mathfrak{L}^\Phi$ at λ_0. Then, thanks to Theorem 5.2.1,

$$N[\mathfrak{M}_{\{i\}}(\lambda_0)] = N[\mathfrak{L}_0^\Phi] \cap \cdots \cap N[\mathfrak{L}_i^\Phi], \qquad 0 \le i \le \nu,$$

and, hence,

$$\chi_j[\mathfrak{L};\lambda_0] = \sum_{i=0}^{j} \dim N[\mathfrak{L}_0^\Phi] \cap \cdots \cap N[\mathfrak{L}_i^\Phi], \qquad 0 \le j \le \nu - 1.$$

Consequently, thanks to Proposition 4.2.2, we find that

$$\chi_j[\mathfrak{L};\lambda_0] = \sum_{i=0}^{j} \sum_{h=i+1}^{\nu} \ell_h = \sum_{h=1}^{\nu} \min\{h, j+1\} \ell_h$$

for each $0 \le j \le \nu - 1$. The invertibility of the matrix A is evident. □

5.5 Removable singularities

This section consists of the proof of Theorem 5.0.1. Let $r \in \mathbb{N} \cup \{\infty\}$ and

$$\mathfrak{L} \in \mathcal{C}^r(\Omega, \mathcal{L}(U,V)),$$

with $\mathfrak{L}_0 \in \mathrm{Fred}_0(U,V)$, be such that $(\lambda-\lambda_0)^\nu \mathfrak{L}(\lambda)^{-1}$ exists and is bounded for λ in $\Omega \setminus \{\lambda_0\}$, for some integer $0 \le \nu \le r$. Then, $\lambda_0 \in \mathrm{Alg}_k(\mathfrak{L})$ for some $k \in \{0, \dots, \nu\}$.

Suppose $k = 0$. Then, $\mathfrak{L}_0 \in \mathrm{Iso}(U,V)$ and, hence, the family $\mathfrak{L}^{-1}$ is of class $\mathcal{C}^r$ in a neighborhood of λ_0. So we are done in this case.

Suppose $k \ge 1$. Then, by Corollary 5.3.2(b), there exist k finite-rank projections

$$P_0, \dots, P_{k-1} \in \mathcal{L}(U) \setminus \{0\}$$

and $\mathfrak{M} \in \mathcal{C}^{r-k}(\Omega, \mathcal{L}(U,V))$ for which $\mathfrak{M}_0 \in \mathrm{Iso}(U,V)$ and (5.36) is satisfied, where the number ν in (5.36) is to be replaced with k. As $\mathfrak{M}_0$ is an isomorphism, there exists $\Omega' \subset \Omega$, with $\lambda_0 \in \Omega'$, such that the family $\mathfrak{M}^{-1}$ is of class $\mathcal{C}^{r-k}$ in Ω'. By (5.36) and (5.34), it is apparent that, for each $\lambda \in \Omega' \setminus \{\lambda_0\}$,

$$\begin{aligned}\mathfrak{L}(\lambda)^{-1} &= \mathfrak{C}_{P_0}(\lambda-\lambda_0)^{-1} \cdots \mathfrak{C}_{P_{k-1}}(\lambda-\lambda_0)^{-1} \mathfrak{M}(\lambda)^{-1} \\ &= (\lambda-\lambda_0)^{-k} \mathfrak{N}_{P_0}(\lambda-\lambda_0) \cdots \mathfrak{N}_{P_{k-1}}(\lambda-\lambda_0) \mathfrak{M}(\lambda)^{-1}\end{aligned}$$

and, consequently,

$$(\lambda-\lambda_0)^k\mathfrak{L}(\lambda)^{-1}=\mathfrak{R}_{P_0}(\lambda-\lambda_0)\cdots\mathfrak{R}_{P_{k-1}}(\lambda-\lambda_0)\mathfrak{M}(\lambda)^{-1}. \tag{5.37}$$

As the right-hand side of (5.37) is of class $\mathcal{C}^{r-k}$ in Ω', it indeed provides us with an extension of class $\mathcal{C}^{r-k}$ of the function

$$\lambda\mapsto(\lambda-\lambda_0)^k\mathfrak{L}(\lambda)^{-1}$$

to Ω', which concludes the proof of Theorem 5.0.1, because $\nu\geq k$.

5.6 The product formula

This section shows that the multiplicity of the product of two families equals the sum of their multiplicities. This property will be referred to as the *product formula.* It is a pivotal property, because it is the basis of all uniqueness results of Chapter 6.

Theorem 5.6.1 (product formula). *Let U,V,W be three $\mathbb{K}$-Banach spaces, $r,s\in\mathbb{N}\cup\{\infty\}$, and*

$$\mathfrak{L}\in\mathcal{C}^r(\Omega,\mathcal{L}(U,V)),\qquad \mathfrak{M}\in\mathcal{C}^s(\Omega,\mathcal{L}(W,U)),$$

with $\mathfrak{L}_0\in\mathrm{Fred}_0(U,V)$ and $\mathfrak{M}_0\in\mathrm{Fred}_0(W,U)$, such that the algebraic multiplicities $\chi[\mathfrak{L};\lambda_0]$ and $\chi[\mathfrak{M};\lambda_0]$ are well defined, and

$$\min\{r,s\}\geq\alpha[\mathfrak{L};\lambda_0]+\alpha[\mathfrak{M};\lambda_0]. \tag{5.38}$$

Then, $\chi[\mathfrak{L}\mathfrak{M};\lambda_0]$ is well defined,

$$\max\{\alpha[\mathfrak{L};\lambda_0],\alpha[\mathfrak{M};\lambda_0]\}\leq\alpha[\mathfrak{L}\mathfrak{M};\lambda_0]\leq\alpha[\mathfrak{L};\lambda_0]+\alpha[\mathfrak{M};\lambda_0], \tag{5.39}$$

and

$$\chi[\mathfrak{L}\mathfrak{M};\lambda_0]=\chi[\mathfrak{L};\lambda_0]+\chi[\mathfrak{M};\lambda_0]. \tag{5.40}$$

Remark 5.6.2. According to Theorem 5.3.1, given $q\in\mathbb{N}\cup\{\infty\}$ and a family $\mathfrak{N}\in\mathcal{C}^q(\Omega,\mathcal{L}(U,V))$ with $\mathfrak{N}_0\in\mathrm{Fred}_0(U,V)$, the algebraic multiplicity $\chi[\mathfrak{N};\lambda_0]$ is well defined if and only if either $q=\infty$, or $q\in\mathbb{N}$ and $\lambda_0\in\mathrm{Alg}_k(\mathfrak{L})$ for some $k\in\{0,\dots,q\}$.

Identity (5.40) will be referred to as the *product formula.* By Corollary 5.3.2, the proof of Theorem 5.6.1 may be based on the special case emphasized by the next result.

Proposition 5.6.3. *Suppose $\mathfrak{L}\in\mathcal{C}^r(\Omega,\mathcal{L}(U,V))$ for some $r\in\mathbb{N}\cup\{\infty\}$, the operator $\mathfrak{L}_0\in\mathrm{Fred}_0(U,V)$, and $\lambda_0\in\mathrm{Alg}_k(\mathfrak{L})$ for some integer $0\leq k\leq r$. Then, for every finite-rank projection $Q\in\mathcal{L}(U)\setminus\{0\}$,*

$$\chi[\mathfrak{C}_Q(\cdot-\lambda_0);\lambda_0]=\operatorname{rank}Q, \tag{5.41}$$

where

$$\mathfrak{C}_Q(t) = tQ + I_U - Q, \qquad t \in \mathbb{K}.$$

If, in addition, the function

$$\lambda \mapsto \mathfrak{L}(\lambda)\mathfrak{C}_Q(\lambda - \lambda_0) \quad \textit{belongs to} \quad \mathcal{C}^{k+1}(\Omega, \mathcal{L}(U, V)), \tag{5.42}$$

then

$$\chi[\mathfrak{L}\mathfrak{C}_Q(\cdot - \lambda_0); \lambda_0] = \chi[\mathfrak{L}; \lambda_0] + \operatorname{rank} Q. \tag{5.43}$$

Remark 5.6.4. The assumption (5.42) is crucial for the validity of this result. In some sense, this condition imposes that $\mathfrak{L}$ is a first-order derived family from $\mathfrak{L}\mathfrak{C}_Q(\cdot - \lambda_0)$ at λ_0 with corresponding projection Q. As λ_0 may be a $(k+1)$-algebraic eigenvalue of $\mathfrak{L}\mathfrak{C}_Q(\cdot - \lambda_0)$, the multiplicity of $\mathfrak{L}\mathfrak{C}_Q(\cdot - \lambda_0)$ at λ_0 might not be defined if (5.42) fails. This explains why we are imposing the restriction (5.38) on the regularity of the families in the statement of Theorem 5.6.1. It is satisfied if $r = s = \infty$, of course.

Proof. To simplify the notation, we simply set

$$\mathfrak{C}(\lambda) := \mathfrak{C}_Q(\lambda - \lambda_0) = (\lambda - \lambda_0)Q + I_U - Q, \qquad \lambda \in \mathbb{K}.$$

Then, $\mathfrak{C}_0 = I_U - Q$ and $\mathfrak{C}_1 = Q$; therefore,

$$N[\mathfrak{C}_0] = R[Q], \qquad R[\mathfrak{C}_0] = N[Q], \qquad N[Q] \oplus R[Q] = U,$$

because $Q^2 = Q$. Moreover,

$$\mathfrak{C}_1(N[\mathfrak{C}_0]) = Q(R[Q]) = R[Q],$$

and, therefore,

$$\mathfrak{C}_1(N[\mathfrak{C}_0]) \oplus R[\mathfrak{C}_0] = R[Q] \oplus N[Q] = U.$$

Consequently, λ_0 is a 1-transversal eigenvalue of $\mathfrak{C}$ and, according to Definition 4.3.3,

$$\chi[\mathfrak{C}; \lambda_0] = \dim \mathfrak{C}_1(N[\mathfrak{C}_0]) = \dim R[Q] = \operatorname{rank} Q.$$

This concludes the proof of (5.41).

The proof of (5.43) will proceed by induction on $k \geq 0$. Suppose $k = 0$. Then, by definition, $\mathfrak{L}_0 \in \operatorname{Iso}(U, V)$. By (5.42), the map $\tilde{\mathfrak{L}}$ defined by

$$\tilde{\mathfrak{L}}(\lambda) := \mathfrak{L}(\lambda)\mathfrak{C}_Q(\lambda - \lambda_0), \qquad \lambda \in \Omega,$$

belongs to $\mathcal{C}^1(\Omega, \mathcal{L}(U, V))$. Moreover, $\lambda_0 \in \operatorname{Alg}_1(\tilde{\mathfrak{L}})$, because of (5.34). Thus, according to Theorem 5.3.1, $\chi[\tilde{\mathfrak{L}}; \lambda_0]$ is well defined and finite. Necessarily,

$$\tilde{\mathfrak{L}}_0 = \mathfrak{L}_0(I_U - Q), \qquad \tilde{\mathfrak{L}}_1 = \mathfrak{L}_0 Q + A(I_U - Q),$$

for a certain operator A, and, since $\mathfrak{L}_0 \in \mathrm{Iso}(U,V)$, we have that

$$N[\tilde{\mathfrak{L}}_0] = R[Q], \qquad R[\tilde{\mathfrak{L}}_0] = \mathfrak{L}_0(N[Q]), \qquad \tilde{\mathfrak{L}}_1(N[\tilde{\mathfrak{L}}_0]) = \mathfrak{L}_0(R[Q]).$$

Consequently, since

$$U = N[Q] \oplus R[Q]$$

and, hence,

$$V = \mathfrak{L}_0(N[Q]) \oplus \mathfrak{L}_0(R[Q]),$$

we find that

$$\tilde{\mathfrak{L}}_1(N[\tilde{\mathfrak{L}}_0]) \oplus R[\tilde{\mathfrak{L}}_0] = V.$$

Therefore,

$$\chi[\tilde{\mathfrak{L}};\lambda_0] = \dim \tilde{\mathfrak{L}}_1(N[\tilde{\mathfrak{L}}_0]) = \dim \mathfrak{L}_0(R[Q]) = \dim R[Q],$$

which concludes the proof of (5.43) in the case $k = 0$.

Suppose $k \geq 1$, fix $0 \leq s < k$, and assume that

$$\chi[\mathfrak{M}\mathfrak{C}_Q(\cdot - \lambda_0);\lambda_0] = \chi[\mathfrak{M};\lambda_0] + \operatorname{rank} Q$$

for all families $\mathfrak{M}$ of class $\mathcal{C}^s$ such that $\lambda_0 \in \mathrm{Alg}_j(\mathfrak{M})$ for some $j \in \{0,\dots,s\}$ and all finite-rank projections $Q \in \mathcal{L}(U) \setminus \{0\}$ for which

$$\mathfrak{M}\mathfrak{C}_Q(\cdot - \lambda_0) \in \mathcal{C}^{s+1}(\Omega, \mathcal{L}(U,V)).$$

Let $\mathfrak{L} \in \mathcal{C}^{s+1}(\Omega, \mathcal{L}(U,V))$ be such that $\lambda_0 \in \mathrm{Alg}_{s+1}(\mathfrak{L})$, and $Q \in \mathcal{L}(U) \setminus \{0\}$ a finite-rank projection such that

$$\mathfrak{L}\mathfrak{C}_Q(\cdot - \lambda_0) \in \mathcal{C}^{s+2}(\Omega, \mathcal{L}(U,V)).$$

Since the spaces $N[Q] \cap N[\mathfrak{L}_0]$ and $R[Q]$ are finite dimensional and their intersection equals $\{0\}$, there exists a projection $P \in \mathcal{L}(U)$ such that

$$R[P] = N[Q] \cap N[\mathfrak{L}_0], \qquad R[Q] \subset N[P].$$

Then,

$$PQ = QP = 0. \tag{5.44}$$

Consequently,

$$(P+Q)^2 = P^2 + Q^2 = P + Q$$

and, therefore, $P + Q$ is a finite-rank projection for which

$$R[P+Q] = R[P] \oplus R[Q], \quad \operatorname{rank}(P+Q) = \operatorname{rank} P + \operatorname{rank} Q. \tag{5.45}$$

Indeed, due to (5.44), for all $x, y \in U$ we have that

$$Px + Qy = P^2x + Q^2y = (P+Q)(Px+Qy) \in R[P+Q]$$

and, hence,

$$R[P] + R[Q] \subset R[P+Q].$$

This shows that

$$R[P+Q] = R[P] + R[Q].$$

Now, let $x \in R[P] \cap R[Q]$. Then,

$$x = Px = Qx$$

and, hence,

$$Px = PQx = 0 \quad \text{and} \quad (I_U - P)x = (I_U - P)Px = 0,$$

which implies $x = 0$. Therefore, (5.45) holds.

It turns out that, in addition,

$$N[\mathfrak{L}(\lambda_0)\mathfrak{C}_Q(0)] = R[P+Q]. \tag{5.46}$$

Indeed, since $R[P] \subset N[\mathfrak{L}_0]$, (5.44) implies

$$\mathfrak{L}(\lambda_0)\mathfrak{C}_Q(0)(P+Q) = \mathfrak{L}_0(I_U - Q)(P+Q) = \mathfrak{L}_0 P = 0$$

and, hence,

$$R[P+Q] \subset N[\mathfrak{L}(\lambda_0)\mathfrak{C}_Q(0)].$$

On the other hand, for any

$$u \in N[\mathfrak{L}(\lambda_0)\mathfrak{C}_Q(0)] = N[\mathfrak{L}_0(I_U - Q)]$$

we have that

$$(I_U - Q)u \in N[\mathfrak{L}_0] \cap N[Q] = R[P]$$

and, hence, using (5.45) gives

$$u \in R[P] + R[Q] = R[P+Q],$$

which concludes the proof of (5.46).

Thanks to (5.46), $P+Q$ is a projection onto $N[\mathfrak{L}(\lambda_0)\mathfrak{C}_Q(0)]$ and, hence, there exists a sequence of derived families $\{\mathfrak{M}_{\{j\}}\}_{j=0}^{s+2}$ from the family

$$\Omega \ni \lambda \mapsto \mathfrak{L}(\lambda)\mathfrak{C}_Q(\lambda - \lambda_0)$$

at λ_0 with corresponding projections $\{P_j\}_{j=0}^{s+2}$ such that

$$P_0 = P + Q.$$

The family $\mathfrak{M}_{\{1\}}$ is of class $\mathcal{C}^{s+1}$ and, according to (5.3), for each $\lambda \in \Omega \setminus \{\lambda_0\}$,

$$\mathfrak{M}_{\{1\}}(\lambda) = (\lambda - \lambda_0)^{-1}\mathfrak{L}(\lambda)\mathfrak{C}_Q(\lambda - \lambda_0)\mathfrak{R}_{P+Q}(\lambda - \lambda_0). \tag{5.47}$$

Moreover, by Definition 5.1.3, Theorem 5.3.1, and (5.45), we can infer from (5.47) that

$$\chi[\mathfrak{L}\mathfrak{C}_Q(\cdot - \lambda_0); \lambda_0] = \chi[\mathfrak{M}_{\{1\}}; \lambda_0] + \operatorname{rank} P + \operatorname{rank} Q. \tag{5.48}$$

On the other hand, due to (5.44),

$$\begin{aligned} \mathfrak{C}_Q(t)\mathfrak{R}_{P+Q}(t) &= (tQ + I_U - Q)[P + Q + t(I_U - P - Q)] \\ &= P + t(I_U - P) = \mathfrak{R}_P(t) \end{aligned}$$

for each $t \in \mathbb{K}$, and, consequently, we can express (5.47) in the form

$$\mathfrak{M}_{\{1\}}(\lambda) = (\lambda - \lambda_0)^{-1}\mathfrak{L}(\lambda)\mathfrak{R}_P(\lambda - \lambda_0), \qquad \lambda \in \Omega \setminus \{\lambda_0\}. \tag{5.49}$$

Since $N[\mathfrak{L}_0]$ is a finite-dimensional space containing $R[P]$, and $N[P]$ is a finite codimensional subspace, because $R[P]$ is finite dimensional and $U = N[P] \oplus R[P]$, then, there exists a projection $\Pi \in \mathcal{L}(U)$ such that

$$R[\Pi] = N[\mathfrak{L}_0], \qquad N[\Pi] \subset N[P].$$

Let $\{\mathfrak{N}_{\{j\}}\}_{j=0}^{s+1}$ be a sequence of derived families from $\mathfrak{L}$ at λ_0 with corresponding projections $\{\Pi_j\}_{j=0}^{s+1}$ for which $\Pi_0 = \Pi$. Then, $\mathfrak{N}_{\{1\}}$ is of class $\mathcal{C}^s$ and, according to (5.3),

$$\mathfrak{N}_{\{1\}}(\lambda) = (\lambda - \lambda_0)^{-1}\mathfrak{L}(\lambda)\mathfrak{R}_\Pi(\lambda - \lambda_0), \qquad \lambda \in \Omega \setminus \{\lambda_0\}. \tag{5.50}$$

Moreover, by Definition 5.1.3 and Theorem 5.3.1,

$$\chi[\mathfrak{L}; \lambda_0] = \chi[\mathfrak{N}_{\{1\}}; \lambda_0] + \operatorname{rank} \Pi, \tag{5.51}$$

and λ_0 is an algebraic value of $\mathfrak{N}_{\{1\}}$ of order

$$\alpha[\mathfrak{N}_{\{1\}}; \lambda_0] = s. \tag{5.52}$$

In addition, substituting (5.50) into (5.49) and using (5.2) show that

$$\mathfrak{M}_{\{1\}}(\lambda) = \mathfrak{N}_{\{1\}}(\lambda)\mathfrak{R}_\Pi\left((\lambda - \lambda_0)^{-1}\right)\mathfrak{R}_P(\lambda - \lambda_0) \tag{5.53}$$

for each $\lambda \in \Omega \setminus \{\lambda_0\}$. To simplify the right-hand side of (5.53), we still need some additional information. As $R[P] \subset R[\Pi]$, we have that $\Pi P = P$. Also, $P\Pi = P$, since $N[\Pi] \subset N[P]$. Thus,

$$P\Pi = \Pi P = P \tag{5.54}$$

and, hence,

$$(\Pi - P)^2 = \Pi - P.$$

Consequently,

$$R[\Pi] = R[\Pi - P] \oplus R[P]$$

and, therefore

$$\operatorname{rank}(\Pi - P) = \operatorname{rank}\Pi - \operatorname{rank} P. \tag{5.55}$$

Also, again by (5.54), we have that, for each $t \in \mathbb{K} \setminus \{0\}$,

$$\begin{aligned}\mathfrak{R}_{\Pi}(t^{-1})\, \mathfrak{R}_P(t) &= [\Pi + t^{-1}(I_U - \Pi)][P + t(I_U - P)] \\ &= t(\Pi - P) + I_U - \Pi + P = \mathfrak{C}_{\Pi - P}(t)\end{aligned}$$

and, hence, (5.53) becomes

$$\mathfrak{M}_{\{1\}}(\lambda) = \mathfrak{N}_{\{1\}}(\lambda)\mathfrak{C}_{\Pi-P}(\lambda - \lambda_0), \qquad \lambda \in \Omega.$$

Note that $\mathfrak{N}_{\{1\}}$ is of class $\mathcal{C}^s$, and that $\mathfrak{M}_{\{1\}}$ is of class $\mathcal{C}^{s+1}$. By (5.52) and (5.55), the induction hypothesis implies

$$\chi[\mathfrak{M}_{\{1\}}; \lambda_0] = \chi[\mathfrak{N}_{\{1\}}; \lambda_0] + \operatorname{rank}\Pi - \operatorname{rank} P. \tag{5.56}$$

Therefore, applying (5.48), (5.56) and (5.51) gives

$$\begin{aligned}\chi[\mathfrak{L}\mathfrak{C}_Q(\cdot - \lambda_0); \lambda_0] &= \chi[\mathfrak{M}_{\{1\}}; \lambda_0] + \operatorname{rank} P + \operatorname{rank} Q \\ &= \chi[\mathfrak{N}_{\{1\}}; \lambda_0] + \operatorname{rank}\Pi + \operatorname{rank} Q \\ &= \chi[\mathfrak{L}; \lambda_0] + \operatorname{rank} Q,\end{aligned}$$

which concludes the proof. □

Proof of Theorem 5.6.1. We set

$$k := \alpha[\mathfrak{L}; \lambda_0], \qquad \ell := \alpha[\mathfrak{M}; \lambda_0].$$

According to Theorem 5.3.1, $k \in \mathbb{N}$ (respectively, $\ell \in \mathbb{N}$) if and only if $\lambda_0 \in \operatorname{Alg}_k(\mathfrak{L})$ (respectively, $\lambda_0 \in \operatorname{Alg}_\ell(\mathfrak{M})$).

Suppose $k, \ell \in \mathbb{N}$. Then, there exists a constant $C > 0$ such that for every λ in some perforated neighborhood of λ_0,

$$\|(\mathfrak{L}\mathfrak{M})^{-1}(\lambda)\| \leq \|\mathfrak{M}(\lambda)^{-1}\| \cdot \|\mathfrak{L}(\lambda)^{-1}\| \leq \frac{C}{|\lambda - \lambda_0|^{k+\ell}}.$$

Thus, $\lambda_0 \in \operatorname{Alg}_\nu(\mathfrak{L}\mathfrak{M})$ for some integer $0 \leq \nu \leq k + \ell$, and, since

$$\mathfrak{L}\mathfrak{M} \in \mathcal{C}^{k+\ell}(\Omega, \mathcal{L}(W, V)),$$

it follows from Theorem 5.3.1 that $\chi[\mathfrak{L}\mathfrak{M}; \lambda_0]$ is well defined and that

$$\alpha[\mathfrak{L}\mathfrak{M}; \lambda_0] = \nu \in \mathbb{N}.$$

There also exists a constant $B > 0$ such that for every λ in some perforated neighborhood of λ_0,

$$\|\mathfrak{L}(\lambda)^{-1}\| \leq \|\mathfrak{M}(\lambda)\| \cdot \|(\mathfrak{L}\mathfrak{M})^{-1}(\lambda)\| \leq \frac{B}{|\lambda - \lambda_0|^{\nu}}. \tag{5.57}$$

Hence, $k \leq \nu$. Similarly, one can conclude that $\ell \leq \nu$ and, therefore,

$$\max\{k, \ell\} \leq \nu \leq k + \ell. \tag{5.58}$$

Therefore, (5.39) holds true if $k, \ell \in \mathbb{N}$.

Now, suppose $k = \infty$. Then, by (5.38), $\mathfrak{L}$ and $\mathfrak{M}$ are of class $\mathcal{C}^\infty$ and, hence, $\mathfrak{L}\mathfrak{M}$ is also of class $\mathcal{C}^\infty$. In particular, $\chi[\mathfrak{L}\mathfrak{M}; \lambda_0]$ is well defined. Moreover, by (5.57), it follows from Theorem 5.3.1 that

$$\alpha[\mathfrak{L}\mathfrak{M}; \lambda_0] = \infty,$$

because $k = \infty$. Therefore, when $k = \infty$,

$$\max\{k, \ell\} = \alpha[\mathfrak{L}\mathfrak{M}; \lambda_0] = k + \ell = \infty.$$

Also, in such a case, by Definition 5.1.3 and Theorem 5.3.1,

$$\chi[\mathfrak{L}\mathfrak{M}; \lambda_0] = \chi[\mathfrak{L}; \lambda_0] + \chi[\mathfrak{M}; \lambda_0] = \infty.$$

By symmetry, the same features hold if $\ell = \infty$.

The previous arguments show that $\chi[\mathfrak{L}\mathfrak{M}; \lambda_0]$ is well defined in all possible cases and that (5.39) is always satisfied. Moreover, (5.40) is satisfied if $k = \infty$ or $\ell = \infty$.

Consequently, to complete the proof it suffices to show (5.40) when $k, \ell \in \mathbb{N}$. In such a case, we already know that

$$\nu := \alpha[\mathfrak{L}\mathfrak{M}; \lambda_0] \in \mathbb{N}.$$

By definition, $\chi[\mathfrak{L}; \lambda_0] = 0$ if $k = 0$, and $\chi[\mathfrak{M}; \lambda_0] = 0$ if $\ell = 0$. Therefore, (5.40) follows straight away from Proposition 5.1.4 and Theorem 5.3.1 if $k = 0$ or $\ell = 0$. It remains to complete the proof in the case

$$1 \leq k \in \mathbb{N}, \qquad 1 \leq \ell \in \mathbb{N}. \tag{5.59}$$

Suppose (5.59). Let $\{\mathfrak{M}_{\{j\}}\}_{j=0}^{\ell}$ be a derived sequence from $\mathfrak{M}$ at λ_0 with corresponding sequence of projections $\{\Pi_j\}_{j=0}^{\ell}$. Then, $\mathfrak{M}_{\{\ell\}}(\lambda_0)$ is an isomorphism, $\mathfrak{M}_{\{0\}} = \mathfrak{M}$, and, for every $j \in \{0, \dots, \ell\}$,

$$\mathfrak{M}_{\{j\}}(\lambda_0) \in \mathrm{Fred}_0(W, U), \qquad \mathfrak{M}_{\{j\}} \in \mathcal{C}^{s-j}(\Omega, \mathcal{L}(W, U)), \tag{5.60}$$

and

$$\mathfrak{M}_{\{j\}}(\lambda) = \mathfrak{M}_{\{j+1\}}(\lambda)\mathfrak{C}_{\Pi_j}(\lambda - \lambda_0), \qquad \lambda \in \Omega.$$

Moreover,

$$\chi[\mathfrak{M}; \lambda_0] = \sum_{j=0}^{\ell-1} \operatorname{rank} \Pi_j. \tag{5.61}$$

In particular, for each $j \in \{0, \dots, \ell - 1\}$,

$$\mathfrak{L}(\lambda)\mathfrak{M}_{\{j\}}(\lambda) = \mathfrak{L}(\lambda)\mathfrak{M}_{\{j+1\}}(\lambda)\mathfrak{C}_{\Pi_j}(\lambda - \lambda_0), \qquad \lambda \in \Omega.$$

Fix a $j \in \{0, \dots, \ell - 1\}$. Then, by (5.38) and (5.60), $\mathfrak{L}\mathfrak{M}_{\{j\}}$ is of class $\mathcal{C}^{k+\ell-j}$ and $\mathfrak{L}\mathfrak{M}_{\{j+1\}}$ is of class $\mathcal{C}^{k+\ell-j-1}$. Moreover, λ_0 is an algebraic eigenvalue of $\mathfrak{L}\mathfrak{M}_{\{j+1\}}$ with order less than or equal to $k+\ell-j-1$. Thus, according to Proposition 5.6.3, we have that

$$\chi[\mathfrak{L}\mathfrak{M}_{\{j\}}; \lambda_0] = \chi[\mathfrak{L}\mathfrak{M}_{\{j+1\}}; \lambda_0] + \operatorname{rank} \Pi_j.$$

Consequently, using (5.61), shows that

$$\chi[\mathfrak{L}\mathfrak{M}; \lambda_0] = \chi[\mathfrak{L}\mathfrak{M}_{\{\ell\}}; \lambda_0] + \sum_{j=0}^{\ell-1} \operatorname{rank} \Pi_j = \chi[\mathfrak{L}\mathfrak{M}_{\{\ell\}}; \lambda_0] + \chi[\mathfrak{M}; \lambda_0].$$

As $\mathfrak{M}_{\{\ell\}}$ is of class $\mathcal{C}^k$ and $\mathfrak{M}_{\{\ell\}}(\lambda_0)$ is an isomorphism, due to Proposition 5.1.4,

$$\chi[\mathfrak{L}\mathfrak{M}_{\{\ell\}}; \lambda_0] = \chi[\mathfrak{L}; \lambda_0]$$

and, therefore,

$$\chi[\mathfrak{L}\mathfrak{M}; \lambda_0] = \chi[\mathfrak{L}; \lambda_0] + \chi[\mathfrak{M}; \lambda_0],$$

which concludes the proof. □

5.7 Perturbation from simple eigenvalues revisited

According to Theorem 5.3.1, the following substantial extension of Theorem 4.4.3 holds. It should be noted that λ_0 might not be a ν-algebraic eigenvalue of $\mathfrak{L}$ for any $1 \leq \nu \leq r$, and that, in such a case, Theorem 4.4.3 cannot be applied.

Theorem 5.7.1. *Suppose U, V are $\mathbb{K}$-Banach spaces such that U is a subspace of V with continuous inclusion, $\Omega \subset \mathbb{K}$ is open, $r \in \mathbb{N}^* \cup \{\infty, \omega\}$, the family $\mathfrak{L} \in \mathcal{C}^r(\Omega, \mathcal{L}(U, V))$ satisfies $\lambda_0 \in \Omega$ and*

$$\dim N[\mathfrak{L}_0] = 1 \quad \textit{and} \quad N[\mathfrak{L}_0] \oplus R[\mathfrak{L}_0] = V. \tag{5.62}$$

Let Y be a closed subspace of U and $\varphi_0 \in N[\mathfrak{L}_0] \setminus \{0\}$ such that

$$N[\mathfrak{L}_0] = \operatorname{span}[\varphi_0] \quad \textit{and} \quad N[\mathfrak{L}_0] \oplus Y = U,$$

and denote by a and φ the perturbed functions from $0 \in \mathbb{K}$ and φ_0, whose existence was guaranteed by Lemma 4.4.1. Then, $\operatorname{ord}_{\lambda_0} a$ is defined and

$$\operatorname{ord}_{\lambda_0} a := \chi[\mathfrak{L}; \lambda_0] = \alpha[\mathfrak{L}; \lambda_0] \tag{5.63}$$

when $\chi[\mathfrak{L}; \lambda_0]$ is defined, while $\operatorname{ord}_{\lambda_0} a$ is undefined or $\operatorname{ord}_{\lambda_0} a > r$ if $\chi[\mathfrak{L}; \lambda_0]$ is undefined.

Proof. By Theorem 5.3.1, we already know that $\chi[\mathfrak{L};\lambda_0]$ is defined if and only if $\lambda_0 \in \mathrm{Alg}_\nu(\mathfrak{L})$ for some integer $1 \le \nu \le r$, or $r = \infty$. Actually, $\chi[\mathfrak{L};\lambda_0] < \infty$ if and only if $\lambda_0 \in \mathrm{Alg}_\nu(\mathfrak{L})$ for some integer $1 \le \nu \le r$ and, in such a case, both $\alpha[\mathfrak{L};\lambda_0]$ and ν equal the order of transversality of every transversalized family from $\mathfrak{L}$ at λ_0. Consequently, (5.63) follows straight ahead from Theorem 4.4.3 if the algebraic multiplicity is finite.

Suppose $\chi[\mathfrak{L};\lambda_0] = \infty$. Then $r = \infty$ and $\lambda_0 \notin \mathrm{Alg}(\mathfrak{L})$. By Lemma 4.4.1, the functions a and φ are of class $\mathcal{C}^\infty$ and, hence, $\mathrm{ord}_{\lambda_0} a \ge 1$ is well defined. We claim that

$$\mathrm{ord}_{\lambda_0} a = \infty.$$

Suppose, looking for a contradiction, that

$$k := \mathrm{ord}_{\lambda_0} a \in \mathbb{N}.$$

By Theorems 4.3.2 and 5.3.1, there exists a polynomial $\Phi \in \mathbb{P}[\mathcal{L}(U)]$ such that $\Phi(\lambda_0) = I_U$ and the product family $\mathfrak{L}^\Phi := \mathfrak{L}\Phi$ satisfies

$$V_k := \bigoplus_{j=1}^{k} \mathfrak{L}_j^\Phi(N[\mathfrak{L}_0^\Phi] \cap \cdots \cap N[\mathfrak{L}_{j-1}^\Phi]) \oplus R[\mathfrak{L}_0^\Phi] \subsetneq V. \tag{5.64}$$

Actually, (5.64) implies

$$\bigoplus_{j=1}^{k} \mathfrak{L}_j^\Phi(N[\mathfrak{L}_0^\Phi] \cap \cdots \cap N[\mathfrak{L}_{j-1}^\Phi]) = \{0\}, \tag{5.65}$$

because $\mathfrak{L}^\Phi = \mathfrak{L}_0$ and $\operatorname{codim} R[\mathfrak{L}_0] = 1$.

Now we set

$$\psi(\lambda) := \Phi(\lambda)^{-1}\varphi(\lambda), \qquad \lambda \sim \lambda_0.$$

Since $\mathfrak{L}\varphi = a\varphi$, we have that

$$\mathfrak{L}^\Phi(\lambda)\psi(\lambda) = a(\lambda)\varphi(\lambda), \qquad \lambda \sim \lambda_0.$$

Thus,

$$\mathfrak{L}^\Phi(\lambda)\psi(\lambda) = o((\lambda-\lambda_0)^{k-1}), \qquad \text{as } \lambda \to \lambda_0, \tag{5.66}$$

and

$$\sum_{i=0}^{k} \mathfrak{L}_i^\Phi \psi_{k-i} = a_k\psi_0 \neq 0. \tag{5.67}$$

Thanks to Lemma 4.1.2, (5.66) implies

$$\psi_0 \in \bigcap_{j=0}^{k-1} N[\mathfrak{L}_j^\Phi], \qquad \psi_0, \ldots, \psi_{k-1} \in N[\mathfrak{L}_0],$$

and, consequently,

$$N[\mathfrak{L}_0] = \bigcap_{j=0}^{k-1} N[\mathfrak{L}_j^\Phi] \qquad \text{and} \qquad \psi_0, \dots, \psi_{k-1} \in \bigcap_{j=0}^{k-1} N[\mathfrak{L}_j^\Phi],$$

because $\dim N[\mathfrak{L}_0] = 1$. Thus, (5.67) becomes

$$\mathfrak{L}_0^\Phi \psi_k + \mathfrak{L}_k^\Phi \psi_0 = a_k \psi_0 \neq 0.$$

By (5.62), we find from this identity that $\mathfrak{L}_k^\Phi \psi_0 \neq 0$ and, therefore,

$$\mathfrak{L}_0^\Phi \Big(\bigcap_{j=0}^{k-1} N[\mathfrak{L}_j^\Phi] \Big) \neq \{0\},$$

which contradicts (5.65). This contradiction shows that $\mathrm{ord}_{\lambda_0} a = \infty$, so concluding the proof when $\chi[\mathfrak{L}; \lambda_0]$ is defined.

Finally, suppose $\chi[\mathfrak{L}; \lambda_0]$ is undefined. Then, there exists an integer $k \geq r$ such that $\mathfrak{L}$ is of class $\mathcal{C}^k$ in a neighborhood of λ_0, but in no neighborhood of λ_0 is it of class $\mathcal{C}^{k+1}$. Consequently, according to Theorems 4.3.2 and 5.3.1, there exists a polynomial $\Phi \in \mathbb{P}[\mathcal{L}(U)]$ such that $\Phi(\lambda_0) = I_U$ and $\mathfrak{L}^\Phi := \mathfrak{L}\Phi$ satisfies (5.64). Arguing as in the previous case when the multiplicity was infinity shows that

$$a_0 = \cdots = a_k = 0. \tag{5.68}$$

As, in general, the perturbed eigenvalue a is of $\mathcal{C}^k$, but it is not necessarily of class $\mathcal{C}^{k+1}$, it is apparent that either $\mathrm{ord}_{\lambda_0} a$ is undefined, or, due to (5.68)

$$\mathrm{ord}_{\lambda_0} a > k \geq r.$$

This concludes the proof. □

5.8 Exercises

1. Suppose $r \geq 1$ and the families $\mathfrak{L} \in \mathcal{C}^r(\Omega, \mathcal{L}(U,V))$ and $\Psi \in \mathcal{C}^r(\Omega, \mathcal{L}(V))$ satisfy $\mathfrak{L}_0 \in \mathrm{Fred}_0(U,V)$ and $\Psi_0 \in \mathrm{Iso}(V)$. Let $\{\mathfrak{L}_{\{j\}}\}_{j=0}^\infty$ be a derived sequence of $\mathfrak{L}$ at λ_0 with corresponding projections $\{\Pi_j\}_{j=0}^\infty$. Prove that $\{\mathfrak{L}_{\{j\}}\}_{j=0}^\infty$ is a derived sequence of $\Psi\mathfrak{L}$ at λ_0 with corresponding projections $\{\Pi_j\}_{j=0}^\infty$. Use this fact to give a proof of the last assertion of Proposition 5.1.4.

2. Prove the result of Exercise 5 of Chapter 4 by using the definition of the μ-multiplicity.

3. Let $r \in \mathbb{N} \cup \{\infty\}$, consider $\lambda_0 \in \Omega$ and $\mathfrak{L} \in \mathcal{C}^r(\Omega, \mathcal{L}(U,V))$ with $\mathfrak{L}_0 \in \mathrm{Fred}_0(U,V)$ such that $\chi[\mathfrak{L}; \lambda_0]$ is defined. Prove that

$$\chi[\mathfrak{L}; \lambda_0] = \dim N[\mathfrak{L}_0]$$

if and only if λ_0 is a k-algebraic value of $\mathfrak{L}$ for some $k \in \{0,1\} \cap [0,r]$. This complements the result proposed in Exercise 6 of Chapter 4.

4. Let $r \in \mathbb{N} \cup \{\infty\}$, consider $\lambda_0 \in \Omega$ and $\mathfrak{L} \in \mathcal{C}^r(\Omega, \mathcal{L}(U,V))$ such that $\mathfrak{L}_0 \in \mathrm{Fred}_0(U,V)$ and $\lambda_0 \in \mathrm{Alg}_k(\mathfrak{L})$ for some $k \in \mathbb{N} \cap [0,r]$. Let $\Phi \in \mathcal{C}^\infty(\Omega, \mathcal{L}(U))$ be such that $\Phi(\lambda_0)$ is an isomorphism, and λ_0 is a k-transversal value of $\mathfrak{L}\Phi$. Prove that, for every $\hat{\mathfrak{L}} \in \mathcal{C}^r(\Omega, \mathcal{L}(U,V))$ satisfying

$$\hat{\mathfrak{L}}_j = \mathfrak{L}_j, \qquad 0 \leq j \leq k,$$

λ_0 is a k-algebraic value of $\hat{\mathfrak{L}}$ and a k-transversal value of $\hat{\mathfrak{L}}\Phi$. Moreover,

$$\chi[\mathfrak{L}; \lambda_0] = \chi[\hat{\mathfrak{L}}; \lambda_0].$$

5. This exercise provides a dual construction of the construction carried out in Section 5.1.

Suppose $\Omega \subset \mathbb{K}$ is open, U and V are $\mathbb{K}$-Banach spaces, $\lambda_0 \in \Omega$, and $\mathfrak{L} \in \mathcal{C}^r(\Omega, \mathcal{L}(U,V))$ satisfies $\mathfrak{L}_0 \in \mathrm{Fred}_0(U,V)$ for some $r \in \mathbb{N} \cup \{\infty\}$. Consider the $r+1$ families $\mathfrak{M}_{\{j\}} \in \mathcal{C}^{r-i}(\Omega, \mathcal{L}(U,V))$, for $0 \leq j < r+1$, defined inductively as follows:

- $\mathfrak{M}_{\{0\}} := \mathfrak{L}$.
- Given $0 \leq i < r$ and $\mathfrak{M}_{\{i\}} \in \mathcal{C}^{r-i}(\Omega, \mathcal{L}(U,V))$, let $\sigma_i \in \mathcal{L}(V)$ be a projection with range $R[\mathfrak{M}_{\{i\}}(\lambda_0)]$ and define

$$\mathfrak{M}_{\{i+1\}}(\lambda) := \begin{cases} [\sigma_i + (\lambda - \lambda_0)^{-1}(I_V - \sigma_i)]\mathfrak{M}_{\{i\}}(\lambda) & \text{if } \lambda \in \Omega \setminus \{\lambda_0\}, \\ \sigma_i \mathfrak{M}_{\{i\}}(\lambda_0) + (I_V - \sigma_i)\mathfrak{M}'_{\{i\}}(\lambda_0) & \text{if } \lambda = \lambda_0. \end{cases}$$

The (finite or infinite) sequence $\{\mathfrak{M}_{\{j\}}\}_{j=0}^r$ is called a sequence of dual derived families from $\mathfrak{L}$ at λ_0. The aim of this exercise is to show that the dual derived families enjoy dual properties of the derived families constructed in Section 5.1. Precisely, it is requested to prove the following:

- $\mathfrak{M}_{\{j\}}(\lambda) \in \mathrm{Fred}_0(U,V)$ for all $0 \leq j < r+1$ and each $\lambda \in \Omega$ satisfying $\mathfrak{L}(\lambda) \in \mathrm{Fred}_0(U,V)$. Moreover, for all $\lambda \in \Omega \setminus \{\lambda_0\}$ and $0 \leq j < r+1$,

$$\mathfrak{M}_{\{j\}}(\lambda) = [\sigma_{j-1} + (\lambda - \lambda_0)^{-1}(I_V - \sigma_{j-1})] \cdots [\sigma_0 + (\lambda - \lambda_0)^{-1}(I_V - \sigma_0)]\mathfrak{L}(\lambda).$$

  In addition,

$$N[\sigma_j] + R[\sigma_{j+1}] = V$$

  and, so,

$$\dim R[\sigma_j] \leq \dim R[\sigma_{j+1}]$$

  for each $0 \leq j < r$.
- Define, for each $0 \leq j < r+1$, the dual partial μ-multiplicity of $\mathfrak{L}$ at λ_0 by

$$\mu_j^*[\mathfrak{L}; \lambda_0] := \sum_{i=0}^{j} \operatorname{codim} R[\sigma_i].$$

  If $\mathfrak{M}_{\{j\}}(\lambda_0)$ is invertible for some integer $0 \leq j \leq r$, then the dual ascent of $\mathfrak{L}$ at λ_0, denoted by $\alpha^*[\mathfrak{L}; \lambda_0]$, is defined as the least integer j for which $\mathfrak{M}_{\{j\}}(\lambda_0)$ is invertible. In such a case, the dual multiplicity of $\mathfrak{L}$ at λ_0, denoted by $\mu^*[\mathfrak{L}; \lambda_0]$, is defined through

$$\mu^*[\mathfrak{L}; \lambda_0] := \mu^*_{\alpha^*[\mathfrak{L}; \lambda_0]}[\mathfrak{L}; \lambda_0].$$

If $r = \infty$ and there does not exist $j \in \mathbb{N}$ for which $\mathfrak{M}_{\{j\}}(\lambda_0)$ is invertible, then the dual ascent and dual multiplicity are defined as

$$\alpha^*[\mathfrak{L}; \lambda_0] = \mu^*[\mathfrak{L}; \lambda_0] = \infty.$$

Prove the consistency of these concepts.

- Prove the dual counterpart of Proposition 5.1.4.
- The reader may try to find the relationships between this dual construction and the construction of Section 5.1; in particular, that the functions μ and μ^* are identical, and so are the functions α and α^*. But an easier proof will be sketched in Exercise 17 of Chapter 7.

6. Let $U = V = \mathbb{K}^2$, identify $\mathcal{L}(U, V)$ with $\mathcal{M}_2(\mathbb{K})$, and consider the families $\mathfrak{L}, \mathfrak{M} \in \mathbb{P}[\mathcal{M}_2(\mathbb{K})]$ defined by

$$\mathfrak{L}(\lambda) := \operatorname{diag}\{\lambda^k, 1\}, \quad \mathfrak{M}(\lambda) := \operatorname{diag}\{\lambda^s, \lambda^\ell\}, \qquad \lambda \in \mathbb{K},$$

for some integers $k, s, \ell \geq 0$. Prove that

$$\begin{aligned} &\alpha[\mathfrak{L}; 0] = \mu[\mathfrak{L}; 0] = k, \qquad \alpha[\mathfrak{M}; 0] = \max\{s, \ell\}, \qquad \mu[\mathfrak{M}; 0] = s + \ell, \\ &\alpha[\mathfrak{L}\mathfrak{M}; 0] = \max\{k + s, \ell\}, \qquad \mu[\mathfrak{L}\mathfrak{M}; 0] = k + s + \ell, \end{aligned}$$

to conclude that the inequalities (5.39) are optimal.

7. This exercise proposes an alternative proof of the main result of Exercise 9 of Chapter 4 by means of the characterization of χ provided by μ through Theorem 5.3.1. Precisely, under the same assumptions as in that exercise, let $\{\mathfrak{L}_{\{j\}}\}_{j=0}^r$ and $\{\mathfrak{M}_{\{j\}}\}_{j=0}^r$ be two sequences of derived families from $\mathfrak{L}$ and $\mathfrak{M}$ at λ_0 with corresponding projections $\{\Pi_j\}_{j=0}^r$ and $\{P_j\}_{j=0}^r$, respectively. Prove the following:

- $\{\mathfrak{L}_{\{j\}} \times \mathfrak{M}_{\{j\}}\}_{j=0}^r$ is a sequence of derived families from $\mathfrak{L} \times \mathfrak{M}$ at λ_0 with corresponding projections $\{\Pi_j \times P_j\}_{j=0}^r$.
- $\mu_j[\mathfrak{L} \times \mathfrak{M}; \lambda_0] = \mu_j[\mathfrak{L}; \lambda_0] + \mu_j[\mathfrak{M}; \lambda_0]$ for all $0 \leq j < r + 1$.
- $\mu[\mathfrak{L} \times \mathfrak{M}; \lambda_0]$ is defined if and only if both $\mu[\mathfrak{L}; \lambda_0]$ and $\mu[\mathfrak{M}; \lambda_0]$ are defined, and, in such a case,
$$\mu[\mathfrak{L} \times \mathfrak{M}; \lambda_0] = \mu[\mathfrak{L}; \lambda_0] + \mu[\mathfrak{M}; \lambda_0]$$
and
$$\alpha[\mathfrak{L} \times \mathfrak{M}; \lambda_0] = \max\{\alpha[\mathfrak{L}; \lambda_0], \alpha[\mathfrak{M}; \lambda_0]\}.$$

8. Let Ω', Ω be two open neighborhoods of $\lambda_0 \in \mathbb{K}$, suppose $r, s \in \mathbb{N} \cup \{\infty\}$, consider $\mathfrak{L} \in \mathcal{C}^r(\Omega, \mathcal{L}(U, V))$ with $\mathfrak{L}_0 \in \operatorname{Fred}_0(U, V)$, and let $f \in \mathcal{C}^s(\Omega')$ satisfy

$$f(\lambda_0) = \lambda_0 \quad \text{and} \quad f(\Omega') \subset \Omega.$$

Let $g \in \mathcal{C}^s(\Omega')$ be the function defined by

$$g(\lambda) := f(\lambda) - \lambda_0, \qquad \lambda_0 \in \Omega'.$$

We ask the following:

- Assume $r = s = \infty$ and, using Theorem 5.6.1, (5.3) and Exercise 2, prove the following identities:

$$\alpha[\mathfrak{L} \circ f; \lambda_0] = \mu[g; \lambda_0]\, \alpha[\mathfrak{L}; \lambda_0], \qquad \mu[\mathfrak{L} \circ f; \lambda_0] = \mu[g; \lambda_0]\, \mu[\mathfrak{L}; \lambda_0]. \tag{5.69}$$

- Assume that the three multiplicities $\mu[\mathfrak{L}; \lambda_0]$, $\mu[g; \lambda_0]$ and $\mu[\mathfrak{L} \circ f; \lambda_0]$ are defined, impose conditions on r, s, and use Theorem 5.6.1, (5.3) and Exercise 2 to obtain (5.69).

9. This exercise shows that Theorem 5.0.1 may fail when the assumption $\mathfrak{L}_0 \in \mathrm{Fred}_0(U, V)$ is removed.
Let $\mathbb{K} = \mathbb{R}$, the spaces $U = V = \mathcal{C}([0,1], \mathbb{C})$, and the operator $T \in \mathcal{L}(U)$ defined by

$$Tf(t) = \int_0^t f(s)\, ds, \qquad t \in [0,1], \quad f \in U.$$

Let $\mathfrak{L} \in \mathbb{P}[\mathcal{L}(U)]$ be the real polynomial defined by

$$\mathfrak{L}(\lambda) = iI_U\lambda - T, \qquad \lambda \in \mathbb{R}.$$

Prove, step by step, the following:

- T is injective but not surjective. Therefore, T is not Fredholm of index zero.
- $\mathfrak{L}(\lambda)$ is invertible for each $\lambda \neq 0$ and, actually, for each $f \in U$,

$$\mathfrak{L}(\lambda)^{-1} f(t) = \frac{1}{i\lambda} f(t) - \frac{1}{\lambda^2} \int_0^t f(s) e^{(t-s)/(i\lambda)}\, ds, \qquad t \in [0,1].$$

- T is not a Fredholm operator.
- $\|\mathfrak{L}(\lambda)^{-1}\| \leq 2|\lambda|^{-2}$ for all $\lambda \in \mathbb{R} \setminus \{0\}$.
- There does not exist any integer n for which the limit

$$\lim_{\lambda \to 0} \lambda^n \mathfrak{L}(\lambda)^{-1}$$

does exist.

5.9 Comments on Chapter 5

The most pioneering materials of this chapter were developed by Gohberg & Sigal [54] and Gohberg [42]. The concept of multiplicity introduced by Definition 5.1.3 goes back to Magnus [92]. It was introduced in the context of nonlinear analysis to generalize some celebrated results by Krasnosel'skiĭ [69] and Rabinowitz [111, 112]. It is extremely significant that Magnus [92] himself asserted

> "It looks as if the calculation of multiplicity might be fraught with difficulties, as our definition is far from transparent."

This sincere declaration might have been provoked by the fact that, except when $U = V = \mathbb{R}^d$, or $U = V$ and $\mathfrak{L}(\lambda) = \lambda I_U - T$ for some $T \in \mathcal{L}(U)$ and every $\lambda \in \mathbb{K}$, Magnus [92] could not ascertain whether or not $\mu[\mathfrak{L}; \lambda_0] < \infty$. Theorem 5.3.1 has actually shown that

$$\mu[\mathfrak{L}; \lambda_0] < \infty \quad \text{if and only if} \quad \lambda_0 \in \mathrm{Alg}(\mathfrak{L})$$

through the identification $\chi = \mu$. Chapter 7 will show that $\chi[\mathfrak{L}; \lambda_0] < \infty$ if λ_0 is an isolated eigenvalue of an analytic family $\mathfrak{L}$. Essentially, Theorem 5.3.1 is attributable to López-Gómez [82]. Besides Definition 5.1.3 and Theorem 4.3.2, the crucial step in the proof of Theorem 5.3.1 relies on Theorem 5.2.1, which goes back to Esquinas & López-Gómez [31]. The key step in proving Theorem 5.2.1 was to show that, for every $1 \leq j \leq k$,

$$\mathfrak{M}_{\{j\}}(\lambda_0) = \mathfrak{L}_0(I_U - \Pi_0) + \sum_{i=1}^{j-1} \mathfrak{L}_i \Pi_{i-1}(I_U - \Pi_i) + \mathfrak{L}_j \Pi_{j-1}$$

if $\lambda_0 \in \mathrm{Alg}_k(\mathfrak{L})$, which expresses $\mathfrak{M}_{\{j\}}(\lambda_0)$ in terms of the classical derivatives of $\mathfrak{L}$ at λ_0. When λ_0 is not a transversal eigenvalue of $\mathfrak{L}$, in order to construct the derived sequence $\{\mathfrak{M}_{\{j\}}\}_{j=0}^{k}$, one should first search for a local transversalizing family of isomorphisms Φ, and then apply the previous formula to

$$\mathfrak{L}^{\Phi} := \mathfrak{L}\Phi.$$

Going back to the proof of Theorem 4.3.2, it is easily realized that such a task might be extremely tedious and hard. Consequently, establishing the identification $\chi = \mu$ involves plenty of serious technical difficulties.

Nevertheless, the approach adopted in this chapter provides an extremely versatile polynomial factorization of $\mathfrak{L}$ through

$$\mathfrak{L}(\lambda) = \mathfrak{M}_{\{k\}}(\lambda)\mathfrak{C}_{\Pi_{k-1}}(\lambda - \lambda_0) \cdots \mathfrak{C}_{\Pi_0}(\lambda - \lambda_0), \qquad \lambda \sim \lambda_0,$$

which is the main technical tool used in the proof of the product formula

$$\chi[\mathfrak{L}\mathfrak{M}; \lambda_0] = \chi[\mathfrak{L}; \lambda_0] + \chi[\mathfrak{M}; \lambda_0],$$

which has been established by Theorem 5.6.1, attributable to Magnus [92]. Originally, Magnus established it for the very special case when $r = s = \infty$, where most of the technical difficulties of the general case covered in this chapter are easily overcome; however, this should not underrate the well-deserved merit of his result. Actually, from Definition 5.1.3 itself, it becomes apparent that μ was introduced having in mind the product formula.

Seemingly, the idea of factorization of the family goes back to Gohberg & Sigal [54] and Gohberg [42]. In these papers, with a similar idea of factorization, but overall different approach, they obtain fundamental properties of the multiplicity, like the stability property (see Chapter 8) and the logarithmic residue theorem.

Rather strikingly, Chapter 6 shows that the product formula together with the property that the multiplicity of some rank-one projection equals one do characterize the algebraic multiplicity $\chi = \mu$, in the sense that all remaining properties must follow from these two.

The importance of all these features greatly escapes from linear to nonlinear analysis, as, in the case $\mathbb{K} = \mathbb{R}$, those polynomial factorizations of $\mathfrak{L}$ considerably facilitate the analysis of the problem of ascertaining whether or not the Poincaré topological index of $\mathfrak{L}(\lambda)$ at $u = 0$, to be introduced in Section 12.2, changes as λ crosses λ_0. It turns out that it changes if and only if

$$\chi[\mathfrak{L}; \lambda_0] \in 2\mathbb{N} + 1,$$

if and only if λ_0 is a nonlinear eigenvalue of $\mathfrak{L}$, as discussed in Chapter 12.

The result of Exercise 8 goes back, essentially, to Magnus [92] (the product formula for the multiplicity to be introduced in Chapter 7 was already proved by Sigal [119]). The counter-example of Exercise 9 was constructed by Magnus & Mora-Corral [94]. We think that Theorem 5.0.1 is a striking property of Fredholm operators that deserves further attention.

Chapter 6

Uniqueness of the Algebraic Multiplicity

Throughout this chapter, given $\mathbb{K} \in \{\mathbb{R}, \mathbb{C}\}$, two non-zero $\mathbb{K}$-Banach spaces U, V, and $\lambda_0 \in \mathbb{K}$, we denote by $\mathcal{S}^\infty_{\lambda_0}(U, V)$ the set of all families $\mathfrak{L}$ of class $\mathcal{C}^\infty$ in a neighborhood of λ_0 with values in $\mathcal{L}(U, V)$ such that

$$\mathfrak{L}_0 := \mathfrak{L}(\lambda_0) \in \mathrm{Fred}_0(U, V);$$

the neighborhood may depend on $\mathfrak{L}$. We also set

$$\mathcal{S}^\infty_{\lambda_0}(U) := \mathcal{S}^\infty_{\lambda_0}(U, U).$$

Clearly, $\mathfrak{L}\mathfrak{M} \in \mathcal{S}^\infty_{\lambda_0}(U, V)$ if $\mathfrak{L} \in \mathcal{S}^\infty_{\lambda_0}(W, V)$ and $\mathfrak{M} \in \mathcal{S}^\infty_{\lambda_0}(U, W)$, where W is another $\mathbb{K}$-Banach space. Moreover, $\mathfrak{L}^{-1} \in \mathcal{S}^\infty_{\lambda_0}(V, U)$ whenever $\mathfrak{L} \in \mathcal{S}^\infty_{\lambda_0}(U, V)$ and $\mathfrak{L}_0 \in \mathrm{Iso}(U, V)$.

So far, in Chapters 4 and 5 we have constructed a function

$$\chi[\,\cdot\,; \lambda_0] \;:\; \mathcal{S}^\infty_{\lambda_0}(U, V) \to \mathbb{N} \cup \{\infty\}$$

that satisfies the following properties:

M1. $\chi[\mathfrak{L}; \lambda_0] = 0$ if and only if $\mathfrak{L}_0 \in \mathrm{Iso}(U, V)$; in this case, $\alpha[\mathfrak{L}; \lambda_0] = 0$.

M2. $1 \leq \chi[\mathfrak{L}; \lambda_0] < \infty$ if and only if $\lambda_0 \in \mathrm{Alg}_\nu(\mathfrak{L})$ for some integer $\nu \geq 1$; in this case, $\alpha[\mathfrak{L}; \lambda_0] = \nu$.

M3. $\chi[\mathfrak{L}; \lambda_0] = \infty$ if and only if $\lambda_0 \notin \mathrm{Alg}(\mathfrak{L})$; in this case, $\alpha[\mathfrak{L}; \lambda_0] = \infty$.

M4. For every $\mathfrak{L} \in \mathcal{S}^\infty_{\lambda_0}(W, V)$ and $\mathfrak{M} \in \mathcal{S}^\infty_{\lambda_0}(U, W)$,

$$\chi[\mathfrak{L}\mathfrak{M}; \lambda_0] = \chi[\mathfrak{L}; \lambda_0] + \chi[\mathfrak{M}; \lambda_0].$$

M5. $\chi[\mathfrak{C}_P(\cdot - \lambda_0); \lambda_0] = \operatorname{rank} P$ for every finite-rank projection $P \in \mathcal{L}(U)$, where

$$\mathfrak{C}_P(t) = t\, P + I_U - P, \qquad t \in \mathbb{K}.$$

Indeed, these properties are consequences of Theorem 5.3.1, Proposition 5.6.3 and Theorem 5.6.1.

The main goal of this chapter is to show that all properties of the algebraic multiplicity χ (equivalently, μ) can be obtained from Properties M4 and M5. For example, in the case $U = V$, the following uniqueness theorem is satisfied.

Theorem 6.0.1 (of uniqueness). *Let U be a non-zero $\mathbb{K}$-Banach space and $\lambda_0 \in \mathbb{K}$. Then, there exists a unique function*

$$\mathfrak{m}[\,\cdot\,; \lambda_0] \;:\; \mathcal{S}^{\infty}_{\lambda_0}(U) \to [0, \infty]$$

satisfying the following properties:

A1. *For all $\mathfrak{L}, \mathfrak{M} \in \mathcal{S}^{\infty}_{\lambda_0}(U)$,*

$$\mathfrak{m}[\mathfrak{L}\mathfrak{M}; \lambda_0] = \mathfrak{m}[\mathfrak{L}; \lambda_0] + \mathfrak{m}[\mathfrak{M}; \lambda_0].$$

A2. *There exists a rank-one projection $P_0 \in \mathcal{L}(U)$ such that*

$$\mathfrak{m}[\mathfrak{C}_{P_0}(\cdot - \lambda_0); \lambda_0] = 1.$$

Moreover,

$$\mathfrak{m}[\mathfrak{L}; \lambda_0] = \chi[\mathfrak{L}; \lambda_0] \qquad \textit{for all } \mathfrak{L} \in \mathcal{S}^{\infty}_{\lambda_0}(U)$$

and, in particular, $\mathfrak{m}$ ranges over $\mathbb{N} \cup \{\infty\}$.

Consequently, all definitions, constructions, properties and relationships established in Chapters 4 and 5 do actually follow from Axioms A1 and A2 if $U = V$. Axiom 2 can be regarded as a *normalization property* of the algebraic multiplicity, in the sense that the algebraic multiplicity is uniquely determined by the product formula (Axiom A1) up to a multiplicative constant, which is determined by fixing the value of

$$\mathfrak{m}[\mathfrak{C}_{P_0}(\cdot - \lambda_0); \lambda_0]$$

for some rank-one projection P_0.

This chapter is organized as follows. Section 6.1 shows the similarity of two arbitrary rank-one projections, which is needed in the proof of the uniqueness theorems. Section 6.2 gives the proof of Theorem 6.0.1 and establishes another version of it. Section 6.3 relaxes the regularity requirements of Theorem 6.0.1. Section 6.4 extends Theorem 6.0.1 to cover the general case when the spaces U, V are different but isomorphic. Finally, Section 6.5 applies these uniqueness theorems to get some explicit formulae for χ.

6.1 Similarity of rank-one projections

The following result will be useful in the proof of Theorem 6.0.1.

Lemma 6.1.1. *Let $P, Q \in \mathcal{L}(U)$ be two rank-one projections. Then, there exists an isomorphism $T \in \mathcal{L}(U)$ such that*

$$P = TQT^{-1}. \tag{6.1}$$

Proof. As $P, Q \in \mathcal{L}(U)$ are projections,

$$U = N[P] \oplus R[P] = N[Q] \oplus R[Q].$$

To construct the isomorphism T we distinguish two cases. First, suppose

$$N[P] \subset N[Q] \quad \text{or} \quad N[Q] \subset N[P]. \tag{6.2}$$

Since

$$\operatorname{codim} N[P] = \operatorname{codim} N[Q] = 1,$$

necessarily $N[P] = N[Q]$. Let

$$p \in R[P] \setminus \{0\} \quad \text{and} \quad q \in R[Q] \setminus \{0\}.$$

Then, the unique operator $T \in \mathcal{L}(U)$ satisfying

$$T|_{N[Q]} = I|_{N[Q]} = I|_{N[P]}, \qquad Tq = p,$$

where I stands for the identity, provides us with an isomorphism for which

$$PT = TQ.$$

Now, suppose that none of the inclusions (6.2) holds. Then,

$$N[P] \cap N[Q] \neq N[P] \quad \text{and} \quad N[P] \cap N[Q] \neq N[Q].$$

Take

$$y \in N[P], \qquad x_0 \in N[P] \setminus N[Q].$$

Then, $Qy \in R[Q]$ and $Qx_0 \in R[Q] \setminus \{0\}$. Since $\operatorname{rank} Q = 1$, we obtain that there exists $\lambda \in \mathbb{K}$ such that $Qy = \lambda Qx_0$. Thus,

$$y - \lambda x_0 \in N[P] \cap N[Q].$$

This proves that

$$N[P] = (N[P] \cap N[Q]) + \operatorname{span}[x_0]$$

and, hence, the codimension of $N[P] \cap N[Q]$ in $N[P]$ is 1. Therefore, $N[P] \cap N[Q]$ is a closed hyperplane of $N[P]$ and admits a one-dimensional topological supplement U_0

$$N[P] = (N[P] \cap N[Q]) \oplus U_0.$$

By symmetry, $N[P] \cap N[Q]$ is a closed hyperplane of $N[Q]$ and there exists a one-dimensional subspace U_1 of U such that

$$N[Q] = (N[P] \cap N[Q]) \oplus U_1.$$

Let

$$u_0 \in U_0 \setminus \{0\}, \quad u_1 \in U_1 \setminus \{0\}, \quad p \in R[P] \setminus \{0\}, \quad q \in R[Q] \setminus \{0\}.$$

Then, it is easy to see that the unique operator $T \in \mathcal{L}(U)$ satisfying

$$T|_{N[P]\cap N[Q]} = I|_{N[P]\cap N[Q]}, \quad Tu_1 = u_0, \quad Tq = p,$$

is an isomorphism that satisfies (6.1). This concludes the proof. □

6.2 Proof of Theorem 6.0.1

This section is devoted to the proof of Theorem 6.0.1.

Let U be a non-zero $\mathbb{K}$-Banach space and $\lambda_0 \in \mathbb{K}$. Then, thanks to the Hahn–Banach theorem, U admits rank-1 projections. Moreover, by Theorem 5.6.1 and Proposition 5.6.3, we already know that the function $\chi[\cdot;\lambda_0]$ satisfies Axioms A1 and A2. It remains to prove the uniqueness.

Let

$$\mathfrak{m}[\cdot;\lambda_0] : \mathcal{S}^\infty_{\lambda_0}(U) \to [0,\infty]$$

satisfy Axioms A1 and A2. Then, the following properties hold:

i) $\mathfrak{m}[\mathfrak{L};\lambda_0] = 0$ if $\mathfrak{L} \in \mathcal{S}^\infty_{\lambda_0}(U)$ and $\mathfrak{L}_0 \in \mathrm{Iso}(U)$.
ii) $\mathfrak{m}[\mathfrak{C}_P(\cdot - \lambda_0);\lambda_0] = 1$ for every rank-1 projection $P \in \mathcal{L}(U)$.
iii) $\mathfrak{m}[\mathfrak{C}_P(\cdot - \lambda_0);\lambda_0] = \operatorname{rank} P$ for every non-zero finite-rank projection $P \in \mathcal{L}(U)$.

Indeed, let $\mathfrak{L} \in \mathcal{S}^\infty_{\lambda_0}(U)$ be such that $\mathfrak{L}(\lambda_0)$ is invertible. Then, by Axioms A1 and A2, we find

$$\begin{aligned} 1 &= \mathfrak{m}[\mathfrak{C}_{P_0}(\cdot - \lambda_0);\lambda_0] = \mathfrak{m}[\mathfrak{C}_{P_0}(\cdot - \lambda_0)\mathfrak{L}\mathfrak{L}^{-1};\lambda_0] \\ &= \mathfrak{m}[\mathfrak{C}_{P_0}(\cdot - \lambda_0);\lambda_0] + \mathfrak{m}[\mathfrak{L};\lambda_0] + \mathfrak{m}[\mathfrak{L}^{-1};\lambda_0] = 1 + \mathfrak{m}[\mathfrak{L};\lambda_0] + \mathfrak{m}[\mathfrak{L}^{-1};\lambda_0]. \end{aligned}$$

Consequently, $\mathfrak{m}[\mathfrak{L};\lambda_0] = 0$, because $\mathfrak{m} \geq 0$. This concludes the proof of Property i).

To prove Property ii), let $P \in \mathcal{L}(U)$ be a rank-1 projection. By Lemma 6.1.1, there exists an invertible operator $T \in \mathcal{L}(U)$ such that $P = TP_0T^{-1}$, where P_0 is the projection of Axiom A2. Thus,

$$\mathfrak{C}_P = T\mathfrak{C}_{P_0}T^{-1}$$

and, therefore, combining Axioms A1 and A2 with Property i) shows that

$$\begin{aligned}\mathfrak{m}[\mathfrak{C}_P(\cdot - \lambda_0); \lambda_0] &= \mathfrak{m}[T; \lambda_0] + \mathfrak{m}[\mathfrak{C}_{P_0}(\cdot - \lambda_0); \lambda_0] + \mathfrak{m}[T^{-1}; \lambda_0] \\ &= \mathfrak{m}[\mathfrak{C}_{P_0}(\cdot - \lambda_0); \lambda_0] = 1,\end{aligned}$$

which concludes the proof of Property ii).

To prove Property iii), let $P \in \mathcal{L}(U)$ be a projection with rank $n \geq 1$. Then, there exist n rank-1 projections $P_1, \dots, P_n \in \mathcal{L}(U)$ such that

$$P = \sum_{i=1}^{n} P_i$$

and

$$P_i P_j = 0, \qquad 1 \leq i, j \leq n, \quad i \neq j.$$

These projections satisfy

$$\mathfrak{C}_P = \mathfrak{C}_{P_1} \cdots \mathfrak{C}_{P_n}$$

and, therefore, by Axiom A1 and Property ii), we find that

$$\mathfrak{m}[\mathfrak{C}_P(\cdot - \lambda_0); \lambda_0] = \sum_{i=1}^{n} \mathfrak{m}[\mathfrak{C}_{P_i}(\cdot - \lambda_0); \lambda_0] = n = \operatorname{rank} P.$$

This concludes the proof of Property iii).

We have all the necessary ingredients to conclude the proof of Theorem 6.0.1. Let $\mathfrak{L} \in \mathcal{S}^{\infty}_{\lambda_0}(U)$. Then, some of the following mutually exclusive options occur: either $\mathfrak{L}(\lambda_0)$ is an isomorphism, or λ_0 is a k-algebraic eigenvalue of $\mathfrak{L}(\lambda_0)$ for some integer $k \geq 1$, or λ_0 is not an algebraic eigenvalue of $\mathfrak{L}$.

Suppose $\mathfrak{L}(\lambda_0)$ is an isomorphism. Then, according to Property i) we have that $\mathfrak{m}[\mathfrak{L}; \lambda_0] = 0$. Suppose λ_0 is a k-algebraic eigenvalue of $\mathfrak{L}(\lambda_0)$ for some integer $k \geq 1$. Then, by Corollary 5.3.2(b), there exist k finite-rank projections

$$\Pi_0, \dots, \Pi_{k-1} \in \mathcal{L}(U) \setminus \{0\}$$

and a family $\mathfrak{M} \in \mathcal{S}^{\infty}_{\lambda_0}(U)$ with $\mathfrak{M}(\lambda_0)$ invertible, such that

$$\mathfrak{L}(\lambda) = \mathfrak{M}(\lambda)\mathfrak{C}_{\Pi_{k-1}}(\lambda - \lambda_0) \cdots \mathfrak{C}_{\Pi_0}(\lambda - \lambda_0), \qquad \lambda \sim \lambda_0.$$

Consequently, by Axiom A1 and Properties iii) and i), we have that

$$\mathfrak{m}[\mathfrak{L}; \lambda_0] = \mathfrak{m}[\mathfrak{M}; \lambda_0] + \sum_{i=0}^{k-1} \mathfrak{m}[\mathfrak{C}_{\Pi_i}(\cdot - \lambda_0); \lambda_0] = \sum_{i=0}^{k-1} \operatorname{rank} \Pi_i.$$

Finally, suppose that λ_0 is not an algebraic eigenvalue of $\mathfrak{L}$. Then, according to Corollary 5.3.2(c), for every integer $\nu \geq 1$ there are ν finite-rank projections

$$\Pi_0, \dots, \Pi_{\nu-1} \in \mathcal{L}(U) \setminus \{0\}$$

and an operator $\mathfrak{M}_\nu \in \mathcal{S}^\infty_{\lambda_0}(U)$ such that

$$\mathfrak{L}(\lambda) = \mathfrak{M}_\nu(\lambda)\mathfrak{C}_{\Pi_{\nu-1}}(\lambda - \lambda_0) \cdots \mathfrak{C}_{\Pi_0}(\lambda - \lambda_0), \qquad \lambda \sim \lambda_0.$$

Therefore, again by Axiom A1 and Property iii), we find that

$$\mathfrak{m}[\mathfrak{L}; \lambda_0] = \mathfrak{m}[\mathfrak{M}_\nu; \lambda_0] + \sum_{i=0}^{\nu-1} \operatorname{rank} \Pi_i \geq \nu, \tag{6.3}$$

because $\mathfrak{m} \geq 0$. As $\nu \geq 1$ is arbitrary, this implies that $\mathfrak{m}[\mathfrak{L}; \lambda_0] = \infty$ and completes the proof of Theorem 6.0.1.

In the previous proof the hypothesis that $\mathfrak{m}$ takes values in $[0, \infty]$ was used exclusively to show that $\mathfrak{m}[\mathfrak{L}; \lambda_0] = 0$ if $\mathfrak{L} \in \mathcal{S}^\infty_{\lambda_0}(U)$ is invertible, and in (6.3). Therefore, the following uniqueness result holds true (see Exercise 4).

Theorem 6.2.1 (of uniqueness). *Let U be a non-zero $\mathbb{K}$-Banach space, $\lambda_0 \in \mathbb{K}$, and denote*

$$\mathcal{A}^\infty_{\lambda_0}(U) := \left\{ \mathfrak{L} \in \mathcal{S}^\infty_{\lambda_0}(U) \ : \ \lambda_0 \in \mathrm{Alg}(\mathfrak{L}) \right\}.$$

Then, there exists a unique function

$$\mathfrak{m}[\,\cdot\,; \lambda_0] \ : \ \mathcal{A}^\infty_{\lambda_0}(U) \to \mathbb{C}$$

satisfying the following properties:

A1. *For all $\mathfrak{L}, \mathfrak{M} \in \mathcal{A}^\infty_{\lambda_0}(U)$,*

$$\mathfrak{m}[\mathfrak{L}\mathfrak{M}; \lambda_0] = \mathfrak{m}[\mathfrak{L}; \lambda_0] + \mathfrak{m}[\mathfrak{M}; \lambda_0].$$

A2. *There exists a rank-1 projection $P_0 \in \mathcal{L}(U)$ such that*

$$\mathfrak{m}[\mathfrak{C}_{P_0}(\cdot - \lambda_0); \lambda_0] = 1.$$

A3. $\mathfrak{m}[\mathfrak{L}; \lambda_0] = 0$ *for all $\mathfrak{L} \in \mathcal{A}^\infty_{\lambda_0}(U)$ such that $\mathfrak{L}(\lambda_0) \in \mathrm{Iso}(U)$.*

Moreover,

$$\mathfrak{m}[\mathfrak{L}; \lambda_0] = \chi[\mathfrak{L}; \lambda_0] \qquad \textit{for all } \mathfrak{L} \in \mathcal{A}^\infty_{\lambda_0}(U)$$

and, in particular, $\mathfrak{m}$ ranges over $\mathbb{N}$.

6.3 Relaxing the regularity requirements

The main difficulty in generalizing Theorem 6.0.1 to a $\mathcal{C}^r$ class regularity setting relies on the troubling fact that $\chi[\mathfrak{L}\mathfrak{M}; \lambda_0]$ may be undefined even if $\chi[\mathfrak{L}; \lambda_0]$ and $\chi[\mathfrak{M}; \lambda_0]$ are defined. In other words, we should overcome the difficulty that the class of families of class $\mathcal{C}^r$ in a neighborhood of λ_0 with values in $\mathcal{L}(U)$ for which $\chi[\mathfrak{L}; \lambda_0]$ is defined is not closed under the composition, which makes it impossible to impose the product formula as an axiom. To overcome such a problem we must reduce the class of admissible families in order to guarantee that $\chi[\mathfrak{L}\mathfrak{M};\lambda_0]$ is defined whenever $\chi[\mathfrak{L};\lambda_0]$ and $\chi[\mathfrak{M};\lambda_0]$ are so. We have seen that algebraic families at λ_0 provide a reasonable class of families for which a uniqueness theorem from the product formula can be given.

Precisely, given a point $\lambda_0 \in \mathbb{K}$ and two non-zero $\mathbb{K}$-Banach spaces U and V, let $\mathcal{M}_{\lambda_0}(U,V)$ denote the set of functions $\mathfrak{L} \in \mathcal{C}^r(\Omega, \mathcal{L}(U,V))$, for some open neighborhood $\Omega \subset \mathbb{K}$ of λ_0, such that $\mathfrak{L}(\lambda_0) \in \mathrm{Fred}_0(U,V)$ and either $r = \infty$, or $r \in \mathbb{N}$ and $\lambda_0 \in \mathrm{Alg}_k(\mathfrak{L})$ for some integer $0 \leq k \leq r$. Note that these are the functions for which the multiplicity $\chi[\mathfrak{L};\lambda_0]$ is well defined. Set also

$$\mathcal{M}_{\lambda_0}(U) := \mathcal{M}_{\lambda_0}(U,U).$$

As $\mathcal{S}^\infty_{\lambda_0}(U) \subset \mathcal{M}_{\lambda_0}(U)$, the next uniqueness theorem provides an extension of Theorem 6.0.1.

Theorem 6.3.1. *Let U be a non-zero $\mathbb{K}$-Banach space and $\lambda_0 \in \mathbb{K}$. Then, there exists a unique function*

$$\mathfrak{m}[\cdot;\lambda_0] \; : \; \mathcal{M}_{\lambda_0}(U) \to [0,\infty]$$

satisfying the following properties:

A1. *For every $\mathfrak{L},\mathfrak{M} \in \mathcal{M}_{\lambda_0}(U)$ such that $\mathfrak{L},\mathfrak{M}$ are of class $\mathcal{C}^{\alpha[\mathfrak{L};\lambda_0]+\alpha[\mathfrak{M};\lambda_0]}$ in a neighborhood of λ_0,*

$$\mathfrak{m}[\mathfrak{L}\mathfrak{M};\lambda_0] = \mathfrak{m}[\mathfrak{L};\lambda_0] + \mathfrak{m}[\mathfrak{M};\lambda_0].$$

A2. *There exists a rank-1 projection $P_0 \in \mathcal{L}(U)$ such that*

$$\mathfrak{m}[\mathfrak{C}_{P_0}(\cdot - \lambda_0);\lambda_0] = 1.$$

Moreover,

$$\mathfrak{m}[\mathfrak{L};\lambda_0] = \chi[\mathfrak{L};\lambda_0] \qquad \text{for all } \mathfrak{L} \in \mathcal{M}_{\lambda_0}(U)$$

and, in particular, $\mathfrak{m}$ ranges over $\mathbb{N} \cup \{\infty\}$.

Proof. Let U be a non-zero $\mathbb{K}$-Banach space and $\lambda_0 \in \mathbb{K}$. Then, thanks to the Hahn–Banach theorem, U admits rank-1 projections. Moreover, thanks to Theorem 5.6.1 and Proposition 5.6.3, we know that the function $\chi[\cdot;\lambda_0]$ satisfies Axioms A1 and A2. It remains to prove the uniqueness.

Let

$$\mathfrak{m}[\cdot;\lambda_0] : \mathcal{M}_{\lambda_0}(U) \to [0,\infty]$$

satisfy Axioms A1 and A2.

Then, the following properties hold:

i) $\mathfrak{m}[\mathfrak{L};\lambda_0]=0$ if $\mathfrak{L}\in\mathcal{M}_{\lambda_0}(U)$ and $\mathfrak{L}(\lambda_0)$ is invertible.

ii) $\mathfrak{m}[\mathfrak{C}_P(\cdot-\lambda_0);\lambda_0]=1$ for every rank-1 projection $P\in\mathcal{L}(U)$.

iii) $\mathfrak{m}[\mathfrak{C}_P(\cdot-\lambda_0);\lambda_0]=\operatorname{rank}P$ for every non-zero finite-rank projection $P\in\mathcal{L}(U)$.

Indeed, let $\mathfrak{L}\in\mathcal{M}_{\lambda_0}(U)$ be such that $\mathfrak{L}(\lambda_0)$ is invertible. Then,

$$\mathfrak{C}_{P_0},\ \mathfrak{L},\ \mathfrak{L}^{-1},\ \mathfrak{C}_{P_0}\mathfrak{L}\mathfrak{L}^{-1},\ \mathfrak{L}\mathfrak{L}^{-1}\in\mathcal{M}_{\lambda_0}(U)$$

and, hence, according to Axioms A1 and A2,

$$1=\mathfrak{m}[\mathfrak{C}_{P_0}(\cdot-\lambda_0);\lambda_0]=1+\mathfrak{m}[\mathfrak{L};\lambda_0]+\mathfrak{m}[\mathfrak{L}^{-1};\lambda_0].$$

Therefore,

$$\mathfrak{m}[\mathfrak{L};\lambda_0]=\mathfrak{m}[\mathfrak{L}^{-1};\lambda_0]=0,$$

because $\mathfrak{m}\geq 0$, which ends the proof of Property i). Properties ii) and iii) have already been shown in the proof of Theorem 6.0.1.

Let $\mathfrak{L}\in\mathcal{M}_{\lambda_0}(U)$. Then, $\mathfrak{L}$ is of class $\mathcal{C}^r$ for some $r\in\mathbb{N}\cup\{\infty\}$, and some of the following mutually exclusive alternatives occur: either $\mathfrak{L}_0\in\operatorname{Iso}(U)$; or $\lambda_0\in\operatorname{Alg}_k(\mathfrak{L})$ for some integer $1\leq k\leq r$; or $r=\infty$ but $\lambda_0\notin\operatorname{Alg}(\mathfrak{L})$.

If $\mathfrak{L}(\lambda_0)$ is an isomorphism, then $\mathfrak{m}[\mathfrak{L};\lambda_0]=0$, by Property i). Suppose λ_0 is a k-algebraic eigenvalue of $\mathfrak{L}(\lambda_0)$ for some integer number $1\leq k\leq r$. By Theorem 5.3.1, $k=\alpha[\mathfrak{L};\lambda_0]$. Let $\{\mathfrak{M}_{\{j\}}\}_{j=0}^{r}$ be a derived family from $\mathfrak{L}$ at λ_0 with associated projections $\{\Pi_j\}_{j=0}^{r}$. Then, $\mathfrak{M}_{\{0\}}=\mathfrak{L}$, the family $\mathfrak{M}_{\{k\}}$ is continuous and take invertible values, and, for every $j\in\{0,\dots,k-1\}$, the family $\mathfrak{M}_{\{j\}}$ is of class $\mathcal{C}^{k-j}$,

$$\mathfrak{M}_{\{j\}}(\lambda)=\mathfrak{M}_{\{j+1\}}(\lambda)\mathfrak{C}_{\Pi_j}(\lambda-\lambda_0),\qquad \lambda\in\Omega,$$

and

$$\alpha[\mathfrak{M}_{\{j+1\}};\lambda_0]=\alpha[\mathfrak{L};\lambda_0]-j-1,\quad \alpha[\mathfrak{C}_{\Pi_j}(\cdot-\lambda_0);\lambda_0]=1.$$

Thus, for each $0\leq j\leq k-1$,

$$\mathfrak{M}_{\{j\}}\ \text{ is of class }\ \mathcal{C}^{\alpha[\mathfrak{M}_{\{j+1\}};\lambda_0]+\alpha[\mathfrak{C}_{\Pi_j}(\cdot-\lambda_0);\lambda_0]}$$

in a neighborhood of λ_0, and, according to Axiom A1 and Properties iii) and i), we have that

$$\mathfrak{m}[\mathfrak{L};\lambda_0]=\mathfrak{m}[\mathfrak{M}_{\{k\}};\lambda_0]+\sum_{i=0}^{k-1}\mathfrak{m}[\mathfrak{C}_{\Pi_i}(\cdot-\lambda_0);\lambda_0]=\sum_{i=0}^{k-1}\operatorname{rank}\Pi_i.$$

Finally, suppose $r=\infty$ and λ_0 is not an algebraic eigenvalue of $\mathfrak{L}$. Let $\{\mathfrak{M}_{\{j\}}\}_{j=0}^{\infty}$ be a derived family from $\mathfrak{L}$ at λ_0 with associated projections $\{\Pi_j\}_{j=0}^{\infty}$. Then, for every integer $\nu\geq 1$, the family $\mathfrak{M}_\nu\in\mathcal{S}^\infty_{\lambda_0}(U)$ satisfies

$$\mathfrak{L}(\lambda)=\mathfrak{M}_\nu(\lambda)\mathfrak{C}_{\Pi_{\nu-1}}(\lambda-\lambda_0)\cdots\mathfrak{C}_{\Pi_0}(\lambda-\lambda_0),\qquad \lambda\sim\lambda_0.$$

Therefore, again by Axiom A1 and Property iii), we find that

$$\mathfrak{m}[\mathfrak{L};\lambda_0] = \mathfrak{m}[\mathfrak{M}_\nu;\lambda_0] + \sum_{i=0}^{\nu-1} \operatorname{rank}\Pi_i \geq \nu,$$

because $\mathfrak{m} \geq 0$. As $\nu \geq 1$ is arbitrary, this implies that $\mathfrak{m}[\mathfrak{L};\lambda_0] = \infty$ and concludes the proof. □

6.4 A general uniqueness theorem

The assumption $U = V$ in the statement of Theorems 6.0.1, 6.2.1 and 6.3.1 is not really restrictive, since U and V must be isomorphic if $\mathrm{Fred}_0(U,V)$ is non-empty. In fact, the following improvement of Theorem 6.3.1 can be obtained.

Theorem 6.4.1. *Let X be a non-zero $\mathbb{K}$-Banach space and $\lambda_0 \in \mathbb{K}$. Let $\mathfrak{J}(X)$ be a set of $\mathbb{K}$-Banach spaces isomorphic to X such that $X \in \mathfrak{J}(X)$. Then, there exists a unique function*

$$\mathfrak{m}[\,\cdot\,;\lambda_0] \; : \; \bigcup_{U,V\in\mathfrak{J}(X)} \mathcal{M}_{\lambda_0}(U,V) \to [0,\infty]$$

satisfying the following properties:

A1. *If $U, V, W \in \mathfrak{J}(X)$ with*

$$\mathfrak{L} \in \mathcal{M}_{\lambda_0}(U,V), \qquad \mathfrak{M} \in \mathcal{M}_{\lambda_0}(W,U),$$

and $\mathfrak{L}, \mathfrak{M}$ are of class $\mathcal{C}^{\alpha[\mathfrak{L};\lambda_0]+\alpha[\mathfrak{M};\lambda_0]}$ in a neighborhood of λ_0, then

$$\mathfrak{m}[\mathfrak{L}\mathfrak{M};\lambda_0] = \mathfrak{m}[\mathfrak{L};\lambda_0] + \mathfrak{m}[\mathfrak{M};\lambda_0].$$

A2. *There exists a rank-1 projection $P_0 \in \mathcal{L}(X)$ such that*

$$\mathfrak{m}[\mathfrak{C}_{P_0}(\cdot - \lambda_0);\lambda_0] = 1.$$

Actually, for every $U, V \in \mathfrak{J}(X)$ and $\mathfrak{L} \in \mathcal{M}_{\lambda_0}(U,V)$,

$$\mathfrak{m}[\mathfrak{L},\lambda_0] = \chi[\mathfrak{L};\lambda_0].$$

In particular, $\mathfrak{m}$ ranges over $\mathbb{N} \cup \{\infty\}$.

Proof. Thanks to Theorem 5.6.1, we know that the algebraic multiplicity χ constructed in Chapter 4, or Chapter 5, satisfies Axioms A1 and A2. It remains to prove its uniqueness.

Let

$$\mathfrak{m}[\,\cdot\,;\lambda_0] : \bigcup_{U,V\in\mathfrak{J}(X)} \mathcal{M}_{\lambda_0}(U,V) \to [0,\infty]$$

be a function satisfying Axioms A1 and A2 of this theorem. Then, the auxiliary function

$$\mathfrak{m}_X[\,\cdot\,;\lambda_0] \;:\; \mathcal{M}_{\lambda_0}(X) \to [0,\infty]$$

defined through

$$\mathfrak{m}_X[\mathfrak{L};\lambda_0] := \mathfrak{m}[\mathfrak{L};\lambda_0], \qquad \mathfrak{L} \in \mathcal{M}_{\lambda_0}(X),$$

satisfies Axioms A1 and A2 of Theorem 6.3.1. Consequently,

$$\mathfrak{m}_X[\mathfrak{L};\lambda_0] = \chi[\mathfrak{L};\lambda_0] \qquad \text{for all } \mathfrak{L} \in \mathcal{M}_{\lambda_0}(X). \tag{6.4}$$

Let $U \in \mathfrak{J}(X)$ and consider any isomorphism $J \in \mathcal{L}(X,U)$. Then, according to Axioms A2 and A1, we have that

$$\begin{aligned} 1 &= \mathfrak{m}[\mathfrak{C}_{P_0}(\cdot - \lambda_0);\lambda_0] = \mathfrak{m}[\mathfrak{C}_{P_0}(\cdot - \lambda_0)J^{-1}J;\lambda_0] \\ &= \mathfrak{m}[\mathfrak{C}_{P_0}(\cdot - \lambda_0);\lambda_0] + \mathfrak{m}[J^{-1};\lambda_0] + \mathfrak{m}[J;\lambda_0] = 1 + \mathfrak{m}[J^{-1};\lambda_0] + \mathfrak{m}[J;\lambda_0] \end{aligned}$$

and, therefore,

$$\mathfrak{m}[J^{-1};\lambda_0] = \mathfrak{m}[J;\lambda_0] = 0, \tag{6.5}$$

because $\mathfrak{m} \geq 0$.

Let $U, V \in \mathfrak{J}(X)$ and $\mathfrak{L} \in \mathcal{M}_{\lambda_0}(U,V)$. Then, there exist isomorphisms $M \in \mathcal{L}(V,X)$ and $J \in \mathcal{L}(X,U)$. Thanks to (6.5),

$$\mathfrak{m}[J;\lambda_0] = \mathfrak{m}[M;\lambda_0] = 0$$

and, hence, by Axiom A1,

$$\mathfrak{m}[\mathfrak{L};\lambda_0] = \mathfrak{m}[M\mathfrak{L}J;\lambda_0].$$

As $M\mathfrak{L}J \in \mathcal{M}_{\lambda_0}(X)$, (6.4) and the product formula for χ show that

$$\mathfrak{m}[\mathfrak{L};\lambda_0] = \mathfrak{m}_X[M\mathfrak{L}J;\lambda_0] = \chi[M\mathfrak{L}J;\lambda_0] = \chi[\mathfrak{L};\lambda_0].$$

This concludes the proof. □

6.5 Applications. Classical multiplicity formulae

This section gives two applications of Theorems 6.0.1 and 6.2.1 to characterize $\chi[\mathfrak{L};\lambda_0]$ in the special case when

$$U = V = \mathbb{K}^N.$$

Corollary 6.5.1. *Let $N \geq 1$ be an integer and choose a basis in $\mathbb{K}^N$ to identify $\mathcal{L}(\mathbb{K}^N)$ with $\mathcal{M}_N(\mathbb{K})$. Then, for every $\lambda_0 \in \mathbb{K}$ and $\mathfrak{L} \in \mathcal{S}^{\infty}_{\lambda_0}(\mathbb{K}^N)$,*

$$\chi[\mathfrak{L};\lambda_0] = \operatorname{ord}_{\lambda_0} \det \mathfrak{L}.$$

Proof. Fix $\lambda_0 \in \mathbb{K}$. Then, the function $\mathfrak{m}[\cdot;\lambda_0] : \mathcal{S}^\infty_{\lambda_0}(\mathbb{K}^N) \to [0,\infty]$ defined by

$$\mathfrak{m}[\mathfrak{L};\lambda_0] := \operatorname{ord}_{\lambda_0} \det \mathfrak{L}, \qquad \mathfrak{L} \in \mathcal{S}^\infty_{\lambda_0}(\mathbb{K}^N),$$

satisfies Axioms A1 and A2 of Theorem 6.0.1. Indeed, $\mathfrak{m}$ is well defined because $\mathfrak{L}$ is of class $\mathcal{C}^\infty$ in a neighborhood of λ_0. Moreover, for every $\mathfrak{L}, \mathfrak{M} \in \mathcal{S}^\infty_{\lambda_0}(\mathbb{K}^N)$, we have that

$$\begin{aligned}\mathfrak{m}[\mathfrak{L}\mathfrak{M};\lambda_0] &= \operatorname{ord}_{\lambda_0} \det(\mathfrak{L}\mathfrak{M}) = \operatorname{ord}_{\lambda_0}[\det \mathfrak{L} \cdot \det \mathfrak{M}] \\ &= \operatorname{ord}_{\lambda_0} \det \mathfrak{L} + \operatorname{ord}_{\lambda_0} \det \mathfrak{M} = \mathfrak{m}[\mathfrak{L};\lambda_0] + \mathfrak{m}[\mathfrak{M};\lambda_0].\end{aligned}$$

Consequently, $\mathfrak{m}$ satisfies Axiom A1.

Let P_0 be the rank-one projection

$$P_0 := \operatorname{diag}\{1, 0, \dots, 0\}.$$

Then, for each $t \in \mathbb{K}$,

$$\mathfrak{C}_{P_0}(t) = t\ P_0 + I_N - P_0 = \operatorname{diag}\{t, 1, \dots, 1\}$$

and, hence,

$$\mathfrak{m}[\mathfrak{C}_{P_0}(\cdot - \lambda_0);\lambda_0] = \operatorname{ord}_{\lambda_0} \det \mathfrak{C}_{P_0}(\cdot - \lambda_0) = \operatorname{ord}_{\lambda_0}(\cdot - \lambda_0) = 1.$$

Consequently, $\mathfrak{m}$ also satisfies Axiom A2. Therefore, by Theorem 6.0.1,

$$\mathfrak{m}[\mathfrak{L};\lambda_0] = \chi[\mathfrak{L};\lambda_0] \quad \text{for all } \ \mathfrak{L} \in \mathcal{S}^\infty_{\lambda_0}(\mathbb{K}^N).$$

This concludes the proof. □

Corollary 6.5.2. *Let $N \geq 1$ be an integer and choose a basis in $\mathbb{C}^N$ to identify $\mathcal{L}(\mathbb{C}^N)$ with $\mathcal{M}_N(\mathbb{C})$. Then, for all $\lambda_0 \in \mathbb{C}$ and $\mathfrak{L} \in \mathcal{A}^\infty_{\lambda_0}(\mathbb{C}^N)$,*

$$\chi[\mathfrak{L};\lambda_0] = \operatorname{tr} \frac{1}{2\pi i} \int_\gamma \mathfrak{L}'(\lambda)\mathfrak{L}(\lambda)^{-1}\, d\lambda, \tag{6.6}$$

where γ is any Cauchy contour whose inner domain D_γ contains λ_0, the family $\mathfrak{L}$ is holomorphic in a neighborhood of $\bar{D}_\gamma$, and $\bar{D}_\gamma \setminus \{\lambda_0\}$ contains no eigenvalue of $\mathfrak{L}$.

Proof. Fix $\lambda_0 \in \mathbb{C}$, and note that $\mathfrak{L}$ is holomorphic at λ_0 if $\mathfrak{L} \in \mathcal{A}^\infty_{\lambda_0}(\mathbb{C}^N)$. By Theorem 6.2.1, the proof will be completed if we show that the function $\mathfrak{m}[\cdot;\lambda_0] : \mathcal{A}^\infty_{\lambda_0}(\mathbb{C}^N) \to \mathbb{C}$ defined through

$$\mathfrak{m}[\mathfrak{L};\lambda_0] := \operatorname{tr} \frac{1}{2\pi i} \int_\gamma \mathfrak{L}'(\lambda)\mathfrak{L}(\lambda)^{-1}\, d\lambda, \qquad \mathfrak{L} \in \mathcal{A}^\infty_{\lambda_0}(\mathbb{C}^N),$$

satisfies Axioms A1–A3 of Theorem 6.2.1, where γ is any Cauchy contour satisfying the requirements of the statement.

We point out that this is a typical example where checking that $\mathfrak{m}$ is non-negative may be rather involved, though Axiom A3 of Theorem 6.2.1 is a direct consequence of the fact that $\mathfrak{L}'\mathfrak{L}^{-1}$ is holomorphic in an open neighborhood of λ_0 if $\mathfrak{L}(\lambda_0)$ is an isomorphism and, therefore, in such a case, the Cauchy theorem guarantees that $\mathfrak{m}[\mathfrak{L};\lambda_0]=0$.

Let $\mathfrak{L},\mathfrak{M}\in\mathcal{A}^\infty_{\lambda_0}(\mathbb{C}^N)$ and consider a Cauchy contour γ with inner domain D_γ containing λ_0 such that both families are holomorphic in $\bar{D}_\gamma$, and $\bar{D}_\gamma\setminus\{\lambda_0\}$ contains no eigenvalue of $\mathfrak{L}$ or of $\mathfrak{M}$. Then,

$$(\mathfrak{L}\mathfrak{M})'(\mathfrak{L}\mathfrak{M})^{-1}=\mathfrak{L}'\mathfrak{L}^{-1}+\mathfrak{L}\mathfrak{M}'\mathfrak{M}^{-1}\mathfrak{L}^{-1}$$

and, hence,

$$\begin{aligned}\mathfrak{m}[\mathfrak{L}\mathfrak{M};\lambda_0]&=\operatorname{tr}\frac{1}{2\pi i}\int_\gamma(\mathfrak{L}\mathfrak{M})'(\lambda)(\mathfrak{L}\mathfrak{M})(\lambda)^{-1}\,d\lambda\\&=\operatorname{tr}\frac{1}{2\pi i}\int_\gamma\mathfrak{L}'(\lambda)\mathfrak{L}(\lambda)^{-1}\,d\lambda+\operatorname{tr}\frac{1}{2\pi i}\int_\gamma\mathfrak{L}(\lambda)\mathfrak{M}'(\lambda)\mathfrak{M}(\lambda)^{-1}\mathfrak{L}(\lambda)^{-1}\,d\lambda\\&=\mathfrak{m}[\mathfrak{L};\lambda_0]+\operatorname{tr}\frac{1}{2\pi i}\int_\gamma\mathfrak{M}'(\lambda)\mathfrak{M}(\lambda)^{-1}\,d\lambda=\mathfrak{m}[\mathfrak{L};\lambda_0]+\mathfrak{m}[\mathfrak{M};\lambda_0],\end{aligned}$$

because the trace is invariant by similarity. Therefore, $\mathfrak{m}$ satisfies Axiom A1 of Theorem 6.2.1.

Finally, let $P\in\mathcal{M}(\mathbb{C})$ be a rank-1 projection. Then,

$$\mathfrak{C}_P{}'(t)\mathfrak{C}_P(t)^{-1}=P\left(t^{-1}P+I_{\mathbb{C}^N}-P\right)=t^{-1}P,\qquad t\in\mathbb{C}\setminus\{0\},$$

and, hence,

$$\mathfrak{m}[\mathfrak{C}_P(\cdot-\lambda_0);\lambda_0]=\frac{1}{2\pi i}\int_\gamma(\lambda-\lambda_0)^{-1}\,d\lambda\ \operatorname{tr}P=1.$$

Consequently, Axiom A2 of Theorem 6.2.1 is satisfied as well. Therefore, the result is a direct consequence of Theorem 6.2.1. □

Remark 6.5.3. Actually, as an easy consequence of Proposition 3.1.1, it readily follows that every isolated eigenvalue λ_0 of a holomorphic family must be an algebraic eigenvalue (see Exercise 3). Therefore, $\mathcal{A}^\infty_{\lambda_0}(\mathbb{C}^N)$ consists of all $\mathfrak{L}\in\mathcal{S}^\infty_{\lambda_0}(\mathbb{C}^N)$ for which either $\mathfrak{L}(\lambda_0)$ is an isomorphism, or λ_0 is an isolated eigenvalue of $\mathfrak{L}$.

The following result is an immediate consequence of Corollary 6.5.2.

Corollary 6.5.4. *Let $N\geq 1$ be an integer and choose a basis in $\mathbb{C}^N$ to identify $\mathcal{L}(\mathbb{C}^N)$ with $\mathcal{M}_N(\mathbb{C})$. Then, for all $A\in\mathcal{M}_N(\mathbb{C})$ and $\lambda_0\in\sigma(A)$,*

$$\chi[\cdot I_{\mathbb{C}^N}-A;\lambda_0]=m_a(\lambda_0),$$

where $m_a(\lambda_0)$ denotes the algebraic multiplicity of λ_0 as an eigenvalue of the matrix A.

Proof. According to (6.6), we have that, for sufficiently small $\varepsilon > 0$,

$$\chi[\cdot I_{\mathbb{C}^N} - A; \lambda_0] = \operatorname{tr} \frac{1}{2\pi i} \int_{C_\varepsilon(\lambda_0)} (\lambda I_{\mathbb{C}^N} - A)^{-1} \, d\lambda = \operatorname{tr} \frac{1}{2\pi i} \int_{C_\varepsilon(\lambda_0)} \mathcal{R}(\lambda; A) \, d\lambda.$$

On the other hand, by Theorem 3.4.1, the operator

$$P_{\lambda_0} := \frac{1}{2\pi i} \int_{C_\varepsilon(\lambda_0)} \mathcal{R}(\lambda; A) \, d\lambda$$

projects $\mathbb{C}^N$ onto $N[(\lambda_0 I_{\mathbb{C}^N} - A)^{\nu(\lambda_0)}]$, where $\nu(\lambda_0)$ is the algebraic ascent of λ_0 as an eigenvalue of A. Therefore, we conclude from Theorem 1.2.1 that

$$\chi[\cdot I_{\mathbb{C}^N} - A; \lambda_0] = \operatorname{tr} P_{\lambda_0} = \operatorname{rank} P_{\lambda_0} = \dim N[(\lambda_0 I_{\mathbb{C}^N} - A)^{\nu(\lambda_0)}] = m_a(\lambda_0).$$

The fact that the trace of $\mathcal{P}_{\lambda_0}$ equals its range follows very easily from the fact that

$$\mathcal{P}_{\lambda_0}|_{N[(\lambda_0 I_{\mathbb{C}^N} - A)^{\nu(\lambda_0)}]} = I_{N[(\lambda_0 I_{\mathbb{C}^N} - A)^{\nu(\lambda_0)}]};$$

see Proposition 9.2.2 for a general proof of this feature. This concludes the proof. □

In Chapters 7 and 9, all the results of this section will be generalized to the infinite-dimensional setting.

6.6 Exercises

1. Prove successively the following:

- If $T \in \mathcal{L}(U)$ is an operator of finite rank, then there exist a finite-dimensional subspace $X \subset U$ and a closed subspace $Y \subset U$ such that

$$U = X \oplus Y, \qquad T(X) \subset X \quad \text{and} \quad Y \subset N[T].$$

- Use this feature to obtain the following generalization of Lemma 6.1.1: If $P, Q \in \mathcal{L}(U)$ are two projections of the same finite rank then there exists an isomorphism $T \in \mathcal{L}(U)$ such that $P = TQT^{-1}$.

2. Use Lemma 6.1.1 and induction to give an alternative proof of the second result of Exercise 1: If $P, Q \in \mathcal{L}(U)$ are two projections of the same finite rank then there exists an isomorphism $T \in \mathcal{L}(U)$ such that $P = TQT^{-1}$.

3. Use Proposition 3.1.1 to prove that every isolated eigenvalue λ_0 of a holomorphic family with values in $\mathcal{M}_N(\mathbb{C})$ must be algebraic.

4. The following is requested to be done:

- Prove that $\mathfrak{L}\mathfrak{M} \in \mathcal{A}^{\infty}_{\lambda_0}(U)$ if $\mathfrak{L}, \mathfrak{M} \in \mathcal{A}^{\infty}_{\lambda_0}(U)$.
- Given a semigroup $(G, +)$ containing $(\mathbb{N}, +)$ as a sub-semigroup, prove that there exists a unique function
$$\mathfrak{m}[\,\cdot\,; \lambda_0] \;:\; \mathcal{A}^{\infty}_{\lambda_0}(U) \to G$$
satisfying the following properties:
A1. $\mathfrak{m}[\mathfrak{L}\mathfrak{M}; \lambda_0] = \mathfrak{m}[\mathfrak{L}; \lambda_0] + \mathfrak{m}[\mathfrak{M}; \lambda_0]$ for all $\mathfrak{L}, \mathfrak{M} \in \mathcal{A}^{\infty}_{\lambda_0}(U)$.
A2. $\mathfrak{m}[\mathfrak{C}_{P_0}(\cdot - \lambda_0); \lambda_0] = 1$ for some rank-1 projection $P_0 \in \mathcal{L}(U)$.
A3. $\mathfrak{m}[\mathfrak{L}; \lambda_0] = 0$ whenever $\mathfrak{L} \in \mathcal{A}^{\infty}_{\lambda_0}(U)$ and $\mathfrak{L}(\lambda_0)$ is an isomorphism.

5. Prove that there are two different functions
$$\mathfrak{m}[\,\cdot\,; \lambda_0] \;:\; \mathcal{S}^{\infty}_{\lambda_0}(U) \to \mathbb{Z} \cup \{\infty\}$$
satisfying Axioms A1 and A2 of the statement of Theorem 6.0.1.

6. Prove that Axioms A1–A3 of Exercise 4 are independent. In other words, none of them is inferable from the remaining two.

7. State and prove the version of the result of Exercise 4 for the general case when $U \neq V$.

8. Let U, V be non-zero spaces with the same finite dimension, and choose bases so that the determinant can be defined in $\mathcal{L}(U, V)$. Prove that
$$\chi[\mathfrak{L}; \lambda_0] = \operatorname{ord}_{\lambda_0} \det \mathfrak{L} \quad \text{for all} \quad \mathfrak{L} \in \mathcal{S}^{\infty}_{\lambda_0}(U, V).$$

9. Generalize the result of Corollary 6.5.2 to cover the general case when U and V are isomorphic to $\mathbb{C}^N$, for some $N \geq 1$.

10. Adapt the proof of Theorem 6.3.1 to prove the following result: Let U be a Banach space, and $\lambda_0 \in \mathbb{K}$. Then there exists at most one function
$$\mathfrak{m}[\cdot; \lambda_0] : \mathcal{M}_{\lambda_0}(U) \to [0, \infty]$$
satisfying the following properties:

- For all $\mathfrak{L}, \mathfrak{M} \in \mathcal{M}_{\lambda_0}(U)$ such that $\mathfrak{L}\mathfrak{M} \in \mathcal{M}_{\lambda_0}(U)$,
$$\mathfrak{m}[\mathfrak{L}\mathfrak{M}; \lambda_0] = \mathfrak{m}[\mathfrak{L}; \lambda_0] + \mathfrak{m}[\mathfrak{M}; \lambda_0].$$
- A projection $P_0 \in \mathcal{L}(U)$ of rank-one exists such that $\mathfrak{m}[\mathfrak{C}_{P_0}(\cdot - \lambda_0); \lambda_0] = 1$.

6.7 Comments on Chapter 6

This chapter is based on Mora-Corral [99]. Actually, all the uniqueness theorems studied in this chapter go back to Mora-Corral [100, 99]. More general uniqueness theorems can be found in [99] and in the list of exercises of Chapter 7.

The results of Section 6.5 reveal that the concept of multiplicity introduced in Chapter 4 equals the classic concepts of algebraic multiplicity in Euclidean spaces, which might have been expected from the very beginning of Chapter 4. All those results will be generalized in Chapters 7 and 9 through the local Smith form. So deeper discussions are postponed.

Chapter 7

Algebraic Multiplicity Through Jordan Chains. Smith Form

This chapter studies the algebraic multiplicity according to the theory of *Jordan chains*. The concept of Jordan chain extends the classic concept of chain of generalized eigenvectors, already studied in Section 1.3. It will provide us with a further approach to the algebraic multiplicities χ and μ introduced and analyzed in Chapters 4 and 5, respectively, whose axiomatization has already been accomplished through the uniqueness theorems included in Chapter 6.

More precisely, the new concept of multiplicity to be introduced in this chapter is based upon the number and lengths of all Jordan chains of the family $\mathfrak{L}$ at the value λ_0, through the concept of *canonical set of Jordan chains*. Not surprisingly, it equals the multiplicities $\chi[\mathfrak{L}; \lambda_0]$ and $\mu[\mathfrak{L}; \lambda_0]$, because it satisfies the product formula and, consequently, it inherits the axiomatic of Chapter 6.

Actually, establishing the hidden inter-relations connecting the associated invariants of all those different approaches to the algebraic multiplicity will extraordinarily facilitate the derivation of a series of extremely sharp properties of the canonical sets of Jordan chains. For example, it will reveal that, for a given family $\mathfrak{L} \in \mathcal{C}^r(\Omega, \mathcal{L}(U, V))$ with $r \in \mathbb{N} \cup \{\infty\}$ (not necessarily analytic) with $\mathfrak{L}_0 \in \mathrm{Fred}_0(U, V)$, the lengths of all Jordan chains of $\mathfrak{L}$ at $\lambda_0 \in \mathrm{Eig}(\mathfrak{L})$ are uniformly bounded above if and only if λ_0 is an algebraic eigenvalue of $\mathfrak{L}$. As a by-product, this pivotal property will allow us to characterize whether or not the family $\mathfrak{L}$ admits a *local Smith form* at λ_0. Basically, the local Smith form of $\mathfrak{L}$ at λ_0 is the simplest canonical form that can be obtained through pre- and post-multiplication of $\mathfrak{L}$ by two families of isomorphisms. Among the main results of this chapter, it will be shown that $\mathfrak{L}$ *possesses a local Smith form of at* λ_0 *if and only if* $\lambda_0 \in \mathrm{Alg}(\mathfrak{L})$. In other words, if and only if

$$\chi[\mathfrak{L}; \lambda_0] = \mu[\mathfrak{L}; \lambda_0] < \infty.$$

The local Smith form tremendously facilitates the proof of some of the most useful properties of the algebraic multiplicity that has been axiomatized in Chapter 6, such as the stability theorem for holomorphic families (see Section 8.4), and the product formula in its more general version (see next Exercise 19). In addition to that, another important interest relies on the fact that it provides the most versatile way to characterize whether two given families are locally equivalent. Two families are said to be locally equivalent when one of them can be obtained from the other through pre- and post-multiplication by two isomorphism families. It turns out that *two families are locally equivalent at* λ_0 *if and only if they possess the same local Smith form at* λ_0.

The general plan of this chapter is as follows. Section 7.1 introduces the concept of Jordan chain and uses the theory of Chapter 4 for characterizing the Jordan chains at any transversal eigenvalue. Section 7.2 fixes the concepts of canonical set of Jordan chains and of partial and total κ-multiplicities, denoted by $\kappa_j[\mathfrak{L};\lambda_0]$ and $\kappa[\mathfrak{L};\lambda_0]$, through the number of Jordan chains of a given length in any canonical set. Section 7.3 shows the invariance of the partial κ-multiplicities under pre- and post-multiplication by continuous isomorphism families. Section 7.4 introduces the concept of local Smith form for $\mathfrak{L} \in \mathcal{C}^r(\Omega, \mathcal{L}(U,V))$ at any $\lambda_0 \in \mathrm{Eig}(\mathfrak{L})$ and establishes the existence of a local Smith form when $r=\infty$ and no Jordan chain of $\mathfrak{L}$ can be continued indefinitely in the special case when U and V are isomorphic to $\mathbb{K}^m$, for some $m \geq 1$. Section 7.5 combines the transversalization theorem of Chapter 4 with the invariance theorem of Section 7.3 to obtain a finite scheme for ascertaining the partial κ-multiplicities and, consequently, constructing canonical sets of Jordan chains systematically.

The remaining sections contain some of the main consequences of the basic analysis carried out in the previous ones. Namely, in Section 7.6 it is proved that the multiplicity $\kappa[\mathfrak{L};\lambda_0]$ coincides with the multiplicities $\chi[\mathfrak{L};\lambda_0]$ and $\mu[\mathfrak{L};\lambda_0]$, and that $\kappa[\mathfrak{L};\lambda_0] < \infty$ if and only if the length of every Jordan chain of $\mathfrak{L}$ at λ_0 is bounded above by a $\nu < r+1$. According to the results of Chapters 4 and 5, it turns out that $\kappa[\mathfrak{L};\lambda_0] < \infty$ if and only if $\lambda_0 \in \mathrm{Alg}_\nu(\mathfrak{L})$ for some $0 \leq \nu < r+1$. Incidentally, much like in the context of classic spectral theory, ν can be characterized as the lowest natural number k for which

$$\dim N\left[\operatorname{trng}\{\mathfrak{L}_0,\dots,\mathfrak{L}_{k-1}\}\right] = \dim N\left[\operatorname{trng}\{\mathfrak{L}_0,\dots,\mathfrak{L}_k\}\right],$$

though this point will not be completely clarified until Chapter 8. Section 7.7 labels the vectors of a canonical set of Jordan chains in an appropriate way useful in Chapter 10. All previous information will be brought together in Section 7.8 to characterize the existence of the local Smith form. Section 7.9 illustrates that characterization by analyzing two simple examples. Finally, Section 7.10 introduces the notion of local equivalence of families and uses the theorem of characterization of the local Smith form to determine whether two given families are locally equivalent at a point.

As in Chapters 4 and 5, throughout this chapter we consider $\mathbb{K} \in \{\mathbb{R}, \mathbb{C}\}$, two $\mathbb{K}$-Banach spaces U and V, an open subset $\Omega \subset \mathbb{K}$, a point $\lambda_0 \in \Omega$, and a family $\mathfrak{L} \in \mathcal{C}^r(\Omega, \mathcal{L}(U, V))$, for some $r \in \mathbb{N} \cup \{\infty, \omega\}$, with $\mathfrak{L}_0 \in \mathrm{Fred}_0(U, V)$, even if not mentioned. As $\mathrm{Fred}_0(U, V)$ is an open set of $\mathcal{L}(U, V)$, we actually have that $\mathfrak{L}(\lambda) \in \mathrm{Fred}_0(U, V)$ for λ sufficiently close to λ_0.

Throughout the rest of this book, for any family $\mathfrak{L} \in \mathcal{C}^r(\Omega, \mathcal{L}(U, V))$ and any integer $0 \leq s \leq r$, we will denote

$$\mathfrak{L}_{[s]} := \mathrm{trng}\{\mathfrak{L}_0, \dots, \mathfrak{L}_s\} = \begin{pmatrix} \mathfrak{L}_0 & 0 & \cdots & 0 \\ \mathfrak{L}_1 & \mathfrak{L}_0 & \cdots & 0 \\ \vdots & \vdots & \ddots & \vdots \\ \mathfrak{L}_s & \mathfrak{L}_{s-1} & \cdots & \mathfrak{L}_0 \end{pmatrix}, \tag{7.1}$$

where, as usual in this book,

$$j!\mathfrak{L}_j = \mathfrak{L}^{j)}(\lambda_0), \qquad 0 \leq j < r+1.$$

These conventions will substantially simplify most of the subsequent notation and statements.

7.1 The concept of Jordan chain

The next definition introduces the concept of Jordan chain.

Definition 7.1.1. Let $r \in \mathbb{N} \cup \{\infty\}$ and $\mathfrak{L} \in \mathcal{C}^r(\Omega, \mathcal{L}(U, V))$. Then, for a given integer $0 \leq s \leq r$, a vector $(u_0, \dots, u_s) \in U^{s+1}$ with $u_0 \neq 0$ is said to be a Jordan chain of length $s+1$ of $\mathfrak{L}$ at λ_0 starting at u_0 if

$$\sum_{i=0}^{j} \mathfrak{L}_i u_{j-i} = 0, \qquad 0 \leq j \leq s. \tag{7.2}$$

Remark 7.1.2. Using (7.1), the identities (7.2) can be equivalently expressed as

$$\mathfrak{L}_{[s]} \operatorname{col}[u_0, \dots, u_s] = 0$$

and, consequently, the set of Jordan chains of length $s+1$ of $\mathfrak{L}$ at λ_0 is given through the column vectors

$$\operatorname{col}[u_0, \dots, u_s] \in N[\mathfrak{L}_{[s]}]$$

with $u_0 \neq 0$. Therefore, the following properties are satisfied:

- If $(u_0, \dots, u_s)$ is a Jordan chain of $\mathfrak{L}$ at λ_0, then, so is $(u_0, \dots, u_p)$ for each integer $0 \leq p \leq s$.
- If $(u_0, \dots, u_s)$ and $(v_0, \dots, v_s)$ are two Jordan chains of length $s+1$ of $\mathfrak{L}$ at λ_0, then, so is $(\lambda u_0 + \mu v_0, \dots, \lambda u_s + \mu v_s)$ for every $\lambda, \mu \in \mathbb{K}$ with $\lambda u_0 + \mu v_0 \neq 0$.

As (7.2) imposes $\mathfrak{L}_0 u_0 = 0$, the family $\mathfrak{L}$ admits a Jordan chain at λ_0 if and only if $\lambda_0 \in \operatorname{Eig}(\mathfrak{L})$. In such a case, any vector $u_0 \in N[\mathfrak{L}_0] \setminus \{0\}$ is said to be an *eigenvector* of $\mathfrak{L}$ at λ_0; the vector (u_0) itself provides a Jordan chain of length 1 of $\mathfrak{L}$ at λ_0 starting at u_0, of course.

Suppose $U \subset V$ with continuous embedding $J : U \hookrightarrow V$, and consider the associated operator family

$$\mathfrak{L}(\lambda) = \lambda J - T, \qquad \lambda \in \mathbb{K},$$

for a given $T \in \mathcal{L}(U,V)$. Then, $\mathfrak{L} \in \mathbb{P}[\mathcal{L}(U,V)]$ and for any $\lambda_0 \in \mathbb{K}$,

$$\mathfrak{L}_0 = \lambda_0 J - T, \qquad \mathfrak{L}_1 = J, \qquad \mathfrak{L}_h = 0 \quad \text{for all } h \geq 2.$$

Thus, in this special case, system (7.2) becomes

$$(T - \lambda_0 J)u_0 = 0, \quad (T - \lambda_0 J)u_1 = u_0, \quad \ldots, \quad (T - \lambda_0 J)u_s = u_{s-1},$$

and, hence, the vector $(u_0, \ldots, u_s)$ is a Jordan chain of length $s+1$ of $\mathfrak{L}$ at λ_0 starting at u_0 if and only if it equals

$$\big((T - \lambda_0 J)^s u_s, (T - \lambda_0 J)^{s-1} u_s, \ldots, (T - \lambda_0 J)^2 u_s, (T - \lambda_0 J)u_s, u_s\big)$$

with

$$u_s \in N[(T - \lambda_0 J)^{s+1}] \setminus N[(T - \lambda_0 J)^s].$$

Consequently, in the classical setting, the Jordan chains provide all possible chains of generalized eigenvectors forming the Jordan basis, as discussed in Section 1.3. So the concept of Jordan chain introduced by Definition 7.1.1 is a natural generalization of the concept of generalized eigenvector chain associated with an eigenvalue λ_0 of a square matrix.

Systems of the type (7.2) were already introduced in Section 4.1 to motivate the concept of transversal eigenvalue. Actually, the concept of transversal eigenvalue was introduced to get system (7.2) uncoupled and, consequently, at a transversal eigenvalue the structure of the Jordan chains should be extremely simple. In fact, the following result holds.

Proposition 7.1.3. *Suppose $r \in \mathbb{N}^* \cup \{\infty\}$, the family $\mathfrak{L} \in \mathcal{C}^r(\Omega, \mathcal{L}(U,V))$ satisfies $\mathfrak{L}_0 \in \operatorname{Fred}_0(U,V)$, and, for some integer $1 \leq \nu \leq r$, the subspaces*

$$R[\mathfrak{L}_0], \ \mathfrak{L}_1(N[\mathfrak{L}_0]), \ \ldots, \ \mathfrak{L}_\nu\Big(\bigcap_{j=0}^{\nu-1} N[\mathfrak{L}_j]\Big)$$

are in direct sum. Then, for each integer $0 \leq s \leq \nu$, the vector $(u_0, \ldots, u_s) \in U^{s+1}$ is a Jordan chain of $\mathfrak{L}$ at λ_0 if and only if $u_0 \neq 0$ and

$$u_j \in \bigcap_{i=0}^{s-j} N[\mathfrak{L}_i], \qquad 0 \leq j \leq s. \tag{7.3}$$

In other words,

$$N[\mathfrak{L}_{[s]}] = \left(\bigcap_{j=0}^{s} N[\mathfrak{L}_j]\right) \times \left(\bigcap_{j=0}^{s-1} N[\mathfrak{L}_j]\right) \times \cdots \times N[\mathfrak{L}_0].$$

Consequently, $s+1 \leq \nu$ when λ_0 is a ν-transversal eigenvalue of $\mathfrak{L}$ and there are Jordan chains of $\mathfrak{L}$ at λ_0 of length $s+1$.

Proof. Let s be an integer such that $0 \leq s \leq \nu$. According to Lemma 4.1.1, it is obvious that $(u_0, \ldots, u_s) \in U^{s+1}$ is a Jordan chain of $\mathfrak{L}$ at λ_0 if and only if $u_0 \neq 0$ and (7.3) is satisfied.

Suppose λ_0 is a ν-transversal eigenvalue of $\mathfrak{L}$ for some integer $1 \leq \nu \leq r$. Then, by Proposition 4.2.2,

$$\bigcap_{j=0}^{\nu} N[\mathfrak{L}_j] = \{0\}$$

and, therefore, $s \leq \nu - 1$ for all Jordan chains of length $s+1$, which concludes the proof. □

7.2 Canonical sets and κ-multiplicity

The next definition introduces the concept of *rank map* associated with an operator family at a given $\lambda_0 \in \mathrm{Eig}(\mathfrak{L})$.

Definition 7.2.1. Let $r \in \mathbb{N} \cup \{\infty\}$ and $\mathfrak{L} \in \mathcal{C}^r(\Omega, \mathcal{L}(U,V))$ with $\lambda_0 \in \mathrm{Eig}(\mathfrak{L})$ and $\mathfrak{L}_0 \in \mathrm{Fred}_0(U,V)$. Let L_{r+1} denote either the empty set, if $r = \infty$, or the set of eigenvectors $u_0 \in N[\mathfrak{L}_0] \setminus \{0\}$ for which $\mathfrak{L}$ admits a Jordan chain of length $r+1$ at λ_0 starting at u_0, if $r \in \mathbb{N}$. Then, the rank map associated with $\mathfrak{L}$ at λ_0 is the map

$$\mathrm{rank} \;:\; N[\mathfrak{L}_0] \setminus (L_{r+1} \cup \{0\}) \to \mathbb{N} \cup \{\infty\}$$

defined, for every

$$u_0 \in N[\mathfrak{L}_0] \setminus (L_{r+1} \cup \{0\}),$$

as follows:

- $\mathrm{rank}(u_0) = k \in \mathbb{N}$ if $k < r+1$ and $\mathfrak{L}$ admits a Jordan chain of length k at λ_0 starting at u_0, but not a Jordan chain of length $k+1$.
- $\mathrm{rank}(u_0) = \infty$ when, for every integer $k \geq 1$, the family $\mathfrak{L}$ possesses a Jordan chain of length k at λ_0 starting at u_0.

In the case $r = \infty$, the rank map has been defined over all $N[\mathfrak{L}_0] \setminus \{0\}$. Another closely related concept of infinity rank might be given through the following definition.

Definition 7.2.2. Suppose $r = \infty$ and $(u_0, \dots, u_s)$ is a Jordan chain of $\mathfrak{L}$ at λ_0. It is said that the Jordan chain $(u_0, \dots, u_s)$ can be continued indefinitely if there exists a sequence $\{u_n\}_{n \geq s+1}$ in U such that $(u_0, \dots, u_n)$ is a Jordan chain of $\mathfrak{L}$ at λ_0 for all $n \geq s+1$.

According to Definition 7.2.1, $\operatorname{rank} u_0 = \infty$ if the Jordan chain (u_0) can be continued indefinitely; however, in general there might exist some $u_0 \in N[\mathfrak{L}_0] \setminus \{0\}$ with $\operatorname{rank} u_0 = \infty$ for which (u_0) cannot be continued indefinitely. Actually, Exercise 15 shows that this can indeed occur if $\mathfrak{L}_0 \notin \mathrm{Fred}_0(U, V)$. Nevertheless, in the case $\mathfrak{L}_0 \in \mathrm{Fred}_0(U, V)$, it will become apparent from Theorem 7.8.3 that $\operatorname{rank} u_0 = \infty$ if and only if the chain (u_0) can be continued indefinitely.

Simple examples show the existence of eigenvectors with infinity rank. Indeed, suppose $U = V = \mathbb{K} = \mathbb{R}$, the point λ_0 is 0, and $\mathfrak{L}$ is the family defined by (4.20). Then, $\mathfrak{L}_j = 0$ for all $j \geq 0$, and, hence,

$$\operatorname{rank} u_0 = \infty \qquad \text{for all } \; u_0 \in \mathbb{R} \setminus \{0\},$$

because, for any $n \geq 1$, the vector $(u_0, \dots, u_0) \in \mathbb{R}^n$ is a Jordan chain of length n.

Now we describe the concept of *canonical set of Jordan chains*. To this aim, we consider $r \in \mathbb{N} \cup \{\infty\}$ and $\mathfrak{L} \in \mathcal{C}^r(\Omega, \mathcal{L}(U, V))$ with

$$\mathfrak{L}_0 = \mathfrak{L}(\lambda_0) \in \mathrm{Fred}_0(U, V) \setminus \mathrm{Iso}(U, V), \tag{7.4}$$

for which the length of all Jordan chains of $\mathfrak{L}$ at λ_0 is uniformly bounded above by some integer $k \leq r$. Then, the rank map is defined over the whole $N[\mathfrak{L}_0] \setminus \{0\}$. Let $\kappa_1 \leq k$ satisfy that $\mathfrak{L}$ possesses a Jordan chain of length κ_1, but it does not admit a Jordan chain of length $\kappa_1 + 1$. Then, there exists

$$u_{0,1} \in N[\mathfrak{L}_0] \setminus \{0\} \quad \text{with} \quad \operatorname{rank} u_{0,1} = \kappa_1,$$

and the length of all Jordan chains of $\mathfrak{L}$ at λ_0 is bounded above by κ_1. Define

$$n := \dim N[\mathfrak{L}_0],$$

and proceeding inductively, suppose that for some $2 \leq i \leq n$ we have chosen $i-1$ linearly independent vectors

$$u_{0,1}, \dots, u_{0,i-1} \in N[\mathfrak{L}_0],$$

and take

$$u_{0,i} \in C_i := N[\mathfrak{L}_0] \setminus \operatorname{span}[u_{0,1}, \dots, u_{0,i-1}]$$

such that

$$\operatorname{rank} u_{0,i} = \kappa_i := \max\{\operatorname{rank} u : u \in C_i\}.$$

This process can be repeated until n linearly independent vectors

$$u_{0,1}, \dots, u_{0,n} \in N[\mathfrak{L}_0]$$

have been selected. They form a basis of $N[\mathfrak{L}_0]$, since $n = \dim N[\mathfrak{L}_0]$. Finally, for each $i \in \{1, \dots, n\}$, let $(u_{0,i}, u_{1,i}, \dots, u_{\kappa_i - 1,i})$ be any Jordan chain of length κ_i of $\mathfrak{L}$ at λ_0 starting at $u_{0,i}$.

Definition 7.2.3. Any ordered family

$$\left[u_{0,1},\ldots,u_{\kappa_1-1,1};\cdots;u_{0,i},\ldots,u_{\kappa_i-1,i};\cdots;u_{0,n},\ldots,u_{\kappa_n-1,n}\right] \qquad (7.5)$$

constructed through the previous scheme is said to be a canonical set of Jordan chains of $\mathfrak{L}$ at λ_0.

By construction, $\kappa_1 \geq \cdots \geq \kappa_n \geq 1$ and the following result is obtained.

Proposition 7.2.4. *Let $r \in \mathbb{N} \cup \{\infty\}$ and $\mathfrak{L} \in \mathcal{C}^r(\Omega, \mathcal{L}(U,V))$ satisfy (7.4) for which the length of all Jordan chains of $\mathfrak{L}$ at λ_0 is uniformly bounded above by some integer $k \leq r$. Then, the integers $\kappa_1,\ldots,\kappa_n$ of the construction of any canonical set of Jordan chains are the same for all canonical sets of Jordan chains.*

Proof. Clearly, κ_1 is the same for every canonical set of Jordan chains. Set

$$K_1 := \{u \in N[\mathfrak{L}_0] \setminus \{0\} : \operatorname{rank} u = \kappa_1\}.$$

By Remark 7.1.2, $E_1 := K_1 \cup \{0\}$ is the linear subspace of $N[\mathfrak{L}_0]$ generated by K_1. Set $\ell := \dim E_1$ and let $\{u_{0,1},\ldots,u_{0,n}\}$ be a basis of $N[\mathfrak{L}_0]$ selected from a canonical set of Jordan chains. Then,

$$E_1 = \operatorname{span}[u_{0,1},\ldots,u_{0,\ell}],$$

because these vectors are linearly independent, by construction. Thus, $\kappa_1 = \cdots = \kappa_\ell$ are independent of the canonical set. This concludes the proof if $\ell = n$.

Suppose $\ell < n$. Clearly, $\kappa_{\ell+1}$ is independent of the canonical set of Jordan chains. Then, reasoning as before, we consider

$$K_2 := \{u \in N[\mathfrak{L}_0] \setminus \{0\} : \operatorname{rank} u \geq \kappa_{\ell+1}\}.$$

By Remark 7.1.2, $E_2 := K_2 \cup \{0\}$ is the subspace of $N[\mathfrak{L}_0]$ generated by K_2. Moreover, since $\kappa_1 \geq \kappa_{\ell+1}$, we have that

$$E_1 \subset E_2 = \operatorname{span}[u_{0,1},\ldots,u_{0,\dim E_2}].$$

Consequently, $\kappa_{\ell+1} = \cdots = \kappa_{\dim E_2}$ are independent of the canonical set of Jordan chains. Repeating this scheme as many times as necessary concludes the proof. □

Proposition 7.2.4 actually shows the consistency of the following concept of algebraic κ-multiplicity.

Definition 7.2.5. Let $r \in \mathbb{N} \cup \{\infty\}$ and $\mathfrak{L} \in \mathcal{C}^r(\Omega, \mathcal{L}(U,V))$ be with $\mathfrak{L}_0 \in \operatorname{Fred}_0(U,V)$. Then, the algebraic κ-multiplicity of $\mathfrak{L}$ at λ_0, denoted by $\kappa[\mathfrak{L};\lambda_0]$, is defined through:

- $\kappa[\mathfrak{L};\lambda_0]=0$, if $\mathfrak{L}_0\in \mathrm{Iso}(U,V)$.
- $\kappa[\mathfrak{L};\lambda_0]:=\sum_{j=1}^n \kappa_j$, if $n:=\dim N[\mathfrak{L}_0]\geq 1$ and the length of all Jordan chains of $\mathfrak{L}$ at λ_0 is uniformly bounded above by some integer $k\leq r$. In such a case, for each $1\leq j\leq n$, the integer
$$\kappa_j[\mathfrak{L};\lambda_0]:=\kappa_j$$
is said to be the jth partial κ-multiplicity of $\mathfrak{L}$ at λ_0.
- $\kappa[\mathfrak{L};\lambda_0]=\infty$, if $r=\infty$ and $\mathfrak{L}$ possesses Jordan chains at λ_0 with arbitrarily large length.

The multiplicity $\kappa[\mathfrak{L};\lambda_0]$, as well as the partial κ-multiplicities, is left undefined when $r\in\mathbb{N}$, the family $\mathfrak{L}$ is not of class $\mathcal{C}^{r+1}$ in any neighborhood of λ_0, and $\mathfrak{L}$ possesses a Jordan chain of length $r+1$ at λ_0. The extension of the concept of partial κ-multiplicities to the case when $r=\infty$ and the lengths of the Jordan chains of $\mathfrak{L}$ are unbounded has been postponed until Exercise 5.

Theorem 7.6.3 establishes that κ coincides with the multiplicity χ introduced in Chapter 4. Consequently, according to Theorem 5.3.1, it must also coincide with the multiplicity μ introduced in Chapter 5.

Example. Suppose $U=V=\mathbb{K}^m$ for some integer $m\geq 1$, and
$$\mathfrak{L}(\lambda)=\mathrm{diag}\{(\lambda-\lambda_0)^{p_1},\dots,(\lambda-\lambda_0)^{p_m}\},\qquad \lambda\in\mathbb{K},$$
for some $p_1,\dots,p_m\in\mathbb{N}$. Then,
$$n:=\dim N[\mathfrak{L}_0]=\mathrm{Card}\{i\in\{1,\dots,m\}\;:\;p_i\geq 1\}$$
and, according to Definition 7.2.5,
$$\kappa[\mathfrak{L};\lambda_0]=\sum_{j=1}^m p_j.$$
Moreover, in the case $n\geq 1$,
$$\kappa_1:=\kappa_1[\mathfrak{L};\lambda_0]=\max\{p_1,\dots,p_m\}$$
and, in general, when $n\geq 2$,
$$\kappa_j[\mathfrak{L};\lambda_0]=\max\left(\{p_1,\dots,p_m\}\setminus\{\kappa_1,\dots,\kappa_{j-1}\}\right),\qquad 2\leq j\leq n.$$
In particular, if
$$\mathfrak{L}(\lambda)=\mathrm{diag}\left\{(\lambda-\lambda_0)^{p_1},\dots,(\lambda-\lambda_0)^{p_n},\underbrace{1,\dots,1}_{m-n}\right\}$$
with
$$p_1\geq\cdots\geq p_n\geq 1,$$
then, $n=\dim N[\mathfrak{L}_0]$,
$$\kappa[\mathfrak{L};\lambda_0]=\sum_{j=1}^n p_j,\quad\text{and}\quad \kappa_j[\mathfrak{L};\lambda_0]=p_j,\qquad 1\leq j\leq n.$$

7.3 Invariance by continuous families of isomorphisms

The next result shows the invariance of the partial multiplicities under multiplication by continuous families of isomorphisms.

Theorem 7.3.1 (of invariance). *Let $r \in \mathbb{N} \cup \{\infty\}$, and let there be four $\mathbb{K}$-Banach spaces U, V, $\hat{U}$, $\hat{V}$, and four operator families*

$$\mathfrak{L} \in \mathcal{C}^r(\Omega, \mathcal{L}(U,V)), \qquad \mathfrak{M} \in \mathcal{C}^r(\Omega, \mathcal{L}(\hat{U},\hat{V})),$$
$$\mathfrak{E} \in \mathcal{C}(\Omega, \mathcal{L}(\hat{V},V)), \qquad \mathfrak{F} \in \mathcal{C}(\Omega, \mathcal{L}(U,\hat{U})),$$

such that

$$\mathfrak{E}_0 := \mathfrak{E}(\lambda_0) \in \mathrm{Iso}(\hat{V},V), \quad \mathfrak{F}_0 := \mathfrak{F}(\lambda_0) \in \mathrm{Iso}(U,\hat{U}), \tag{7.6}$$

and

$$\mathfrak{L} = \mathfrak{E}\mathfrak{M}\mathfrak{F}. \tag{7.7}$$

Then, $\mathfrak{L}_0 \in \mathrm{Fred}_0(U,V)$ if and only if $\mathfrak{M}_0 \in \mathrm{Fred}_0(\hat{U},\hat{V})$, and, in such a case, for every integer $0 \leq s \leq r$ there is a linear bijection between the sets of Jordan chains of length $s+1$ of $\mathfrak{L}$ and $\mathfrak{M}$ at λ_0. In particular, the length of all Jordan chains of $\mathfrak{L}$ at λ_0 is uniformly bounded above by some integer $k \leq r$ if and only if so is the length of all Jordan chains of $\mathfrak{M}$. Moreover, in such a circumstance,

$$\kappa[\mathfrak{L};\lambda_0] = \kappa[\mathfrak{M};\lambda_0],$$

and, actually, when $\mathfrak{L}_0 \notin \mathrm{Iso}(U,V)$,

$$\kappa_j[\mathfrak{L};\lambda_0] = \kappa_j[\mathfrak{M};\lambda_0], \quad 1 \leq j \leq \dim N[\mathfrak{L}_0] = \dim N[\mathfrak{M}_0].$$

Proof. By (7.6) and (7.7), it is clear that

$$\mathfrak{L}_0 \in \mathrm{Fred}_0(U,V) \quad \text{if and only if} \quad \mathfrak{M}_0 \in \mathrm{Fred}_0(\hat{U},\hat{V}).$$

Throughout the rest of the proof we will work under this assumption.

Again by (7.6) and (7.7), $\mathfrak{L}_0 \in \mathrm{Iso}(U,V)$ if and only if $\mathfrak{M}_0 \in \mathrm{Iso}(\hat{U},\hat{V})$. Moreover, in such a case, according to Definition 7.2.5, we have that

$$\kappa[\mathfrak{L};\lambda_0] = \kappa[\mathfrak{M};\lambda_0] = 0,$$

which concludes the proof. So, we will subsequently assume that

$$n = \dim N[\mathfrak{L}_0] = \dim N[\mathfrak{M}_0] \geq 1.$$

First, we will prove the theorem in the special case when

$$\mathfrak{E} \in \mathcal{C}^r(\Omega, \mathcal{L}(\hat{V},V)) \quad \text{and} \quad \mathfrak{F} \in \mathcal{C}^r(\Omega, \mathcal{L}(U,\hat{U})).$$

Fix an integer $0 \leq s \leq r$. Then, applying the Leibniz rule to (7.7) gives

$$\mathfrak{L}_{[s]} = \mathfrak{E}_{[s]}\mathfrak{M}_{[s]}\mathfrak{F}_{[s]}.$$

By Remark 7.1.2, the spaces $N[\mathfrak{M}_{[s]}]$ and $N[\mathfrak{L}_{[s]}]$ provide all Jordan chains of length $s+1$ of $\mathfrak{M}$ and $\mathfrak{L}$ at λ_0, respectively. Moreover, by (7.6),

$$\mathfrak{E}_{[s]} \in \mathrm{Iso}(\hat{V}^{s+1}, V^{s+1}), \qquad \mathfrak{F}_{[s]} \in \mathrm{Iso}(U^{s+1}, \hat{U}^{s+1}).$$

Therefore, the map

$$\begin{array}{rcl} \mathfrak{I}_s : N[\mathfrak{L}_{[s]}] & \to & N[\mathfrak{M}_{[s]}] \\ \mathrm{col}[u_0, \dots, u_s] & \mapsto & \mathfrak{F}_{[s]} \,\mathrm{col}[u_0, \dots, u_s] \end{array}$$

establishes an isomorphism. Moreover, setting

$$\mathrm{col}[\hat{u}_0, \dots, \hat{u}_s] := \mathfrak{F}_{[s]} \,\mathrm{col}[u_0, \dots, u_s]$$

shows that $\hat{u}_0 \neq 0$ if and only if $u_0 \neq 0$, since $\hat{u}_0 = \mathfrak{F}_0 u_0$. Consequently, the map $\mathfrak{I}_s$ establishes a linear bijection between the Jordan chains of length $s+1$ of $\mathfrak{M}$ and those of $\mathfrak{L}$. The remaining assertions of the theorem are straightforward consequences of this fact.

The same idea can be adapted to cover the general case when $\mathfrak{E}$ and $\mathfrak{F}$ are continuous, although carrying out all technical details in the general case is more involved, as identity (7.7) cannot be differentiated.

Fix an integer $0 \leq s \leq r$. By Theorem 4.3.2, there exists $\Phi \in \mathbb{P}[\mathcal{L}(U)]$ with $\Phi_0 = I_U$ for which the transformed family

$$\mathfrak{L}^{\Phi} := \mathfrak{L}\Phi \tag{7.8}$$

satisfies

$$V_s := \bigoplus_{j=1}^{s} \mathfrak{L}_j^{\Phi}(N[\mathfrak{L}_0^{\Phi}] \cap \dots \cap N[\mathfrak{L}_{j-1}^{\Phi}]) \oplus R[\mathfrak{L}_0^{\Phi}] \subset V.$$

Then, setting

$$\mathfrak{F}^{\Phi} := \mathfrak{F}\Phi,$$

we have

$$\mathfrak{F}_0^{\Phi} = \mathfrak{F}_0 \quad \text{and} \quad \mathfrak{L}^{\Phi} = \mathfrak{E}\mathfrak{M}\mathfrak{F}^{\Phi}. \tag{7.9}$$

By (7.8), according to the result already established in the regular case, it is apparent that the map

$$\begin{array}{rcl} N[\mathfrak{L}_{[s]}^{\Phi}] & \to & N[\mathfrak{L}_{[s]}] \\ \mathrm{col}[u_0, \dots, u_s] & \mapsto & \Phi_{[s]} \,\mathrm{col}[u_0, \dots, u_s] \end{array}$$

establishes a linear bijection between the Jordan chains of length $s+1$ of $\mathfrak{L}^{\Phi}$ and those of $\mathfrak{L}$. Consequently, to complete the proof it suffices to establish a linear bijection between the Jordan chains of length $s+1$ of $\mathfrak{L}^{\Phi}$ and those of $\mathfrak{M}$.

According to (7.6) and (7.9), we have that

$$\mathfrak{M}_0\mathfrak{F}_0 u_0 = 0 \quad \text{if and only if} \quad u_0 \in N[\mathfrak{L}_0^{\Phi}] \tag{7.10}$$

and, hence, in the case $s = 0$, the map $u_0 \mapsto \mathfrak{F}_0 u_0$ indeed establishes a linear bijection between the Jordan chains of length 1 of $\mathfrak{L}^\Phi$ and those of $\mathfrak{M}$.

Suppose $1 \leq s < r+1$, and let $(u_0, \dots, u_s)$ be a Jordan chain of $\mathfrak{L}^\Phi$. Then, by Proposition 7.1.3,

$$u_0 \neq 0 \quad \text{and} \quad u_j \in \bigcap_{i=0}^{s-j} N[\mathfrak{L}_i^\Phi], \qquad 0 \leq j \leq s.$$

In particular,

$$\dim \bigcap_{i=0}^{s} N[\mathfrak{L}_i^\Phi] \geq 1$$

and there exist $s+1$ natural numbers

$$n_s \leq n_{s-1} \leq \cdots \leq n_1 \leq n_0,$$

and n_0 linearly independent vectors $u^1, \dots, u^{n_0}$ in U such that

$$\bigcap_{i=0}^{j} N[\mathfrak{L}_i^\Phi] = \operatorname{span}[u^1, \dots, u^{n_j}], \qquad 0 \leq j \leq s.$$

Clearly,

$$N[\mathfrak{L}_{[s]}^\Phi] = \operatorname{span}\Big\{ \operatorname{col}[\underbrace{0, \dots, 0}_{j}, u^i, \underbrace{0, \dots, 0}_{s-j}] \;:\; 1 \leq i \leq n_{s-j}, \;\; 0 \leq j \leq s \Big\}.$$

On the other hand, for each integer $0 \leq \ell \leq s$,

$$0 = \sum_{i=0}^{\ell} \mathfrak{L}_i^\Phi u_{\ell-i} = \mathfrak{L}_0^\Phi u_\ell + \sum_{i=1}^{\ell} \lim_{\lambda \to \lambda_0} \frac{\mathfrak{L}^\Phi(\lambda) - \sum_{j=0}^{i-1} (\lambda - \lambda_0)^j \mathfrak{L}_j^\Phi}{(\lambda - \lambda_0)^i} u_{\ell-i}.$$

Moreover, owing to Remark 7.1.2, we have that, for any $\lambda \neq \lambda_0$,

$$\sum_{i=1}^{\ell} \sum_{j=0}^{i-1} (\lambda - \lambda_0)^{j-i} \mathfrak{L}_j^\Phi u_{\ell-i} = \sum_{k=-\ell}^{-1} (\lambda - \lambda_0)^k \sum_{j=0}^{\ell+k} \mathfrak{L}_j^\Phi u_{\ell+k-j} = 0.$$

Thus,

$$0 = \lim_{\lambda \to \lambda_0} \sum_{i=0}^{\ell} (\lambda - \lambda_0)^{-i} \mathfrak{L}^\Phi(\lambda) u_{\ell-i} = \lim_{\lambda \to \lambda_0} \sum_{i=0}^{\ell} (\lambda - \lambda_0)^{-i} \mathfrak{E}(\lambda) \mathfrak{M}(\lambda) \mathfrak{F}^\Phi(\lambda) u_{\ell-i}.$$

As $\mathfrak{E}_0 \in \operatorname{Iso}(\hat{V}, V)$, the inverse operator $\mathfrak{E}^{-1}$ is continuous in a neighborhood of λ_0 and, hence, for each integer $0 \leq \ell \leq s$,

$$\lim_{\lambda \to \lambda_0} \sum_{i=0}^{\ell} (\lambda - \lambda_0)^{-i} \mathfrak{M}(\lambda) \mathfrak{F}^\Phi(\lambda) u_{\ell-i} = 0.$$

For each $\lambda \neq \lambda_0$ and $0 \leq \ell \leq s$, we also have

$$\mathfrak{M}(\lambda) = \sum_{j=0}^{\ell} (\lambda - \lambda_0)^j \mathfrak{M}_j + o\left((\lambda - \lambda_0)^\ell\right)$$

and, consequently,

$$\lim_{\lambda \to \lambda_0} \sum_{i=0}^{\ell} \sum_{j=0}^{i} \mathfrak{M}_j (\lambda - \lambda_0)^{j-i} \mathfrak{F}^\Phi(\lambda) u_{\ell-i} = 0, \qquad 0 \leq \ell \leq s.$$

Therefore, for every $0 \leq \ell \leq s$,

$$\lim_{\lambda \to \lambda_0} \mathfrak{M}_{[\ell]} \mathfrak{F}^\Phi_{[[\ell]]}(\lambda) \operatorname{col}[u_0, \dots, u_\ell] = 0, \tag{7.11}$$

where

$$\mathfrak{F}^\Phi_{[[\ell]]} : \Omega \setminus \{\lambda_0\} \to \mathcal{L}(U^{\ell+1}, \hat{U}^{\ell+1})$$

stands for the family defined through

$$\mathfrak{F}^\Phi_{[[\ell]]}(\lambda) := \operatorname{trng}\left\{\mathfrak{F}^\Phi(\lambda), (\lambda - \lambda_0)^{-1}\mathfrak{F}^\Phi(\lambda), \dots, (\lambda - \lambda_0)^{-\ell}\mathfrak{F}^\Phi(\lambda)\right\}$$

for all $\lambda \in \Omega \setminus \{\lambda_0\}$.

Take $1 \leq \ell \leq s$ and

$$\operatorname{col}[u_0, \dots, u_\ell] \in N[\mathfrak{L}^\Phi_{[\ell]}].$$

Then, taking into account that $\mathfrak{F}^\Phi_0 = \mathfrak{F}_0$, we find from (7.11) that

$$\begin{aligned}
0 &= \lim_{\lambda \to \lambda_0} \mathfrak{M}_{[\ell]} \mathfrak{F}^\Phi_{[[\ell]]}(\lambda) \operatorname{col}[u_0, \dots, u_\ell] \\
&= \mathfrak{M}_{[\ell]} \operatorname{diag}\{\mathfrak{F}_0, \dots, \mathfrak{F}_0\} \operatorname{col}[u_0, \dots, u_\ell] \\
&\quad + \lim_{\lambda \to \lambda_0} \mathfrak{M}_{[\ell]} \operatorname{trng}\{0, (\lambda - \lambda_0)^{-1}\mathfrak{F}^\Phi(\lambda), \dots, (\lambda - \lambda_0)^{-\ell}\mathfrak{F}^\Phi(\lambda)\} \operatorname{col}[u_0, \dots, u_\ell] \\
&= \operatorname{col}[\mathfrak{M}_0 \mathfrak{F}_0 u_0, \dots, \mathfrak{M}_\ell \mathfrak{F}_0 u_0] + \begin{pmatrix} 0 \\ \mathfrak{M}_{[\ell-1]} \operatorname{col}[\mathfrak{F}_0 u_1, \dots, \mathfrak{F}_0 u_\ell] \end{pmatrix} \\
&\quad + \lim_{\lambda \to \lambda_0} \mathfrak{M}_{[\ell]} \begin{pmatrix} 0 \\ \operatorname{trng}\{(\lambda - \lambda_0)^{-1}\mathfrak{F}^\Phi(\lambda), \dots, (\lambda - \lambda_0)^{-\ell}\mathfrak{F}^\Phi(\lambda)\} \operatorname{col}[u_0, \dots, u_{\ell-1}] \end{pmatrix} \\
&= \operatorname{col}[\mathfrak{M}_0 \mathfrak{F}_0 u_0, \dots, \mathfrak{M}_\ell \mathfrak{F}_0 u_0] + \begin{pmatrix} 0 \\ \mathfrak{M}_{[\ell-1]} \operatorname{col}[\mathfrak{F}_0 u_1, \dots, \mathfrak{F}_0 u_\ell] \end{pmatrix} \\
&\quad + \lim_{\lambda \to \lambda_0} \begin{pmatrix} 0 \\ (\lambda - \lambda_0)^{-1} \mathfrak{M}_{[\ell-1]} \mathfrak{F}^\Phi_{[[\ell-1]]}(\lambda) \operatorname{col}[u_0, \dots, u_{\ell-1}] \end{pmatrix}.
\end{aligned}$$

Therefore, identifying the last ℓ rows shows that

$$\begin{aligned}\lim_{\lambda\to\lambda_0}&(\lambda-\lambda_0)^{-1}\mathfrak{M}_{[\ell-1]}\mathfrak{F}^{\Phi}_{[[\ell-1]]}(\lambda)\operatorname{col}[u_0,\dots,u_{\ell-1}]\\&=-\operatorname{col}[\mathfrak{M}_1\mathfrak{F}_0u_0,\dots,\mathfrak{M}_\ell\mathfrak{F}_0u_0]-\mathfrak{M}_{[\ell-1]}\operatorname{col}[\mathfrak{F}_0u_1,\dots,\mathfrak{F}_0u_\ell].\end{aligned}$$

By the result of Exercise 1,

$$\mathfrak{M}_{[\ell-1]}\in\operatorname{Fred}_0(\hat{U}^\ell,\hat{V}^\ell),$$

so the space $R[\mathfrak{M}_{[\ell-1]}]$ is closed, and, hence, we find from this identity that there exists

$$\operatorname{col}[\hat{u}_0,\dots,\hat{u}_{\ell-1}]\in\hat{U}^\ell$$

such that

$$\begin{aligned}&\mathfrak{M}_{[\ell-1]}\operatorname{col}[\hat{u}_0,\dots,\hat{u}_{\ell-1}]\\&\quad=-\operatorname{col}[\mathfrak{M}_1\mathfrak{F}_0u_0,\dots,\mathfrak{M}_\ell\mathfrak{F}_0u_0]-\mathfrak{M}_{[\ell-1]}\operatorname{col}[\mathfrak{F}_0u_1,\dots,\mathfrak{F}_0u_\ell].\end{aligned}\tag{7.12}$$

To summarize, for each $1\le\ell\le s$ and

$$\operatorname{col}[u_0,\dots,u_\ell]\in N[\mathfrak{L}^{\Phi}_{[\ell]}]$$

there exists

$$\operatorname{col}[\hat{u}_0,\dots,\hat{u}_{\ell-1}]\in\hat{U}^\ell$$

satisfying (7.12).

For every $1\le i\le n_s$, we have that

$$\operatorname{col}[u^i,\underbrace{0,\dots,0}_{s}]\in N[\mathfrak{L}^{\Phi}_{[s]}]$$

and, hence, due to (7.12), there exists

$$\operatorname{col}[\hat{u}^i_0,\dots,\hat{u}^i_{s-1}]\in\hat{U}^s$$

such that

$$\mathfrak{M}_{[s-1]}\operatorname{col}[\hat{u}^i_0,\dots,\hat{u}^i_{s-1}]=-\operatorname{col}[\mathfrak{M}_1\mathfrak{F}_0u^i,\dots,\mathfrak{M}_s\mathfrak{F}_0u^i].\tag{7.13}$$

Now, let $1\le j\le s$ and $n_j<i\le n_{j-1}$. Then,

$$u^i\in\operatorname{span}[u^1,\dots,u^{n_{j-1}}]=\bigcap_{i=0}^{j-1}N[\mathfrak{L}^{\Phi}_i]$$

and, hence,

$$\operatorname{col}[u^i,\underbrace{0,\dots,0}_{j-1}]\in N[\mathfrak{L}^{\Phi}_{[j-1]}].$$

Thus, by (7.12), there exists

$$\operatorname{col}[\hat{u}_0^i, \dots, \hat{u}_{j-1}^i] \in \hat{U}^j$$

such that

$$\mathfrak{M}_{[j-1]} \operatorname{col}[\hat{u}_0^i, \dots, \hat{u}_{j-1}^i] = -\operatorname{col}[\mathfrak{M}_1 \mathfrak{F}_0 u^i, \dots, \mathfrak{M}_j \mathfrak{F}_0 u^i]. \tag{7.14}$$

Moreover, if $1 \le i \le n_0$, then $u^i \in N[\mathfrak{L}_0^\Phi]$, and (7.10) implies

$$\mathfrak{M}_0 \mathfrak{F}_0 u^i = 0. \tag{7.15}$$

As the vectors $u^1, \dots, u^{n_0}$ are linearly independent, there exist operators

$$F_1, \dots, F_s \in \mathcal{L}(U, \hat{U})$$

such that

$$F_j u^i = \hat{u}_{j-1}^i, \qquad 1 \le j \le s, \quad 1 \le i \le n_j.$$

Since $\mathfrak{F}_0 \in \operatorname{Iso}(U, \hat{U})$, we have that

$$\mathfrak{F}_{\{s\}} := \operatorname{trng}\{\mathfrak{F}_0, F_1, \dots, F_s\} \in \operatorname{Iso}(U^{s+1}, \hat{U}^{s+1}).$$

Moreover,

$$\mathfrak{F}_{\{s\}} \left(N[\mathfrak{L}_{[s]}^\Phi] \right) \subset N[\mathfrak{M}_{[s]}].$$

Indeed, by (7.13), (7.14), and (7.15), we have that, for every $0 \le j \le s-1$ and $1 \le i \le n_{s-j}$,

$$\begin{aligned}
\mathfrak{M}_{[s]} \mathfrak{F}_{\{s\}} & \operatorname{col}[\underbrace{0, \dots, 0}_{j}, u^i, \underbrace{0, \dots, 0}_{s-j}] \\
&= \mathfrak{M}_{[s]} \operatorname{col}[\underbrace{0, \dots, 0}_{j}, \mathfrak{F}_0 u^i, F_1 u^i, \dots, F_{s-j} u^i] \\
&= \mathfrak{M}_{[s]} \operatorname{col}[\underbrace{0, \dots, 0}_{j}, \mathfrak{F}_0 u^i, \hat{u}_0^i, \dots, \hat{u}_{s-1-j}^i] \\
&= \operatorname{col}[\underbrace{0, \dots, 0}_{j}, \mathfrak{M}_0 \mathfrak{F}_0 u^i, \dots, \mathfrak{M}_{s-j} \mathfrak{F}_0 u^i] \\
&\quad + \operatorname{col}\Big[\underbrace{0, \dots, 0}_{j+1}, \mathfrak{M}_{[s-1-j]} \operatorname{col}[\hat{u}_0^i, \dots, \hat{u}_{s-1-j}^i]\Big] \\
&= \operatorname{col}[\underbrace{0, \dots, 0}_{j+1}, \mathfrak{M}_1 \mathfrak{F}_0 u^i, \dots, \mathfrak{M}_{s-j} \mathfrak{F}_0 u^i] \\
&\quad + \operatorname{col}\Big[\underbrace{0, \dots, 0}_{j+1}, \mathfrak{M}_{[s-1-j]} \operatorname{col}[\hat{u}_0^i, \dots, \hat{u}_{s-1-j}^i]\Big] \\
&= 0.
\end{aligned}$$

Moreover, for every $1 \le i \le n_0$,

$$\mathfrak{M}_{[s]}\mathfrak{F}_{\{s\}} \operatorname{col}[\underbrace{0,\dots,0}_{s},u^i] = \operatorname{col}[\underbrace{0,\dots,0}_{s},\mathfrak{M}_0\mathfrak{F}_0 u^i] = 0,$$

because of (7.15). Therefore, the map

$$\mathfrak{F}_{\{s\}} \;:\; N[\mathfrak{L}^{\Phi}_{[s]}] \to N[\mathfrak{M}_{[s]}]$$

is well defined, linear, continuous and injective. Moreover, it sends Jordan chains of $\mathfrak{L}^\Phi$ into Jordan chains of $\mathfrak{M}$, because if $u_0 \in U \setminus \{0\}$ then

$$\mathfrak{F}^{\Phi}_0 u_0 = \mathfrak{F}_0 u_0 \neq 0.$$

Therefore, $\mathfrak{F}_{[s]}$ establishes a linear injection from the set of Jordan chains of length $s+1$ of $\mathfrak{L}^\Phi$ at λ_0 to the set of Jordan chains of length $s+1$ of $\mathfrak{M}$. As we have already constructed a linear bijection between the Jordan chains of length $s+1$ of $\mathfrak{L}^\Phi$ and those of $\mathfrak{L}$, we have actually established a linear injection from the set of Jordan chains of length $s+1$ of $\mathfrak{L}$ to the set of Jordan chains of length $s+1$ of $\mathfrak{M}$. By the symmetric nature of the problem, transversalizing in this occasion $\mathfrak{M}$ instead of $\mathfrak{L}$, it is easily realized that there is another linear injection from the set of Jordan chains of length $s+1$ of $\mathfrak{M}$ to the set of Jordan chains of length $s+1$ of $\mathfrak{L}$. Consequently, we actually have established a linear bijection between both sets of Jordan chains, which concludes the proof. □

7.4 Local Smith form for $\mathcal{C}^\infty$ matrix families

As usual, for any integer $m \ge 1$, we identify the space of $m \times m$ square matrices $\mathcal{M}_m(\mathbb{K})$ with $\mathcal{L}(\mathbb{K}^m)$. The following result establishes the importance of the family considered in the Example of Section 7.2.

Theorem 7.4.1. *Let $m \ge 1$ be an integer, $r \in \{\infty, \omega\}$ and $\lambda_0 \in \Omega$. Consider an operator family $\mathfrak{L} \in \mathcal{C}^r(\Omega, \mathcal{M}_m(\mathbb{K}))$ such that*

$$n := \dim N[\mathfrak{L}_0] \ge 1$$

and no Jordan chain of $\mathfrak{L}$ at λ_0 can be continued indefinitely. Then, there exist an open subset $\Omega' \subset \Omega$ with $\lambda_0 \in \Omega'$, and two families $\mathfrak{E}, \mathfrak{F} \in \mathcal{C}^r(\Omega', \mathcal{M}_m(\mathbb{K}))$ such that $\mathfrak{E}_0, \mathfrak{F}_0 \in \operatorname{Iso}(\mathbb{K}^m)$ and

$$\mathfrak{L}(\lambda) = \mathfrak{E}(\lambda)\ \mathfrak{D}(\lambda)\ \mathfrak{F}(\lambda), \qquad \lambda \in \Omega', \tag{7.16}$$

where $\mathfrak{D} \in \mathbb{P}[\mathcal{M}_m(\mathbb{K})]$ is the family defined, for each $\lambda \in \mathbb{K}$, by

$$\mathfrak{D}(\lambda) := \operatorname{diag}\big\{(\lambda-\lambda_0)^{\kappa_1[\mathfrak{L};\lambda_0]}, \dots, (\lambda-\lambda_0)^{\kappa_n[\mathfrak{L};\lambda_0]}, \underbrace{1,\dots,1}_{m-n}\big\}. \tag{7.17}$$

Remark 7.4.2. If $n = m$, then no entry 1 appears in the expression of $\mathfrak{D}$ in (7.17). In general, $\mathfrak{D}$ exhibits as many 1's as indicated by the difference $m - n$. Obviously, the result is also satisfied if $n = 0$. In such a case, $\mathfrak{D} = I_m$ and one can make the choice $\mathfrak{E} = I_m$ and $\mathfrak{F} = \mathfrak{L}$.

Proof of Theorem 7.4.1. It is based on the Gaussian elimination method, and proceeds by induction on m.

Suppose $m = 1$. Then, $\mathfrak{L}$ is a $\mathbb{K}$-valued function and $\mathfrak{L}_0 = 0$. Moreover, since no Jordan chain of $\mathfrak{L}$ at λ_0 can be continued indefinitely,

$$1 \leq \kappa := \kappa[\mathfrak{L}; \lambda_0] = \kappa_1[\mathfrak{L}; \lambda_0] = \operatorname{ord}_{\lambda_0} \mathfrak{L} < \infty$$

and, hence, there exists $\mathfrak{F} \in \mathcal{C}^r(\Omega)$ such that $\mathfrak{F}_0 \neq 0$ and

$$\mathfrak{L}(\lambda) = (\lambda - \lambda_0)^\kappa \mathfrak{F}(\lambda), \qquad \lambda \in \Omega.$$

Therefore, (7.16) holds true by making the choice $\mathfrak{E}(\lambda) = 1$ and $\mathfrak{D}(\lambda) = (\lambda - \lambda_0)^\kappa$ for each $\lambda \in \Omega$.

Suppose $m \geq 2$ and the result to be true for every family in $\mathcal{C}^r(\Omega, \mathcal{M}_{m-1}(\mathbb{K}))$ satisfying the requirements of the theorem.

Let $\mathfrak{L} \in \mathcal{C}^r(\Omega, \mathcal{M}_m(\mathbb{K}))$ satisfy the requirements of the theorem. If $\operatorname{ord}_{\lambda_0} \mathfrak{L} = \infty$, then $\mathfrak{L}_j = 0$ for all $j \geq 0$ and, hence, every Jordan chain of $\mathfrak{L}$ at λ_0 can be continued indefinitely. Thus,

$$\operatorname{ord}_{\lambda_0} \mathfrak{L} < \infty. \tag{7.18}$$

Now, for each $1 \leq i, j \leq m$, we denote by $\ell_{(i,j)}$ the (i,j)th entry of $\mathfrak{L}$. Let (i_0, j_0) be such that

$$0 \leq h := \operatorname{ord}_{\lambda_0} \ell_{(i_0,j_0)} = \min_{1 \leq i,j \leq m} \operatorname{ord}_{\lambda_0} \ell_{(i,j)}.$$

By (7.18), $h < \infty$. Let $P, Q \in \mathcal{M}_m(\mathbb{K})$ be two permutation matrices (constant and invertible) for which the family $\tilde{\mathfrak{L}}$ defined by

$$\tilde{\mathfrak{L}}(\lambda) := P\mathfrak{L}(\lambda)Q, \qquad \lambda \in \Omega, \tag{7.19}$$

satisfies

$$h = \operatorname{ord}_{\lambda_0} \tilde{\ell}_{(m,m)} = \min_{1 \leq i,j \leq m} \operatorname{ord}_{\lambda_0} \tilde{\ell}_{(i,j)},$$

where we have denoted

$$\tilde{\mathfrak{L}} = \left(\tilde{\ell}_{(i,j)}\right)_{1 \leq i,j \leq m}.$$

By the choice of h, for each $(i,j) \in \{1, \ldots, m\}^2$ there exists $b_{(i,j)} \in \mathcal{C}^r(\Omega)$ such that

$$\tilde{\ell}_{(i,j)}(\lambda) = (\lambda - \lambda_0)^h b_{(i,j)}(\lambda), \qquad \lambda \in \Omega.$$

Moreover, since $b_{(m,m)}(\lambda_0) \neq 0$, there exists an open neighborhood $\tilde{\Omega} \subset \Omega$ of λ_0 such that

$$b_{(m,m)}^{-1}(0) \cap \tilde{\Omega} = \emptyset.$$

Now we consider the matrix families $\tilde{\mathfrak{E}}, \tilde{\mathfrak{F}} \in \mathcal{C}^r(\tilde{\Omega}, \mathcal{M}_m(\mathbb{K}))$ defined by

$$\tilde{\mathfrak{E}} := \left(\begin{array}{ccc|c} & & & -\frac{b_{(1,m)}}{b_{(m,m)}} \\ & I_{m-1} & & \vdots \\ & & & -\frac{b_{(m-1,m)}}{b_{(m,m)}} \\ \hline 0 & \cdots & 0 & 1 \end{array}\right),$$

$$\tilde{\mathfrak{F}} := \left(\begin{array}{ccc|c} & & & 0 \\ & I_{m-1} & & \vdots \\ & & & 0 \\ \hline -\frac{b_{(m,1)}}{b_{(m,m)}} & \cdots & -\frac{b_{(m,m-1)}}{b_{(m,m)}} & 1 \end{array}\right).$$

Clearly, $\tilde{\mathfrak{E}}$ and $\tilde{\mathfrak{F}}$ take invertible values. Moreover, for each $\lambda \in \tilde{\Omega}$, we have that

$$\begin{aligned} \tilde{\mathfrak{E}}(\lambda)\tilde{\mathfrak{L}}(\lambda)\tilde{\mathfrak{F}}(\lambda) &= \operatorname{diag}\{\tilde{\mathfrak{M}}(\lambda), \tilde{\ell}_{(m,m)}(\lambda)\} \\ &= \operatorname{diag}\{\tilde{\mathfrak{M}}(\lambda), (\lambda-\lambda_0)^h\} \operatorname{diag}\{I_{m-1}, b_{(m,m)}(\lambda)\}, \end{aligned} \tag{7.20}$$

where $\tilde{\mathfrak{M}} \in \mathcal{C}^r(\tilde{\Omega}, \mathcal{M}_{m-1}(\mathbb{K}))$ stands for the matrix whose (i,j)th entry is given by

$$\tilde{m}_{(i,j)} := \tilde{\ell}_{(i,j)} - \tilde{\ell}_{(m,j)} \frac{b_{(i,m)}}{b_{(m,m)}}, \qquad 1 \le i, j \le m-1.$$

Since

$$\operatorname{diag}\{I_{m-1}, b_{(m,m)}(\lambda_0)\} \in \operatorname{Iso}(\mathbb{K}^m),$$

by (7.19) and (7.20), we find from Theorem 7.3.1 that no Jordan chain of

$$\operatorname{diag}\{\tilde{\mathfrak{M}}, \tilde{\ell}_{mm}\}$$

can be continued indefinitely, as it was assumed to occur for $\mathfrak{L}$. On the other hand, a vector

$$\left(\begin{pmatrix} u_0 \\ 0 \end{pmatrix}, \dots, \begin{pmatrix} u_s \\ 0 \end{pmatrix}\right) \in (\mathbb{K}^m)^{s+1}$$

is a Jordan chain of $\operatorname{diag}\{\tilde{\mathfrak{M}}, \tilde{\ell}_{mm}\}$ at λ_0 whenever the vector

$$(u_0, \dots, u_s) \in (\mathbb{K}^{m-1})^{s+1}$$

is a Jordan chain of $\tilde{\mathfrak{M}}$ at λ_0. Therefore, no Jordan chain of $\tilde{\mathfrak{M}}$ can be continued indefinitely. Consequently, by the induction hypothesis, there exist an open neighborhood $\hat{\Omega} \subset \tilde{\Omega}$ of λ_0 and two families

$$\hat{\mathfrak{E}}, \hat{\mathfrak{F}} \in \mathcal{C}^r(\hat{\Omega}, \mathcal{M}_{m-1}(\mathbb{K}))$$

such that $\hat{\mathfrak{E}}_0, \hat{\mathfrak{F}}_0 \in \mathrm{Iso}(\mathbb{K}^{m-1})$ and

$$\tilde{\mathfrak{M}}(\lambda) = \hat{\mathfrak{E}}(\lambda)\tilde{\mathfrak{D}}(\lambda)\hat{\mathfrak{F}}(\lambda), \qquad \lambda \in \hat{\Omega}, \tag{7.21}$$

where $\tilde{\mathfrak{D}}$ is given through

$$\tilde{\mathfrak{D}}(\lambda) := \mathrm{diag}\left\{(\lambda-\lambda_0)^{\kappa_1[\tilde{\mathfrak{M}};\lambda_0]}, \dots, (\lambda-\lambda_0)^{\kappa_{\tilde{n}}[\tilde{\mathfrak{M}};\lambda_0]}, \underbrace{1,\dots,1}_{m-1-\tilde{n}}\right\}$$

and $\tilde{n} := \dim N[\tilde{\mathfrak{M}}_0]$. By (7.21), we have that, for each $\lambda \in \hat{\Omega}$,

$$\mathrm{diag}\{\tilde{\mathfrak{M}}(\lambda), (\lambda-\lambda_0)^h\} = \mathrm{diag}\{\hat{\mathfrak{E}}(\lambda), 1\} \cdot \mathrm{diag}\{\tilde{\mathfrak{D}}(\lambda), (\lambda-\lambda_0)^h\} \cdot \mathrm{diag}\{\hat{\mathfrak{F}}(\lambda), 1\}.$$

Therefore, substituting this identity into (7.20), we find from (7.19) that there exists an open neighborhood $\Omega' \subset \hat{\Omega}$ of λ_0 such that

$$\mathfrak{L}(\lambda) = \mathfrak{E}'(\lambda)\mathfrak{D}'(\lambda)\mathfrak{F}'(\lambda), \qquad \lambda \in \Omega',$$

where, for each $\lambda \in \Omega'$,

$$\begin{aligned}
\mathfrak{E}'(\lambda) &:= P^{-1}\tilde{\mathfrak{E}}(\lambda)^{-1}\,\mathrm{diag}\{\hat{\mathfrak{E}}(\lambda), 1\},\\
\mathfrak{D}'(\lambda) &:= \mathrm{diag}\{\tilde{\mathfrak{D}}(\lambda), (\lambda-\lambda_0)^h\},\\
\mathfrak{F}'(\lambda) &:= \mathrm{diag}\{\hat{\mathfrak{F}}(\lambda), 1\}\,\mathrm{diag}\{I_{m-1}, b_{(m,m)}(\lambda)\}\tilde{\mathfrak{F}}(\lambda)^{-1}Q^{-1}.
\end{aligned}$$

According to Theorem 7.3.1,

$$n = \dim N[\mathfrak{L}_0] = \dim N[\mathfrak{D}'_0]$$

and

$$\kappa_j[\mathfrak{L};\lambda_0] = \kappa_j[\mathfrak{D}';\lambda_0], \qquad 1 \le j \le n.$$

Note that $\tilde{n} = n$ if $h = 0$, while $\tilde{n} = n-1$ if $h \ge 1$. Moreover, according to the Example of Section 7.2, there exist two permutation matrices P' and Q' such that

$$\mathfrak{D}'(\lambda) := P'\,\mathrm{diag}\left\{(\lambda-\lambda_0)^{\kappa_1[\mathfrak{L};\lambda_0]}, \dots, (\lambda-\lambda_0)^{\kappa_n[\mathfrak{L};\lambda_0]}, \underbrace{1,\dots,1}_{m-n}\right\}Q'.$$

Consequently, making the choice

$$\mathfrak{E}(\lambda) := \mathfrak{E}'(\lambda)P', \quad \mathfrak{F}(\lambda) := Q'\mathfrak{F}'(\lambda), \qquad \lambda \in \Omega',$$

concludes the proof. □

Based on the previous result, it becomes rather natural to introduce the following concept of *local Smith form*.

Definition 7.4.3. Let $r \in \mathbb{N} \cup \{\infty\}$ and $\mathfrak{L} \in \mathcal{C}^r(\Omega, \mathcal{L}(U,V))$ be with

$$\mathfrak{L}_0 \in \operatorname{Fred}_0(U,V) \quad \text{and} \quad n := \dim N[\mathfrak{L}_0] \geq 1.$$

The operator family $\mathfrak{L}$ is said to admit a local Smith form at λ_0 when there exist an open neighborhood $\Omega' \subset \Omega$ of λ_0, two families

$$\mathfrak{E} \in \mathcal{C}(\Omega', \mathcal{L}(U,V)) \quad \text{and} \quad \mathfrak{F} \in \mathcal{C}(\Omega', \mathcal{L}(U))$$

with $\mathfrak{E}_0 \in \operatorname{Iso}(U,V)$ and $\mathfrak{F}_0 \in \operatorname{Iso}(U)$, a direct sum topological decomposition

$$U = U_0 \oplus U_1,$$

with $n = \dim U_0$, and n integer positive numbers $k_1 \geq \cdots \geq k_n$ such that

$$\mathfrak{L}(\lambda) = \mathfrak{E}(\lambda)\; [\mathfrak{S}(\lambda) \oplus I_{U_1}]\; \mathfrak{F}(\lambda), \qquad \lambda \in \Omega',$$

where

$$\mathfrak{S}(\lambda) = \operatorname{diag}\left\{(\lambda - \lambda_0)^{k_1}, \dots, (\lambda - \lambda_0)^{k_n}\right\}, \qquad \lambda \in \mathbb{K}.$$

In such a case, $\mathfrak{S}$ is said to be a local Smith form of $\mathfrak{L}$ at λ_0.

According to Theorem 7.3.1, if the lengths of all Jordan chains of $\mathfrak{L}$ at λ_0 are uniformly bounded above by some integer $k \leq r$, necessarily

$$k_j = \kappa_j[\mathfrak{L}; \lambda_0], \qquad 1 \leq j \leq n,$$

and, consequently, the local Smith form should be unique, if it exists, but the rigorous proof of this feature is postponed until Proposition 7.10.2. Theorem 7.4.1 established that every $\mathcal{C}^\infty$ family of square matrices for which no Jordan chain can be continued indefinitely indeed possesses a local Smith form. Theorem 7.8.1 will show that this condition on the Jordan chains of $\mathfrak{L}$ is not only sufficient for the existence of the local Smith form but also necessary. Necessary and sufficient conditions for the existence of the local Smith form for general operator-valued families of class $\mathcal{C}^r$ will be established by Theorem 7.8.3. The Smith form of a family is the simplest canonical form that can be obtained from it through pre- and post-multiplication by isomorphism families, of course.

7.5 Canonical sets at transversal eigenvalues

In this section we combine the transversalization Theorem 4.3.2 with the invariance Theorem 7.3.1 to provide a scheme to calculate the partial κ-multiplicities and, thus, building up, in a rather systematic way, canonical sets of Jordan chains.

Proposition 7.5.1. *Let $r \in \mathbb{N} \cup \{\infty\}$ and $\mathfrak{L} \in \mathcal{C}^r(\Omega, \mathcal{L}(U,V))$ satisfy that $\mathfrak{L}_0 \in \operatorname{Fred}_0(U,V)$ and λ_0 is a ν-transversal eigenvalue for some integer $1 \leq \nu \leq r$. Set*

$$\ell_j := \dim \mathfrak{L}_j(N[\mathfrak{L}_0] \cap \cdots \cap N[\mathfrak{L}_{j-1}]), \qquad 1 \leq j \leq \nu. \tag{7.22}$$

Then, every canonical set of Jordan chains of $\mathfrak{L}$ *at* λ_0 *consists of exactly* ℓ_j *Jordan chains of length* j *for all* $j \in \{1, \dots, \nu\}$. *Consequently, the partial* κ*-multiplicities of* $\mathfrak{L}$ *at* λ_0 *are given through*

$$\underbrace{\nu = \cdots = \nu}_{\ell_\nu} > \underbrace{\nu - 1 = \cdots = \nu - 1}_{\ell_{\nu-1}} > \cdots > \underbrace{1 = \cdots = 1}_{\ell_1} \tag{7.23}$$

and, therefore, $\kappa_1[\mathfrak{L}; \lambda_0] = \nu$ *and*

$$\kappa[\mathfrak{L}; \lambda_0] = \sum_{j=1}^{n} \kappa_j[\mathfrak{L}; \lambda_0] = \sum_{i=1}^{\nu} i\ell_i = \chi[\mathfrak{L}; \lambda_0], \tag{7.24}$$

where χ *is the algebraic multiplicity introduced by Definition* 4.3.3.

Proof. Thanks to Proposition 7.1.3, $\mathfrak{L}$ cannot admit a Jordan chain of length greater than ν. Thus, the length of all Jordan chains of $\mathfrak{L}$ is bounded above by ν. Set $n := \dim N[\mathfrak{L}_0]$. According to Proposition 4.2.2, we have

$$\bigcap_{i=0}^{\nu} N[\mathfrak{L}_i] = \{0\}$$

and

$$\dim \bigcap_{i=0}^{j} N[\mathfrak{L}_i] = \sum_{i=j+1}^{\nu} \ell_i, \qquad 0 \le j \le \nu - 1.$$

Moreover,

$$\bigcap_{i=0}^{j} N[\mathfrak{L}_i] \subset \bigcap_{i=0}^{j-1} N[\mathfrak{L}_i] \subset N[\mathfrak{L}_0], \qquad 1 \le j \le \nu.$$

Therefore, there exists a basis $\{u_1, \dots, u_{n_0}\}$ of $N[\mathfrak{L}_0]$ such that

$$\bigcap_{i=0}^{j} N[\mathfrak{L}_i] = \operatorname{span}\left[u_1, \dots, u_{\sum_{i=j+1}^{\nu} \ell_i}\right], \qquad 0 \le j \le \nu - 1.$$

Thanks to Proposition 7.1.3, it is easy to check that the following ordered set provides us with a canonical set of Jordan chains of $\mathfrak{L}$ at λ_0:

$$\begin{aligned} &\Big[\underbrace{u_1, 0, \dots, 0}_{\nu}; \underbrace{u_2, 0, \dots, 0}_{\nu}; \cdots ; \underbrace{u_{\ell_\nu}, 0, \dots, 0}_{\nu}; \\ &\underbrace{u_{\ell_\nu+1}, 0, \dots, 0}_{\nu-1}; \underbrace{u_{\ell_\nu+2}, 0, \dots, 0}_{\nu-1}; \cdots ; \underbrace{u_{\ell_\nu+\ell_{\nu-1}}, 0, \dots, 0}_{\nu-1}; \cdots ; \\ &\underbrace{u_{\ell_\nu+\cdots+\ell_3+1}, 0}_{2}; \underbrace{u_{\ell_\nu+\cdots+\ell_3+2}, 0}_{2}; \cdots ; \underbrace{u_{\ell_\nu+\cdots+\ell_3+\ell_2}, 0}_{2}; \\ &\underbrace{u_{\ell_\nu+\cdots+\ell_3+\ell_2+1}}_{1}; \underbrace{u_{\ell_\nu+\cdots+\ell_3+\ell_2+2}}_{1}; \cdots ; \underbrace{u_{\ell_\nu+\cdots+\ell_2+\ell_1}}_{1}\Big]. \end{aligned}$$

It consists of ℓ_j Jordan chains of length j for all $1 \leq j \leq \nu$. As (7.23) and (7.24) follow straight away from Definitions 7.2.5 and 4.3.3, this concludes the proof. $\square$

Naturally, Proposition 7.5.1 characterizes the local Smith form given by Theorem 7.4.1.

Corollary 7.5.2. *Consider $m \in \mathbb{N}^*$ and $r \in \{\infty, \omega\}$. Let $\mathfrak{L} \in \mathcal{C}^r(\Omega, \mathcal{M}_m(\mathbb{K}))$ be a family for which $\lambda_0 \in \Omega$ is a ν-transversal eigenvalue for some integer $\nu \geq 1$. Then, the matrix $\mathfrak{D}(\lambda)$ introduced in* (7.17) *equals*

$$\operatorname{diag}\{\underbrace{(\lambda-\lambda_0)^\nu,\dots,(\lambda-\lambda_0)^\nu}_{\ell_\nu},\dots,\underbrace{\lambda-\lambda_0,\dots,\lambda-\lambda_0}_{\ell_1},\underbrace{1,\dots,1}_{m-n}\},$$

where $n = \dim N[\mathfrak{L}_0]$ and the ℓ_j's are given through (7.22). *In particular,*

$$\det \mathfrak{D}(\lambda) = (\lambda-\lambda_0)^{\kappa[\mathfrak{L};\lambda_0]} = (\lambda-\lambda_0)^{\chi[\mathfrak{L};\lambda_0]}, \qquad \lambda \in \mathbb{K}.$$

Actually, according to Theorems 4.3.2 and 7.3.1, the multiplicities χ and κ must coincide at any algebraic eigenvalue of the family, which provides a local Smith form in the finite-dimensional case for $\mathcal{C}^\infty$ families. Precisely, the following result is satisfied.

Theorem 7.5.3. *Let $r \in \mathbb{N}^* \cup \{\infty\}$ and $\mathfrak{L} \in \mathcal{C}^r(\Omega, \mathcal{L}(U,V))$ be with $\mathfrak{L}_0 \in \operatorname{Fred}_0(U,V)$, such that $\lambda_0 \in \operatorname{Alg}_\nu(\mathfrak{L})$ for some $\nu \in \mathbb{N} \cap [0,r]$. Let $\Phi \in \mathcal{C}^r(\Omega, \mathcal{L}(U))$ be such that $\Phi_0 = I_U$ for which λ_0 is a ν-transversal value of $\mathfrak{L}^\Phi := \mathfrak{L}\Phi$, and set*

$$\ell_j := \dim \mathfrak{L}_j^\Phi(N[\mathfrak{L}_0^\Phi] \cap \cdots \cap N[\mathfrak{L}_{j-1}^\Phi]), \qquad 1 \leq j \leq \nu.$$

Then, every canonical set of Jordan chains of $\mathfrak{L}$ at λ_0 possesses exactly ℓ_j chains of length j for all $1 \leq j \leq \nu$. Consequently,

$$\kappa[\mathfrak{L};\lambda_0] = \sum_{j=1}^{\nu} j\ell_j = \chi[\mathfrak{L};\lambda_0].$$

If, in addition, $\dim U = \dim V = m$ for some integer $m \geq 1$ and $r \in \{\infty, \omega\}$, then the family $\mathfrak{S} \in \mathbb{P}[\mathcal{M}_m(\mathbb{K})]$ defined for each $\lambda \in \mathbb{K}$ as

$$\mathfrak{S}(\lambda) := \operatorname{diag}\{\underbrace{(\lambda-\lambda_0)^\nu,\dots,(\lambda-\lambda_0)^\nu}_{\ell_\nu},\dots,\underbrace{\lambda-\lambda_0,\dots,\lambda-\lambda_0}_{\ell_1}\}$$

provides us with a local Smith form of $\mathfrak{L}$ at λ_0.

Proof. The existence of the transversalizing family Φ is guaranteed by Theorem 5.3.1. Thanks to Proposition 7.5.1, every canonical set of Jordan chains of $\mathfrak{L}^\Phi$ at λ_0 consists of ℓ_j Jordan chains of length j for all $1 \leq j \leq \nu$ and, consequently, by Definition 4.3.3,

$$\chi[\mathfrak{L};\lambda_0] = \chi[\mathfrak{L}^\Phi;\lambda_0] = \kappa[\mathfrak{L};\lambda_0] = \sum_{j=1}^{\nu} j\ell_j.$$

The last assertion of the theorem follows straight from Corollary 7.5.2 and Definition 7.4.3. It should be noted that

$$\mathfrak{D}(\lambda) = \mathfrak{S}(\lambda) \oplus I_{m-\dim N[\mathfrak{L}_0]}, \qquad \lambda \in \mathbb{K},$$

where $\mathfrak{D}(\lambda)$ is the matrix introduced in (7.17). This concludes the proof. □

7.6 Coincidence of the multiplicities κ and χ

The next result, of an auxiliary nature, provides us with the dimensions of the kernels of the operators

$$\mathfrak{L}_{[s]} = \operatorname{trng}\{\mathfrak{L}_0, \dots, \mathfrak{L}_s\}$$

for every integer $1 \le s \le r$ and $\mathfrak{L} \in \mathcal{C}^r(\Omega, \mathcal{L}(U,V))$.

Proposition 7.6.1. *Let $r \in \mathbb{N}^* \cup \{\infty\}$ and $\mathfrak{L} \in \mathcal{C}^r(\Omega, \mathcal{L}(U,V))$ be such that $\mathfrak{L}_0 \in \operatorname{Fred}_0(U,V)$ and $\lambda_0 \in \operatorname{Eig}(\mathfrak{L})$. For a given integer $1 \le s \le r$, let $\Phi \in \mathcal{C}^r(\Omega, \mathcal{L}(U))$ be with $\Phi_0 \in \operatorname{Iso}(U)$ for which the family $\mathfrak{L}^\Phi := \mathfrak{L}\Phi$ satisfies*

$$V_s := \bigoplus_{j=1}^{s} \mathfrak{L}_j^\Phi (N[\mathfrak{L}_0^\Phi] \cap \cdots \cap N[\mathfrak{L}_{j-1}^\Phi]) \oplus R[\mathfrak{L}_0^\Phi] \subset V$$

and set

$$\begin{aligned} n_j &:= \dim N[\mathfrak{L}_0^\Phi] \cap \cdots \cap N[\mathfrak{L}_j^\Phi], && 0 \le j \le s, \\ \ell_j &:= \dim \mathfrak{L}_j^\Phi (N[\mathfrak{L}_0^\Phi] \cap \cdots \cap N[\mathfrak{L}_{j-1}^\Phi]), && 1 \le j \le s. \end{aligned}$$

Then (see (7.1)),

$$\dim N[\mathfrak{L}_{[s]}] = \sum_{h=1}^{s} h\ell_h + (s+1)n_s.$$

Proof. Fix an integer $1 \le s \le r$. The existence of Φ is guaranteed by Theorem 4.3.2. According to Lemma 4.1.1,

$$N[\mathfrak{L}_{[s]}^\Phi] = \left(\bigcap_{j=0}^{s} N[\mathfrak{L}_j^\Phi] \right) \times \left(\bigcap_{j=0}^{s-1} N[\mathfrak{L}_j^\Phi] \right) \times \cdots \times N[\mathfrak{L}_0^\Phi],$$

and, hence,

$$\dim N[\mathfrak{L}_{[s]}^\Phi] = \sum_{j=0}^{s} n_j.$$

Owing to Proposition 4.2.2, we have that

$$n_j = n_s + \sum_{h=j+1}^{s} \ell_h, \qquad 0 \le j \le s-1.$$

Thus,

$$\dim N[\mathfrak{L}^{\Phi}_{[s]}] = n_s + \sum_{j=0}^{s-1}\left(n_s + \sum_{h=j+1}^{s} \ell_h\right) = \sum_{h=1}^{s} h\ell_h + (s+1)n_s.$$

Finally, according to the Leibniz rule,

$$\mathfrak{L}^{\Phi}_{[s]} = \mathfrak{L}_{[s]}\Phi_{[s]}$$

and, therefore,

$$\dim N[\mathfrak{L}^{\Phi}_{[s]}] = \dim N[\mathfrak{L}_{[s]}],$$

because $\Phi_{[s]} \in \mathrm{Iso}(U^{s+1})$. This concludes the proof. □

Now, for each integer $1 \leq s \leq r$ we consider the map

$$\begin{array}{rcl} J_s : N[\mathfrak{L}_{[s-1]}] & \to & N[\mathfrak{L}_{[s]}] \\ y & \mapsto & \mathrm{col}[0, y], \end{array}$$

which is a linear injection and, hence,

$$\dim N[\mathfrak{L}_{[s-1]}] \leq \dim N[\mathfrak{L}_{[s]}], \qquad 1 \leq s < r+1.$$

Obviously, the map J_s is an isomorphism if and only if

$$\dim N[\mathfrak{L}_{[s-1]}] = \dim N[\mathfrak{L}_{[s]}].$$

The following result shows that $\lambda_0 \in \mathrm{Alg}_\nu(\mathfrak{L})$ for some integer $1 \leq \nu \leq r$ if and only if ν is the lowest integer k for which

$$J_k\left(N[\mathfrak{L}_{[k-1]}]\right) = N[\mathfrak{L}_{[k]}],$$

much like in the classical case when $U = V = \mathbb{K}^m$ for some integer $m \geq 1$ and

$$\mathfrak{L}(\lambda) = \lambda I_U - T, \qquad \lambda \in \mathbb{K},$$

for a given $T \in \mathcal{M}_m(\mathbb{K})$, where we already know that any eigenvalue $\lambda_0 \in \mathrm{Eig}(\mathfrak{L}) = \sigma(T)$ must be ν-algebraic for some integer $\nu \geq 1$ and that ν is the lowest integer k for which

$$N[(\lambda_0 I_U - T)^k] = N[(\lambda_0 I_U - T)^{k+1}].$$

Such similarities will be completely explained in Chapter 8, where the attention will be focused on these classical families.

Theorem 7.6.2. *Let $r \in \mathbb{N}^* \cup \{\infty\}$ and $\mathfrak{L} \in \mathcal{C}^r(\Omega, \mathcal{L}(U,V))$ be with $\mathfrak{L}_0 \in \mathrm{Fred}_0(U,V)$. Then, $\lambda_0 \in \mathrm{Alg}_\nu(\mathfrak{L})$ for some integer $1 \leq \nu \leq r$ if and only if*

$$\begin{aligned} &\dim N[\mathfrak{L}_{[j-1]}] < \dim N[\mathfrak{L}_{[j]}], \qquad \textit{for all } 1 \leq j \leq \nu - 1, \\ &\dim N[\mathfrak{L}_{[\nu-1]}] = \dim N[\mathfrak{L}_{[\nu]}]. \end{aligned} \tag{7.25}$$

Moreover, in such a case,

$$\dim N[\mathfrak{L}_{[\nu-1]}] = \chi[\mathfrak{L};\lambda_0] = \kappa[\mathfrak{L};\lambda_0] \tag{7.26}$$

and

$$\dim N[\mathfrak{L}_{[\nu-1]}] = \dim N[\mathfrak{L}_{[\ell]}], \qquad \nu \le \ell < r+1. \tag{7.27}$$

Proof. Assume that $\lambda_0 \in \mathrm{Alg}_\nu(\mathfrak{L})$ for some $1 \le \nu < r+1$. Then, according to Theorem 5.3.1, there exists $\Phi \in \mathbb{P}[\mathcal{L}(U)]$ with $\Phi_0 = I_U$ such that λ_0 is a ν-transversal eigenvalue of $\mathfrak{L}^\Phi := \mathfrak{L}\Phi$. Thus,

$$\bigoplus_{j=1}^{\nu} \mathfrak{L}_j^\Phi(N[\mathfrak{L}_0^\Phi] \cap \cdots \cap N[\mathfrak{L}_{j-1}^\Phi]) \oplus R[\mathfrak{L}_0^\Phi] = V,$$

and, setting

$$\begin{aligned} n_j &:= \dim N[\mathfrak{L}_0^\Phi] \cap \cdots \cap N[\mathfrak{L}_j^\Phi], && 0 \le j \le \nu, \\ \ell_j &:= \dim \mathfrak{L}_j^\Phi(N[\mathfrak{L}_0^\Phi] \cap \cdots \cap N[\mathfrak{L}_{j-1}^\Phi]), && 1 \le j \le \nu, \end{aligned} \tag{7.28}$$

we have from Proposition 4.2.2 that

$$n_0 \ge n_1 \ge \cdots \ge n_{\nu-1} \ge 1, \qquad n_\nu = 0.$$

Thus, according to Propositions 7.6.1, 4.2.2 and 7.5.1,

$$\dim N[\mathfrak{L}_{[\nu-1]}] = \sum_{h=1}^{\nu-1} h\ell_h + \nu n_{\nu-1} = \sum_{h=1}^{\nu} h\ell_h = \dim N[\mathfrak{L}_{[\nu]}] = \chi[\mathfrak{L};\lambda_0] = \kappa[\mathfrak{L};\lambda_0],$$

whereas, for every integer $1 \le j \le \nu-1$ (if there is any),

$$\dim N[\mathfrak{L}_{[j-1]}] = \sum_{h=1}^{j-1} h\ell_h + jn_{j-1} = \sum_{h=1}^{j} h\ell_h + jn_j < \sum_{h=1}^{j} h\ell_h + (j+1)n_j = \dim N[\mathfrak{L}_{[j]}],$$

because $n_j \ge 1$. This concludes the proof of (7.25) and (7.26). To show (7.27) we will proceed by contradiction. Suppose it fails. Then, there exists $\nu \le \ell < r$ such that

$$\dim N[\mathfrak{L}_{[\nu]}] = \dim N[\mathfrak{L}_{[\nu+j]}], \qquad 0 \le j \le \ell - \nu,$$

but

$$\dim N[\mathfrak{L}_{[\ell]}] < \dim N[\mathfrak{L}_{[\ell+1]}]. \tag{7.29}$$

Consequently, the map

$$\begin{aligned} J_{\ell+1} : N[\mathfrak{L}_{[\ell]}] &\to N[\mathfrak{L}_{[\ell+1]}] \\ y &\mapsto \mathrm{col}[0, y] \end{aligned}$$

is not surjective and, hence, there exists

$$(u_0, \dots, u_{\ell+1}) \in N[\mathfrak{L}_{[\ell+1]}]$$

with $u_0 \neq 0$. In other words, $\mathfrak{L}$ admits a Jordan chain of length $\ell + 2$ at λ_0. Thus, by Remark 7.1.2, it possesses a Jordan chain of length $\ell + 1$ and, consequently, the map

$$\begin{array}{rcl} J_\ell : N[\mathfrak{L}_{[\ell-1]}] & \rightarrow & N[\mathfrak{L}_{[\ell]}] \\ y & \mapsto & \operatorname{col}[0, y] \end{array}$$

cannot be surjective. Therefore,

$$\dim N[\mathfrak{L}_{[\ell-1]}] < \dim N[\mathfrak{L}_{[\ell]}],$$

which is a contradiction, and concludes the proof of (7.27).

To show the converse implication, suppose (7.25) occurs for some integer $1 \leq \nu \leq r$. By Theorem 4.3.2, there exists $\Phi \in \mathbb{P}[\mathcal{L}(U)]$ with $\Phi_0 = I_U$ such that the subspaces

$$R[\mathfrak{L}_0^\Phi], \; \mathfrak{L}_1^\Phi(N[\mathfrak{L}_0^\Phi]), \; \dots, \; \mathfrak{L}_\nu^\Phi(\bigcap_{j=0}^{\nu-1} N[\mathfrak{L}_j^\Phi])$$

are in direct sum, where

$$\mathfrak{L}^\Phi := \mathfrak{L}\Phi.$$

Setting (7.28), we find from Propositions 7.6.1 and 4.2.2 that

$$\dim N[\mathfrak{L}_{[\nu-1]}] = \sum_{h=1}^{\nu-1} h\ell_h + \nu n_{\nu-1} = \sum_{h=1}^{\nu} h\ell_h + \nu n_\nu$$

and

$$\dim N[\mathfrak{L}_{[\nu]}] = \sum_{h=1}^{\nu} h\ell_h + (\nu+1) n_\nu.$$

Therefore, by (7.25),

$$n_\nu = 0.$$

Similarly,

$$\begin{aligned} \sum_{h-1}^{\nu-1} h\ell_h + (\nu-1) n_{\nu-1} = \sum_{h-1}^{\nu-2} h\ell_h + (\nu-1) n_{\nu-2} &= \dim N[\mathfrak{L}_{[\nu-2]}] \\ < \dim N[\mathfrak{L}_{[\nu-1]}] &= \sum_{h=1}^{\nu-1} h\ell_h + \nu n_{\nu-1} \end{aligned}$$

and, hence,

$$1 \leq n_{\nu-1} = n_\nu + \ell_\nu = \ell_\nu.$$

Therefore, according to Proposition 4.2.2, λ_0 is a ν-transversal eigenvalue of $\mathfrak{L}^\Phi$. Consequently, by Theorem 5.3.1, we conclude that $\lambda_0 \in \operatorname{Alg}_\nu(\mathfrak{L})$. This ends the proof. $\square$

More generally, the following result, establishing the coincidence of χ, μ and κ, is satisfied.

Theorem 7.6.3 (of coincidence of χ and κ). *Let $r \in \mathbb{N}\cup\{\infty\}$ and $\mathfrak{L} \in \mathcal{C}^r(\Omega, \mathcal{L}(U,V))$ be with $\mathfrak{L}_0 \in \mathrm{Fred}_0(U,V)$. Then, $\kappa[\mathfrak{L};\lambda_0]$ is defined if and only if so is $\chi[\mathfrak{L};\lambda_0]$; in such a case,*

$$\kappa[\mathfrak{L};\lambda_0] = \chi[\mathfrak{L};\lambda_0].$$

Moreover, the following conditions are equivalent:

i) $\chi[\mathfrak{L};\lambda_0] < \infty$.

ii) $\lambda_0 \in \mathrm{Alg}_\nu(\mathfrak{L})$ *for some integer* $0 \leq \nu \leq r$.

iii) *The length of any Jordan chain of $\mathfrak{L}$ is bounded above by $\nu < r+1$.*

Proof. Some of the following alternatives occur. Either $\mathfrak{L}_0 \in \mathrm{Iso}(U,V)$, or

$$\lambda_0 \in \mathrm{Alg}_\nu(\mathfrak{L}) \quad \text{for some integer} \quad 1 \leq \nu \leq r,$$

or

$$\lambda_0 \notin \bigcup_{0\leq k<r+1} \mathrm{Alg}_k(\mathfrak{L}).$$

Suppose $\mathfrak{L}_0 \in \mathrm{Iso}(U,V)$. Then, by definition, both multiplicities are defined and

$$\kappa[\mathfrak{L};\lambda_0] = \chi[\mathfrak{L};\lambda_0] = 0.$$

Moreover, in such a case, (ii) holds with $\nu = 0$, and $\mathfrak{L}$ does not admit a Jordan chain. Therefore, the equivalence of (i)–(iii) trivially occurs when $\nu = 0$.

Suppose $\lambda_0 \in \mathrm{Alg}_\nu(\mathfrak{L})$ for some integer $1 \leq \nu \leq r$. Then, according to Theorem 7.6.2, conditions (7.25) and (7.26) are satisfied and, hence,

$$\kappa[\mathfrak{L};\lambda_0] = \chi[\mathfrak{L};\lambda_0] < \infty,$$

by Theorem 5.3.1. In particular, this shows that (ii) implies (i). Actually, the equivalence between (i) and (ii) has already been established by Theorem 5.3.1. In addition, thanks to (7.25) of Theorem 7.6.2, the linear injection

$$\begin{array}{rcl} J_\nu : N[\mathfrak{L}_{[\nu-1]}] & \to & N[\mathfrak{L}_{[\nu]}] \\ y & \mapsto & \mathrm{col}[0,y] \end{array}$$

is an isomorphism and, therefore, $\mathfrak{L}$ cannot admit any Jordan chain at λ_0 of length $\nu+1$. By Remark 7.1.2, the length of any Jordan chain of $\mathfrak{L}$ must be bounded above by ν and, consequently, (ii) also implies (iii).

Now, we will show the validity of the converse implications. Suppose (iii) and let $1 \leq \nu \leq r$ be an integer such that $\mathfrak{L}$ possesses a Jordan chain at λ_0 of length ν, but it does not admit a Jordan chain of length $\nu+1$. Then, the fact

$$(u_0,\ldots,u_\nu) \in N[\mathfrak{L}_{[\nu]}]$$

implies $u_0 = 0$ and, hence, J_ν is an isomorphism. Consequently,

$$\dim N[\mathfrak{L}_{[\nu-1]}] = \dim N[\mathfrak{L}_{[\nu]}]. \tag{7.30}$$

Arguing as in the proof of Theorem 7.6.2, let $\Phi \in \mathbb{P}[\mathcal{L}(U)]$ be such that $\Phi_0 = I_U$ and the subspaces

$$R[\mathfrak{L}_0^\Phi],\ \mathfrak{L}_1^\Phi(N[\mathfrak{L}_0^\Phi]), \ldots,\ \mathfrak{L}_\nu^\Phi(\bigcap_{j=0}^{\nu-1} N[\mathfrak{L}_j^\Phi])$$

are in direct sum, where

$$\mathfrak{L}^\Phi := \mathfrak{L}\Phi.$$

Setting (7.28), we find from Propositions 7.6.1 and 4.2.2 that

$$\dim N[\mathfrak{L}_{[\nu-1]}] = \sum_{h=1}^{\nu-1} h\ell_h + \nu n_{\nu-1} = \sum_{h=1}^{\nu} h\ell_h + \nu n_\nu$$

and

$$\dim N[\mathfrak{L}_{[\nu]}] = \sum_{h=1}^{\nu} h\ell_h + (\nu+1)n_\nu.$$

Therefore, by (7.30), we find that

$$n_\nu = \dim N[\mathfrak{L}_0^\Phi] \cap \cdots \cap N[\mathfrak{L}_\nu^\Phi] = 0.$$

Let ℓ denote the lowest integer $0 \le k \le \nu$ for which $n_k = 0$. Due to Proposition 4.2.2, λ_0 is an ℓ-transversal eigenvalue of $\mathfrak{L}^\Phi$ and, hence, by Theorem 5.3.1, $\lambda_0 \in \mathrm{Alg}_\ell(\mathfrak{L})$. Moreover, as we have already seen that (ii) implies (iii), the length of any Jordan chain of $\mathfrak{L}$ must be bounded above by ℓ and, so, $\nu \le \ell$. Therefore, $\nu = \ell$ and $\lambda_0 \in \mathrm{Alg}_\nu(\mathfrak{L})$, which shows that (iii) indeed implies (ii). Consequently, we have established the equivalence between (i), (ii), and (iii).

Throughout the rest of the proof we suppose

$$\lambda_0 \notin \bigcup_{0 \le k < r+1} \mathrm{Alg}_k(\mathfrak{L}).$$

Either $r = \infty$ or $r \in \mathbb{N}$. Suppose $r = \infty$. Then, according to Theorem 5.3.1, we have that

$$\chi[\mathfrak{L}; \lambda_0] = \infty.$$

Moreover, by Theorem 7.6.2, for every $k \ge 1$,

$$1 \le \dim N[\mathfrak{L}_{[k-1]}] < \dim N[\mathfrak{L}_{[k]}] \tag{7.31}$$

and, therefore, the map J_k cannot be surjective. Consequently, $\mathfrak{L}$ possesses a Jordan chain of length $k+1$ for all $k \ge 1$ and, by Definition 7.2.5,

$$\kappa[\mathfrak{L}; \lambda_0] = \infty = \chi[\mathfrak{L}; \lambda_0].$$

In the case $r \in \mathbb{N}$, (7.31) is still valid for each $1 \leq k \leq r$ and, therefore, $\mathfrak{L}$ admits a Jordan chain of length $r+1$. Consequently, $\kappa[\mathfrak{L};\lambda_0]$ is not defined, neither is $\chi[\mathfrak{L};\lambda_0]$, which concludes the proof. □

7.7 Labeling the vectors of the canonical sets

For later purposes, it is sometimes appropriate to change the notation introduced for the canonical sets of Jordan chains. Suppose $\lambda_0 \in \mathrm{Alg}_\nu(\mathfrak{L})$ for some integer $1 \leq \nu \leq r$ and let (7.5) be a canonical set of Jordan chains of $\mathfrak{L}$ at λ_0, as described in Section 7.2. Then, the partial multiplicities of $\mathfrak{L}$ at λ_0 are

$$\kappa_1 \geq \cdots \geq \kappa_n,$$

with $\kappa_n \geq 1$ and $n := \dim N[\mathfrak{L}_0]$. By Theorem 7.6.2, (7.25) is satisfied and, therefore, ν equals the maximum length of the Jordan chains of $\mathfrak{L}$ at λ_0, in other words,

$$\kappa_1 = \nu.$$

Let $\ell_1, \dots, \ell_\nu \geq 0$ be the integers introduced through Theorem 7.5.3 so that the partial multiplicities of $\mathfrak{L}$ at λ_0 are (7.23). Now, we choose the vectors

$$x_{i,j}^h \in U, \qquad 1 \leq j \leq \nu, \quad 1 \leq h \leq \ell_j, \quad 0 \leq i \leq j-1,$$

in such a way that

$$\begin{aligned} &\big[x_{0,\nu}^1, \dots, x_{\nu-1,\nu}^1; \cdots; x_{0,\nu}^{\ell_\nu}, \dots, x_{\nu-1,\nu}^{\ell_\nu};\\ &\ x_{0,\nu-1}^1, \dots, x_{\nu-2,\nu-1}^1; \cdots; x_{0,\nu-1}^{\ell_{\nu-1}}, \dots, x_{\nu-2,\nu-1}^{\ell_{\nu-1}}; \cdots;\\ &\ x_{0,j}^1, \dots, x_{j-1,j}^1; \cdots; x_{0,j}^{\ell_j}, \dots, x_{j-1,j}^{\ell_j}; \cdots; x_{0,1}^1; \cdots; x_{0,1}^{\ell_1}\big] \end{aligned} \tag{7.32}$$

is a canonical set of Jordan chains of $\mathfrak{L}$ at λ_0. The next result constructs a basis of $N[\mathfrak{L}_{[\nu-1]}]$ from this canonical set of Jordan chains.

Proposition 7.7.1. *Let $r \in \mathbb{N}^* \cup \{\infty\}$ and $\mathfrak{L} \in \mathcal{C}^r(\Omega, \mathcal{L}(U,V))$ be with $\mathfrak{L}_0 \in \mathrm{Fred}_0(U,V)$ such that $\lambda_0 \in \mathrm{Alg}_\nu(\mathfrak{L})$ for some integer $0 \leq \nu \leq r$. Let (7.32) be a canonical set of Jordan chains of $\mathfrak{L}$ at λ_0. Then,*

$$\mathcal{B} := \left\{ \mathrm{col}[\underbrace{0,\dots,0}_{\nu-i-1}, x_{0,j}^h, \dots, x_{i,j}^h] \in U^\nu : \begin{array}{c} 1 \leq h \leq \ell_j \\ 0 \leq i \leq j-1 \\ 1 \leq j \leq \nu \end{array} \right\}$$

is a basis of $N[\mathfrak{L}_{[\nu-1]}]$.

Proof. By construction, the set $\mathcal{B}$ is contained in $N[\mathfrak{L}_{[\nu-1]}]$ and has $\sum_{j=1}^{\nu} j\ell_j$ vectors. Thus, by (7.26), it suffices to show the linear independence of the vectors of $\mathcal{B}$. Let

$$\alpha_{i,j}^h \in \mathbb{K}, \qquad 1 \le h \le \ell_j, \quad 0 \le i \le j-1, \quad 1 \le j \le \nu,$$

be such that

$$\sum_{j=1}^{\nu} \sum_{i=0}^{j-1} \sum_{h=1}^{\ell_j} \alpha_{i,j}^h \operatorname{col}[\underbrace{0, \dots, 0}_{\nu-i-1}, x_{0,j}^h, \dots, x_{i,j}^h] = 0. \tag{7.33}$$

Focusing our attention on the first component of (7.33), we find that

$$\alpha_{\nu-1,\nu}^1 x_{0,\nu}^1 + \cdots + \alpha_{\nu-1,\nu}^{\ell_\nu} x_{0,\nu}^{\ell_\nu} = 0.$$

As (7.32) is a canonical set of Jordan chains, this implies

$$\alpha_{\nu-1,\nu}^h = 0, \qquad 1 \le h \le \ell_\nu.$$

Now, using induction on $r \ge 1$, we will show that

$$\alpha_{\nu-r,j}^h = 0, \qquad 1 \le h \le \ell_j, \quad \nu - r + 1 \le j \le \nu, \quad 1 \le r \le \nu. \tag{7.34}$$

We have just shown the result for $r = 1$. Let $2 \le \rho \le \nu$ and suppose (7.34) is satisfied for all $r \in \{1, \dots, \rho - 1\}$. Then, the ρth component of (7.33) provides

$$0 = \sum_{j=\nu-\rho+1}^{\nu} \sum_{i=\nu-\rho}^{j-1} \sum_{h=1}^{\ell_j} \alpha_{i,j}^h x_{i-(\nu-\rho),j}^h = \sum_{j=\nu-\rho+1}^{\nu} \sum_{h=1}^{\ell_j} \alpha_{\nu-\rho,j}^h x_{0,j}^h,$$

because, due to (7.34), $\alpha_{i,j}^h = 0$ if $\nu - \rho + 1 \le i < j$ and $1 \le h \le \ell_j$. As (7.32) is a canonical set of Jordan chains, we obtain that

$$\alpha_{\nu-\rho,j}^h = 0, \qquad 1 \le h \le \ell_j, \quad \nu - \rho + 1 \le j \le \nu,$$

which completes the proof. □

7.8 Characterizing the existence of the Smith form

This section uses the algebraic multiplicities χ and κ to characterize the existence of the local Smith form of $\mathfrak{L}$ at λ_0 through their associated algebraic invariants. The main finite-dimensional result reads as follows.

Theorem 7.8.1 (of characterization of the existence of the Smith form). *Let $m \ge 1$ be an integer, $r \in \mathbb{N}^* \cup \{\infty, \omega\}$ and $\lambda_0 \in \Omega$. Consider a matrix operator family $\mathfrak{L} \in \mathcal{C}^r(\Omega, \mathcal{L}(\mathbb{K}^m))$ with $n := \dim N[\mathfrak{L}_0] \ge 1$. Then, the following conditions are equivalent:*

i) $\mathfrak{L}$ *possesses a local Smith form at* λ_0, *in the sense of Definition* 7.4.3, *and the number* k_1 *appearing in that definition satisfies* $k_1 \leq r$.

ii) $\lambda_0 \in \mathrm{Alg}_\nu(\mathfrak{L})$ *for some integer* $1 \leq \nu \leq r$.

If, in addition, $r \in \{\infty, \omega\}$, *then these conditions are also equivalent to each of the next two:*

iii) $\mathrm{ord}_{\lambda_0} \det \mathfrak{L} < \infty$.

iv) *No Jordan chain of* $\mathfrak{L}$ *at* λ_0 *can be continued indefinitely.*

Moreover, in such a case, there exist an open neighborhood $\Omega' \subset \Omega$ *of* λ_0, *families* $\mathfrak{E} \in \mathcal{C}^{r-\nu}(\Omega', \mathcal{M}_m(\mathbb{K}))$, *and* $\mathfrak{F} \in \mathcal{C}^{\omega}(\Omega', \mathcal{M}_m(\mathbb{K}))$, *with* $\mathfrak{E}_0, \mathfrak{F}_0 \in \mathrm{Iso}(\mathbb{K}^m)$, *such that*

$$\mathfrak{L}(\lambda) = \mathfrak{E}(\lambda)\ \mathfrak{D}(\lambda)\ \mathfrak{F}(\lambda), \qquad \lambda \in \Omega', \tag{7.35}$$

where

$$\mathfrak{D}(\lambda) = \mathrm{diag}\left\{(\lambda - \lambda_0)^{\kappa_1[\mathfrak{L};\lambda_0]}, \dots, (\lambda - \lambda_0)^{\kappa_n[\mathfrak{L};\lambda_0]}, \underbrace{1, \dots, 1}_{m-n}\right\}, \quad \lambda \in \mathbb{K}.$$

Furthermore, the previous characterizations can be formulated through conditions (i) *and* (iii) *of Theorem* 7.6.3, *as well as through* (iii) *and* (iv) *of Theorem* 5.3.1.

Proof. Suppose (i). Then, there exist an open neighborhood $\Omega' \subset \Omega$ of λ_0, families $\mathfrak{E} \in \mathcal{C}(\Omega', \mathcal{L}(\mathbb{K}^m))$ and $\mathfrak{F} \in \mathcal{C}(\Omega', \mathcal{L}(\mathbb{K}^m))$, with $\mathfrak{E}_0 \in \mathrm{Iso}(\mathbb{K}^m)$ and $\mathfrak{F}_0 \in \mathrm{Iso}(\mathbb{K}^m)$, such that

$$\mathfrak{L}(\lambda) = \mathfrak{E}(\lambda)\ \mathfrak{D}(\lambda)\ \mathfrak{F}(\lambda), \qquad \lambda \in \Omega',$$

where

$$\mathfrak{D}(\lambda) = \mathrm{diag}\left\{(\lambda - \lambda_0)^{k_1}, \dots, (\lambda - \lambda_0)^{k_n}, \underbrace{1, \dots, 1}_{m-n}\right\}, \quad \lambda \in \mathbb{K},$$

for some integers $k_1 \geq \cdots \geq k_n$ such that $k_1 \leq r$ and $k_n \geq 1$. In particular,

$$\mathrm{ord}_{\lambda_0} \det \mathfrak{L} = \sum_{j=1}^{n} k_j$$

and, hence, (i) implies (iii). Moreover,

$$\mathfrak{L}(\lambda)^{-1} = \mathfrak{F}(\lambda)^{-1}\ \mathfrak{D}(\lambda)^{-1}\ \mathfrak{E}(\lambda)^{-1}, \qquad \lambda \sim \lambda_0,\ \lambda \neq \lambda_0,$$

and, since

$$\mathfrak{D}(\lambda)^{-1} = \mathrm{diag}\left\{(\lambda - \lambda_0)^{-k_1}, \dots, (\lambda - \lambda_0)^{-k_n}, 1, \dots, 1\right\}, \qquad \lambda \in \mathbb{K} \setminus \{\lambda_0\},$$

there exists a constant $C > 0$ such that

$$\|\mathfrak{L}(\lambda)^{-1}\| \leq \frac{C}{|\lambda - \lambda_0|^{k_1}}, \qquad \lambda \sim \lambda_0,\ \lambda \neq \lambda_0.$$

Therefore, $\lambda_0 \in \mathrm{Alg}_\nu(\mathfrak{L})$ for some integer ν with $1 \leq \nu \leq k_1 \leq r$, and, consequently, (i) implies (ii).

Suppose (ii) and consider the auxiliary families $\tilde{\mathfrak{L}}$ and $\mathfrak{F}$ defined through

$$\tilde{\mathfrak{L}}(\lambda) := \sum_{i=0}^{\nu} (\lambda - \lambda_0)^i \mathfrak{L}_i, \quad \mathfrak{F}(\lambda) := \tilde{\mathfrak{L}}(\lambda)\mathfrak{L}(\lambda)^{-1}, \qquad \lambda \sim \lambda_0.$$

Clearly, $\tilde{\mathfrak{L}} - \mathfrak{L} \in \mathcal{C}^r(\Omega, \mathcal{L}(\mathbb{K}^m))$ and all its derivatives up to order ν vanish at λ_0. Thus, the auxiliary family $\mathfrak{R}$, defined by

$$\mathfrak{R}(\lambda) := (\lambda - \lambda_0)^{-\nu} \left[\tilde{\mathfrak{L}}(\lambda) - \mathfrak{L}(\lambda)\right], \qquad \lambda \sim \lambda_0, \quad \lambda \neq \lambda_0,$$

and $\mathfrak{R}(\lambda_0) := 0$, is of class $\mathcal{C}^{r-\nu}$. Moreover, by construction, we have that

$$\begin{aligned} \mathfrak{F}(\lambda) &= [\mathfrak{L}(\lambda) + (\lambda - \lambda_0)^\nu \,\mathfrak{R}(\lambda)]\, \mathfrak{L}(\lambda)^{-1} \\ &= I_V + (\lambda - \lambda_0)^\nu \,\mathfrak{R}(\lambda)\mathfrak{L}(\lambda)^{-1}, \qquad \lambda \sim \lambda_0, \quad \lambda \neq \lambda_0. \end{aligned}$$

As $\lambda_0 \in \mathrm{Alg}_\nu(\mathfrak{L})$, due to Theorem 5.0.1, the map

$$\lambda \mapsto (\lambda - \lambda_0)^\nu \mathfrak{L}(\lambda)^{-1}$$

is of class $\mathcal{C}^{r-\nu}$ in a neighborhood of λ_0, as is $\mathfrak{R}$. Therefore, so is $\mathfrak{F}$. Moreover, $\mathfrak{F}(\lambda_0) = I_V$, because $\mathfrak{R}(\lambda_0) = 0$. Thus, there exists a constant $C > 0$ such that, for each λ sufficiently close to λ_0 with $\lambda \neq \lambda_0$,

$$\|\tilde{\mathfrak{L}}(\lambda)^{-1}\| = \|\mathfrak{L}(\lambda)^{-1}\mathfrak{F}(\lambda)^{-1}\| = \frac{\|(\lambda - \lambda_0)^\nu \mathfrak{L}(\lambda)^{-1}\mathfrak{F}(\lambda)^{-1}\|}{|\lambda - \lambda_0|^\nu} \leq \frac{C}{|\lambda - \lambda_0|^\nu}.$$

Consequently, $\lambda_0 \in \mathrm{Alg}_k(\tilde{\mathfrak{L}})$ for some $1 \leq k \leq \nu$. As $\tilde{\mathfrak{L}} \in \mathbb{P}[\mathcal{L}(\mathbb{K}^m)]$, we have that $\tilde{\mathfrak{L}}$ is analytic. Moreover, according to Theorem 7.6.3, the length of all Jordan chains of $\tilde{\mathfrak{L}}$ is bounded above by $k \leq \nu < r+1$. In particular, no Jordan chain of $\tilde{\mathfrak{L}}$ can be continued indefinitely. Thus, thanks to Theorem 7.4.1, there exist an open subset $\tilde{\Omega} \subset \Omega$ with $\lambda_0 \in \tilde{\Omega}$, and two families $\tilde{\mathfrak{E}}, \tilde{\mathfrak{F}} \in \mathcal{C}^\omega(\tilde{\Omega}, \mathcal{M}_m(\mathbb{K}))$ such that $\tilde{\mathfrak{E}}_0, \tilde{\mathfrak{F}}_0 \in \mathrm{Iso}(\mathbb{K}^m)$ and

$$\tilde{\mathfrak{L}}(\lambda) = \tilde{\mathfrak{E}}(\lambda)\; \mathfrak{D}(\lambda)\; \tilde{\mathfrak{F}}(\lambda), \qquad \lambda \in \tilde{\Omega},$$

where, for every $\lambda \in \mathbb{K}$,

$$\mathfrak{D}(\lambda) := \mathrm{diag}\left\{(\lambda - \lambda_0)^{\kappa_1[\tilde{\mathfrak{L}};\lambda_0]}, \ldots, (\lambda - \lambda_0)^{\kappa_n[\tilde{\mathfrak{L}};\lambda_0]}, \underbrace{1, \ldots, 1}_{m-n}\right\}.$$

Note that $n = \dim N[\tilde{\mathfrak{L}}_0]$, since $\mathfrak{L}_0 = \tilde{\mathfrak{L}}_0$. Moreover,

$$\mathfrak{L}(\lambda) = \mathfrak{F}(\lambda)^{-1}\; \tilde{\mathfrak{L}}(\lambda) = \mathfrak{F}(\lambda)^{-1}\; \tilde{\mathfrak{E}}(\lambda)\; \mathfrak{D}(\lambda)\; \tilde{\mathfrak{F}}(\lambda), \qquad \lambda \sim \lambda_0.$$

Thus, setting $\mathfrak{E} := \mathfrak{F}^{-1}\tilde{\mathfrak{E}}$ and $\mathfrak{F} := \tilde{\mathfrak{F}}$, we have that $\mathfrak{E}$ is of class $\mathcal{C}^{r-\nu}$ and $\mathfrak{F}$ is of class $\mathcal{C}^{\omega}$ in a neighborhood of λ_0 and that (7.35) holds with the required regularity properties. In particular, it becomes apparent that (ii) implies (i). Note that, due to Theorem 7.3.1 and the Example of Section 7.2,

$$\kappa_j[\mathfrak{L};\lambda_0] = \kappa_j[\mathfrak{D};\lambda_0] = \kappa_j[\tilde{\mathfrak{L}};\lambda_0], \qquad 1 \leq j \leq n.$$

Moreover,

$$\kappa_1[\mathfrak{L};\lambda_0] = \kappa_1[\mathfrak{D};\lambda_0] = \kappa_1[\tilde{\mathfrak{L}};\lambda_0] = \nu.$$

The previous analysis concludes the proof when $r \in \mathbb{N}$.

From now on, we restrict ourselves to considering the case $r \in \{\infty, \omega\}$. We have already seen that (i) implies (iii). Suppose (iii). Then, thanks to Proposition 3.1.1, it is easy to see that $\lambda_0 \in \mathrm{Alg}_\nu(\mathfrak{L})$ for some integer

$$1 \leq \nu \leq \mathrm{ord}_{\lambda_0} \det \mathfrak{L}$$

and, consequently, (iii) implies (ii). Therefore, (i), (ii) and (iii) are indeed equivalent.

Suppose (ii). Then, according to Theorem 7.6.3, the length of any Jordan chain of $\mathfrak{L}$ is bounded above by ν and, hence, (iv) holds.

Suppose (iv). Then, (i) follows straight away from Theorem 7.4.1. This concludes the proof. □

The following result provides us with the basic technical tool for obtaining the infinite-dimensional counterpart of Theorem 7.8.1.

Lemma 7.8.2. *Let $r \in \mathbb{N} \cup \{\infty, \omega\}$ and $\mathfrak{L} \in \mathcal{C}^r(\Omega, \mathcal{L}(U,V))$ be with $\mathfrak{L}_0 \in \mathrm{Fred}_0(U,V)$. Then, there exist an open neighborhood $\tilde{\Omega} \subset \Omega$ of λ_0, a topological direct sum decomposition $U = U_0 \oplus U_1$, with*

$$\dim U_0 = \dim N[\mathfrak{L}_0] < \infty,$$

and three operator families

$$\mathfrak{E} \in \mathcal{C}^r(\tilde{\Omega}, \mathcal{L}(U,V)), \quad \mathfrak{F} \in \mathcal{C}^r(\tilde{\Omega}, \mathcal{L}(U)), \quad \mathfrak{M} \in \mathcal{C}^r(\tilde{\Omega}, \mathcal{L}(U_0)) \tag{7.36}$$

with $\mathfrak{E}_0 \in \mathrm{Iso}(U,V)$ and $\mathfrak{F}_0 \in \mathrm{Iso}(U)$, such that

$$\mathfrak{L}(\lambda) = \mathfrak{E}(\lambda)\,[\mathfrak{M}(\lambda) \oplus I_{U_1}]\,\mathfrak{F}(\lambda), \qquad \lambda \in \tilde{\Omega}. \tag{7.37}$$

Proof. Suppose $\mathfrak{L}_0 \in \mathrm{Iso}(U,V)$. Then, the result holds by making the choices

$$U_0 = \{0\}, \quad U_1 = U, \quad \mathfrak{F} = I_U, \quad \mathfrak{M} = I_{U_0}, \quad \mathfrak{E} = \mathfrak{L}.$$

So, from now on, we will assume that $\dim N[\mathfrak{L}_0] \geq 1$. Since $\mathfrak{L}_0 \in \mathrm{Fred}_0(U,V)$, there is an operator $F \in \mathcal{L}(U,V)$ with

$$\dim R[F] = \mathrm{codim}\, R[\mathfrak{L}_0] = \dim N[\mathfrak{L}_0], \tag{7.38}$$

such that

$$\mathfrak{L}_0 + F \in \mathrm{Iso}(U, V).$$

Thus, there exists an open neighborhood $\tilde{\Omega} \subset \Omega$ of λ_0 for which the family $\mathfrak{E}$ defined by

$$\mathfrak{E}(\lambda) := \mathfrak{L}(\lambda) + F, \qquad \lambda \in \tilde{\Omega},$$

satisfies $\mathfrak{E}(\lambda) \in \mathrm{Iso}(U, V)$ for each $\lambda \in \tilde{\Omega}$, and $\mathfrak{E}^{-1} \in \mathcal{C}^r(\tilde{\Omega}, \mathcal{L}(V, U))$.

As $\dim R[F] < \infty$, the space $N[F]$ admits a topological finite-dimensional complement U_0 in U, so

$$U = U_0 \oplus U_1, \qquad U_1 := N[F].$$

Let $P \in \mathcal{L}(U)$ be the projection satisfying

$$R[P] = U_0 \quad \text{and} \quad N[P] = N[F].$$

Then, P has finite rank and

$$FP = F,$$

because $I_U - P$ is a projection onto $N[F]$. Thus, for each $\lambda \in \tilde{\Omega}$,

$$I_U - \mathfrak{E}(\lambda)^{-1}F = \left[I_U - P\mathfrak{E}(\lambda)^{-1}FP\right]\left[I_U - (I_U - P)\,\mathfrak{E}(\lambda)^{-1}FP\right],$$

and, setting

$$\mathfrak{F}(\lambda) := I_U - (I_U - P)\mathfrak{E}(\lambda)^{-1}FP, \qquad \lambda \in \tilde{\Omega},$$

we have that

$$I_U - \mathfrak{E}(\lambda)^{-1}F = \left[I_U - P\mathfrak{E}(\lambda)^{-1}FP\right]\mathfrak{F}(\lambda), \qquad \lambda \in \tilde{\Omega}.$$

Consequently, for each $\lambda \in \tilde{\Omega}$,

$$\mathfrak{L}(\lambda) = \mathfrak{E}(\lambda) - F = \mathfrak{E}(\lambda)[I_U - \mathfrak{E}(\lambda)^{-1}F] = \mathfrak{E}(\lambda)\left[I_U - P\mathfrak{E}(\lambda)^{-1}FP\right]\mathfrak{F}(\lambda).$$

Clearly, $\mathfrak{F} \in \mathcal{C}^r(\tilde{\Omega}, \mathcal{L}(U))$ and, for each $\lambda \in \tilde{\Omega}$, the operator $\mathfrak{F}(\lambda)$ is invertible. In fact,

$$\mathfrak{F}(\lambda)^{-1} = I_U + (I_U - P)\mathfrak{E}(\lambda)^{-1}FP, \qquad \lambda \in \tilde{\Omega}.$$

Moreover, setting

$$\mathfrak{D}(\lambda) := I_U - P\mathfrak{E}(\lambda)^{-1}FP, \qquad \lambda \in \tilde{\Omega},$$

we have that

$$\mathfrak{L}(\lambda) = \mathfrak{E}(\lambda)\;\mathfrak{D}(\lambda)\;\mathfrak{F}(\lambda), \qquad \lambda \in \tilde{\Omega}.$$

Furthermore, with respect to the decomposition

$$U = U_0 \oplus U_1 = R[P] \oplus N[P],$$

the family $\mathfrak{D}$ can be expressed as

$$\mathfrak{D}(\lambda) = \operatorname{diag}\{\mathfrak{D}(\lambda)|_{U_0}, I_{N[F]}\} = \mathfrak{D}(\lambda)|_{U_0} \oplus I_{N[F]}, \qquad \lambda \in \tilde{\Omega},$$

since

$$\mathfrak{D}(\lambda)(U_0) \subset U_0 \quad \text{and} \quad \mathfrak{D}(\lambda)|_{N[F]} = I|_{N[F]}, \qquad \lambda \in \tilde{\Omega}.$$

Therefore, (7.37) holds true with

$$\mathfrak{M}(\lambda) := \mathfrak{D}(\lambda)|_{U_0}, \qquad \lambda \in \tilde{\Omega}.$$

Note that, (7.38) indeed implies

$$\dim U_0 = \dim N[\mathfrak{L}_0],$$

which concludes the proof. □

The following substantial extension of Theorem 7.8.1 is satisfied.

Theorem 7.8.3 (of characterization of the existence of the Smith form). *Let $r \in \mathbb{N}^* \cup \{\infty, \omega\}$. Consider $\lambda_0 \in \Omega$ and an operator family $\mathfrak{L} \in \mathcal{C}^r(\Omega, \mathcal{L}(U, V))$ with $\mathfrak{L}_0 \in \mathrm{Fred}_0(U, V)$ and $n := \dim N[\mathfrak{L}_0] \geq 1$. Then, the following conditions are equivalent:*

i) *$\mathfrak{L}$ possesses a local Smith form at λ_0, in the sense of Definition 7.4.3, and the number k_1 appearing in that definition satisfies $k_1 \leq r$.*

ii) *$\lambda_0 \in \mathrm{Alg}_\nu(\mathfrak{L})$ for some integer $1 \leq \nu \leq r$.*

If, in addition, $r \in \{\infty, \omega\}$, then these conditions are also equivalent to each of the next two:

iii) *For every neighborhood Ω' of λ_0, any $\mathbb{K}$-Banach spaces $\hat{U}$ and $\hat{V}$, any topological decompositions*

$$\hat{U} = \hat{U}_1 \oplus \hat{U}_2 \quad \text{and} \quad \hat{V} = \hat{V}_1 \oplus \hat{V}_2$$

with

$$\dim \hat{U}_1 = \dim \hat{V}_1 < \infty,$$

and any families

$$\mathfrak{E} \in \mathcal{C}^\infty(\Omega', \mathcal{L}(\hat{V}, V)), \qquad \mathfrak{F} \in \mathcal{C}^\infty(\Omega', \mathcal{L}(U, \hat{U})),$$
$$\mathfrak{N} \in \mathcal{C}^\infty(\Omega', \mathcal{L}(\hat{U}_2, \hat{V}_2)), \qquad \mathfrak{M} \in \mathcal{C}^\infty(\Omega', \mathcal{L}(\hat{U}_1, \hat{V}_1)),$$

such that $\mathfrak{E}_0, \mathfrak{F}_0, \mathfrak{N}_0$ are isomorphisms, and

$$\mathfrak{L}(\lambda) = \mathfrak{E}(\lambda)\,[\mathfrak{M}(\lambda) \oplus \mathfrak{N}(\lambda)]\,\mathfrak{F}(\lambda), \qquad \lambda \in \Omega',$$

one has

$$\operatorname{ord}_{\lambda_0} \det \mathfrak{M} < \infty.$$

iv) *No Jordan chain of $\mathfrak{L}$ at λ_0 can be continued indefinitely.*

Moreover, in such a case, there actually exist an open neighborhood $\Omega' \subset \Omega$ *of* λ_0*, two operator families* $\mathfrak{E} \in \mathcal{C}^{r-\nu}(\Omega', \mathcal{L}(U,V))$ *and* $\mathfrak{F} \in \mathcal{C}^r(\Omega', \mathcal{L}(U))$ *with* $\mathfrak{E}_0 \in \operatorname{Iso}(U,V)$ *and* $\mathfrak{F}_0 \in \operatorname{Iso}(U)$*, and a topological direct sum decomposition* $U = U_0 \oplus U_1$*, with* $n = \dim U_0$*, such that*

$$\mathfrak{L}(\lambda) = \mathfrak{E}(\lambda)\ [\mathfrak{S}(\lambda) \oplus I_{U_1}]\ \mathfrak{F}(\lambda), \qquad \lambda \in \Omega', \tag{7.39}$$

where, for each $\lambda \in \mathbb{K}$*,*

$$\mathfrak{S}(\lambda) = \operatorname{diag}\left\{(\lambda-\lambda_0)^{\kappa_1[\mathfrak{L};\lambda_0]}, \dots, (\lambda-\lambda_0)^{\kappa_n[\mathfrak{L};\lambda_0]}\right\}.$$

Furthermore, the previous characterizations can be formulated through conditions (i) *and* (iii) *of Theorem* 7.6.3*, as well as through* (iii) *and* (iv) *of Theorem* 5.3.1*.*

Proof. Suppose (i). Then, there exists an open neighborhood $\Omega' \subset \Omega$ of λ_0, families $\mathfrak{E} \in \mathcal{C}(\Omega', \mathcal{L}(U,V))$ and $\mathfrak{F} \in \mathcal{C}(\Omega', \mathcal{L}(U))$, with $\mathfrak{E}_0 \in \operatorname{Iso}(U,V)$ and $\mathfrak{F}_0 \in \operatorname{Iso}(U)$, a direct sum topological decomposition $U = U_0 \oplus U_1$, with $n = \dim U_0$, and n integer numbers $k_1 \geq \cdots \geq k_n$ with $k_1 \leq r$ and $k_n \geq 1$ such that

$$\mathfrak{L}(\lambda) = \mathfrak{E}(\lambda)\ [\mathfrak{S}(\lambda) \oplus I_{U_1}]\ \mathfrak{F}(\lambda), \qquad \lambda \in \Omega',$$

where

$$\mathfrak{S}(\lambda) = \operatorname{diag}\left\{(\lambda-\lambda_0)^{k_1}, \dots, (\lambda-\lambda_0)^{k_n}\right\}, \qquad \lambda \in \mathbb{K}.$$

Thus, there exists a neighborhood $\tilde{\Omega} \subset \Omega'$ of λ_0 such that

$$\mathfrak{L}(\lambda)^{-1} = \mathfrak{F}(\lambda)^{-1}[\mathfrak{S}(\lambda)^{-1} \oplus I_{U_1}]\mathfrak{E}(\lambda)^{-1}, \qquad \lambda \in \tilde{\Omega} \setminus \{\lambda_0\},$$

and, hence, there exists a constant $C > 0$ such that

$$\|\mathfrak{L}(\lambda)^{-1}\| \leq \frac{C}{|\lambda-\lambda_0|^{k_1}}, \qquad \lambda \in \tilde{\Omega} \setminus \{\lambda_0\}.$$

Therefore, $\lambda_0 \in \operatorname{Alg}_\nu(\mathfrak{L})$ for some integer ν with $1 \leq \nu \leq k_1 \leq r$, and, consequently, (i) implies (ii).

Suppose (ii). Then, by Theorem 7.6.3, the length of all Jordan chains of $\mathfrak{L}$ is bounded above by $\nu < r+1$. On the other hand, owing to Lemma 7.8.2, there exist an open neighborhood $\tilde{\Omega} \subset \Omega$ of λ_0, a topological direct sum decomposition $U = U_0 \oplus U_1$, with $n = \dim U_0$, and three operator families

$$\tilde{\mathfrak{E}} \in \mathcal{C}^r(\tilde{\Omega}, \mathcal{L}(U,V)), \quad \tilde{\mathfrak{F}} \in \mathcal{C}^r(\tilde{\Omega}, \mathcal{L}(U)), \quad \mathfrak{M} \in \mathcal{C}^r(\tilde{\Omega}, \mathcal{L}(U_0)),$$

with $\tilde{\mathfrak{E}}_0 \in \operatorname{Iso}(U,V)$ and $\tilde{\mathfrak{F}}_0 \in \operatorname{Iso}(U)$, such that

$$\mathfrak{L}(\lambda) = \tilde{\mathfrak{E}}(\lambda)\ [\mathfrak{M}(\lambda) \oplus I_{U_1}]\ \tilde{\mathfrak{F}}(\lambda), \qquad \lambda \in \tilde{\Omega}. \tag{7.40}$$

Thanks to Theorem 7.3.1, the length of any Jordan chain of $\mathfrak{M} \oplus I_{U_1}$ is bounded above by $\nu < r+1$. Moreover, any Jordan chain of $\mathfrak{M} \oplus I_{U_1}$ at λ_0 must be of the form

$$\left(\begin{pmatrix} u_0 \\ 0 \end{pmatrix}, \dots, \begin{pmatrix} u_s \\ 0 \end{pmatrix}\right)$$

for some Jordan chain $(u_0, \dots, u_s)$ of $\mathfrak{M}$ at λ_0. Consequently, the lengths of the Jordan chains of $\mathfrak{M}$ at λ_0 are bounded above by $\nu < r+1$. Actually, there is a linear bijection between them and those of $\mathfrak{L}$ and, therefore,

$$\kappa_j[\mathfrak{L}; \lambda_0] = \kappa_j[\mathfrak{M}; \lambda_0], \qquad 1 \le j \le n.$$

By Theorem 7.6.3, $\lambda_0 \in \mathrm{Alg}_k(\mathfrak{M})$ for some integer $1 \le k \le \nu$ and, therefore, according to Theorem 7.8.1, there exist an open neighborhood $\Omega' \subset \tilde{\Omega}$ of λ_0, families $\mathfrak{E}' \in \mathcal{C}^{r-k}(\Omega', \mathcal{L}(U_0))$ and $\mathfrak{F}' \in \mathcal{C}^{\omega}(\Omega', \mathcal{L}(U_0))$ with $\mathfrak{E}'_0, \mathfrak{F}'_0 \in \mathrm{Iso}(U_0)$, such that

$$\mathfrak{M}(\lambda) = \mathfrak{E}'(\lambda)\, \mathfrak{S}(\lambda)\, \mathfrak{F}'(\lambda), \qquad \lambda \in \Omega', \tag{7.41}$$

where

$$\mathfrak{S}(\lambda) = \mathrm{diag}\left\{(\lambda-\lambda_0)^{\kappa_1[\mathfrak{L};\lambda_0]}, \dots, (\lambda-\lambda_0)^{\kappa_n[\mathfrak{L};\lambda_0]}\right\}, \quad \lambda \in \mathbb{K}.$$

According to (7.40) and (7.41), we find that, in Ω',

$$\mathfrak{L} = \tilde{\mathfrak{E}}\,(\mathfrak{E}' \oplus I_{U_1})\,(\mathfrak{S} \oplus I_{U_1})\,(\mathfrak{F}' \oplus I_{U_1})\,\tilde{\mathfrak{F}}.$$

Therefore, making the choices

$$\mathfrak{E} := \tilde{\mathfrak{E}}\,(\mathfrak{E}' \oplus I_{U_1}), \qquad \mathfrak{F} := (\mathfrak{F}' \oplus I_{U_1})\,\tilde{\mathfrak{F}},$$

concludes the proof of (i), and shows the equivalence of (i) and (ii). Note that, since $k \le \nu$,

$$\mathfrak{E} \in \mathcal{C}^{r-\nu}(\Omega', \mathcal{L}(U,V)) \quad \text{and} \quad \mathfrak{F} \in \mathcal{C}^{r}(\Omega', \mathcal{L}(U)).$$

Consequently, we have also proved (7.39) subject to the regularity requirements of the theorem. This concludes the proof when $r \in \mathbb{N}$.

From now on, we assume that $r \in \{\infty, \omega\}$. Suppose (ii). Then, according to Theorem 7.6.3, the lengths of the Jordan chains of $\mathfrak{L}$ are bounded above by $\nu < r+1$ and, therefore, (ii) implies (iv). Moreover, under the circumstances detailed in the statement of (iii), for each $\lambda \neq \lambda_0$ sufficiently close to λ_0 we have that

$$\mathfrak{M}(\lambda)^{-1} \oplus \mathfrak{N}(\lambda)^{-1} = \mathfrak{F}(\lambda)\mathfrak{L}(\lambda)^{-1}\mathfrak{E}(\lambda).$$

Thus,

$$\|\mathfrak{M}(\lambda)^{-1}\| \le C\|\mathfrak{L}(\lambda)^{-1}\|$$

for some constant $C > 0$, and, hence, $\lambda_0 \in \mathrm{Alg}_k(\mathfrak{M})$ for some integer $1 \le k \le \nu$. Therefore, by Theorem 7.8.1, $\det \mathfrak{M}$ has a zero of finite order at λ_0. Consequently, (ii) also implies (iii).

Suppose (iii). Thanks to Lemma 7.8.2, there exist an open neighborhood $\tilde{\Omega} \subset \Omega$ of λ_0, a topological direct sum decomposition $U = U_0 \oplus U_1$ with $n = \dim U_0$, and three operator families (7.36) such that

$$\mathfrak{L}(\lambda) = \mathfrak{E}(\lambda)\,[\mathfrak{M}(\lambda) \oplus I_{U_1}]\,\mathfrak{F}(\lambda), \qquad \lambda \in \tilde{\Omega}. \tag{7.42}$$

As we are assuming that (iii) occurs, $\det \mathfrak{M}$ has a zero of finite order at λ_0. By Theorem 7.8.1, λ_0 is an algebraic eigenvalue of $\mathfrak{M}$ and, therefore, we find from (7.42) that $\lambda_0 \in \mathrm{Alg}_\nu(\mathfrak{L})$ for some integer $\nu \geq 1$. Consequently, (iii) implies (ii).

It remains to prove that (iv) implies (i). Suppose (iv) and consider the decomposition (7.42). By Theorem 7.4.1, $\mathfrak{M}$ admits a local Smith form at λ_0. Arguing as in the final part of the proof of (i) from (ii), the Smith form of $\mathfrak{M}$ provides us with a Smith form for $\mathfrak{L}$. Therefore, (i) is satisfied, which ends the proof. □

7.9 Two illustrative examples

This section works out two simple examples to illustrate the practical use of Theorem 7.8.1.

7.9.1 Example 1

Consider $\mathbb{K} = \mathbb{R}$, the spaces $U = V = \mathbb{R}^2$ and the map $\mathfrak{L} : \mathbb{R} \to \mathcal{L}(\mathbb{R}^2)$ defined by

$$\mathfrak{L}(\lambda) := \begin{pmatrix} -2\lambda^2 \left(1 + \lambda \sin \dfrac{1}{\lambda}\right) & \lambda \left(1 + \lambda^3 \cos \dfrac{1}{\lambda}\right) \\ -3\lambda^2 & 2 \end{pmatrix}, \qquad \lambda \in \mathbb{R} \setminus \{0\},$$

and

$$\mathfrak{L}(0) := \lim_{\lambda \to 0} \mathfrak{L}(\lambda) = \begin{pmatrix} 0 & 0 \\ 0 & 2 \end{pmatrix}.$$

It is easy to see that $\mathfrak{L}$ is of class $\mathcal{C}^2$ and that $\lambda_0 := 0$ is an algebraic eigenvalue of $\mathfrak{L}$ of order $\nu = 2 = r$. Consequently, according to Theorem 7.8.1, $\mathfrak{L}$ possesses a local Smith form at 0, in the sense of Definition 7.4.3, and $k_1 \leq 2$. Note that $\mathfrak{L}$ has a non-vanishing constant entry, and its determinant satisfies

$$\det \mathfrak{L}(\lambda) = -4\lambda^2(1 + o(1)) \qquad \text{as } \lambda \to 0.$$

To calculate the local Smith form of $\mathfrak{L}$ at 0, we use the algorithm described in the proof of Theorem 7.4.1.

For each $i, j \in \{1, 2\}$, let ℓ_{ij} denote the (i, j)th entry of $\mathfrak{L}$. Since $\mathrm{ord}_0\, \ell_{22} = 0$, we can use ℓ_{22} as a pivot element. To make a zero in the entry $(2, 1)$th, we replace, for each $\lambda \in \mathbb{R}$, the column

$$\begin{pmatrix} \ell_{11}(\lambda) \\ \ell_{21}(\lambda) \end{pmatrix} \quad \text{with} \quad \begin{pmatrix} \ell_{11}(\lambda) \\ \ell_{21}(\lambda) \end{pmatrix} + \frac{3}{2}\lambda^2 \begin{pmatrix} \ell_{12}(\lambda) \\ \ell_{22}(\lambda) \end{pmatrix}.$$

This amounts to performing the matrix multiplication

$$\begin{aligned}&\begin{pmatrix}-2\lambda^2+\frac{3}{2}\lambda^3-2\lambda^3\sin\frac{1}{\lambda}+\frac{3}{2}\lambda^6\cos\frac{1}{\lambda} & \lambda+\lambda^4\cos\frac{1}{\lambda}\\ 0 & 2\end{pmatrix}\\ &= \mathfrak{L}(\lambda)\begin{pmatrix}1 & 0\\ \frac{3}{2}\lambda^2 & 1\end{pmatrix} =: \mathfrak{L}^1(\lambda),\end{aligned}$$

for every $\lambda\in\mathbb{R}\setminus\{0\}$, and

$$\mathfrak{L}^1(0) := \lim_{\lambda\to 0}\mathfrak{L}^1(\lambda) = \begin{pmatrix}0 & 0\\ 0 & 2\end{pmatrix}.$$

For each $i,j\in\{1,2\}$, let ℓ^1_{ij} denote the (i,j)th entry of $\mathfrak{L}^1$. To make a zero in the $(1,2)$th entry of $\mathfrak{L}^1$, we replace, for each $\lambda\in\mathbb{R}\setminus\{0\}$, the row

$$\begin{pmatrix}\ell^1_{11}(\lambda) & \ell^1_{12}(\lambda)\end{pmatrix}$$

with

$$\begin{pmatrix}\ell^1_{11}(\lambda) & \ell^1_{12}(\lambda)\end{pmatrix} - \frac{1}{2}\left(\lambda+\lambda^4\cos\frac{1}{\lambda}\right)\begin{pmatrix}\ell^1_{21}(\lambda) & \ell^1_{22}(\lambda)\end{pmatrix}.$$

This amounts to performing the matrix multiplication

$$\begin{aligned}&\begin{pmatrix}-2\lambda^2+\frac{3}{2}\lambda^3-2\lambda^3\sin\frac{1}{\lambda}+\frac{3}{2}\lambda^6\cos\frac{1}{\lambda} & 0\\ 0 & 2\end{pmatrix}\\ &= \begin{pmatrix}1 & -\frac{1}{2}\left(\lambda+\lambda^4\cos\frac{1}{\lambda}\right)\\ 0 & 1\end{pmatrix}\mathfrak{L}^1(\lambda) =: \mathfrak{L}^2(\lambda),\end{aligned}$$

for every $\lambda\in\mathbb{R}\setminus\{0\}$, and

$$\mathfrak{L}^2(0) := \lim_{\lambda\to 0}\mathfrak{L}^2(\lambda) = \begin{pmatrix}0 & 0\\ 0 & 2\end{pmatrix}.$$

Finally, factorizing

$$\begin{aligned}&\operatorname{diag}\left\{-2\lambda^2+\frac{3}{2}\lambda^3-2\lambda^3\sin\frac{1}{\lambda}+\frac{3}{2}\lambda^6\cos\frac{1}{\lambda},2\right\}\\ &= \operatorname{diag}\{\lambda^2,1\}\cdot\operatorname{diag}\left\{-2+\frac{3}{2}\lambda-2\lambda\sin\frac{1}{\lambda}+\frac{3}{2}\lambda^4\cos\frac{1}{\lambda},2\right\}\end{aligned}$$

and tracing all the previous identities, we conclude that $\operatorname{diag}\{\lambda^2, 1\}$ is the local Smith form of $\mathfrak{L}$ at 0, and that

$$\mathfrak{L}(\lambda) = \begin{pmatrix} 1 & \frac{1}{2}\left(\lambda + \lambda^4 \cos\frac{1}{\lambda}\right) \\ 0 & 1 \end{pmatrix} \begin{pmatrix} \lambda^2 & 0 \\ 0 & 1 \end{pmatrix}$$
$$\times \begin{pmatrix} -2 + \frac{3}{2}\lambda - 2\lambda \sin\frac{1}{\lambda} + \frac{3}{2}\lambda^4 \cos\frac{1}{\lambda} & 0 \\ -3\lambda^2 & 2 \end{pmatrix}$$

is the factorization of Definition 7.4.3. The value of the matrices above at $\lambda = 0$ is defined by continuity. Finally, note that, according to Theorem 7.6.3,

$$\chi[\mathfrak{L}; 0] = \kappa[\mathfrak{L}; 0] = 2 = \operatorname{ord}_0 \det \mathfrak{L}.$$

7.9.2 Example 2

Consider the map $\mathfrak{L} : \mathbb{R} \to \mathcal{L}(\mathbb{R}^2)$ defined by

$$\mathfrak{L}(\lambda) := \begin{pmatrix} -\lambda^2 \operatorname{sign}\lambda & 4\lambda(1+\lambda) - \lambda^3 \operatorname{sign}\lambda \\ (-1+\lambda)\lambda^2 \operatorname{sign}\lambda & \lambda[-4 + (-1+\lambda)\lambda^2 \operatorname{sign}\lambda] \end{pmatrix}, \qquad \lambda \in \mathbb{R},$$

where $\operatorname{sign} : \mathbb{R} \to \{-1, 0, 1\}$ is the map defined through

$$\operatorname{sign} x = \begin{cases} x/|x|, & \text{if } x \neq 0, \\ 0, & \text{if } x = 0. \end{cases}$$

As the map $\ell : \mathbb{R} \to \mathbb{R}$ defined by

$$\ell(\lambda) := \lambda^2 \operatorname{sign}\lambda, \qquad \lambda \in \mathbb{R},$$

satisfies

$$\frac{d\ell}{d\lambda}(\lambda) = 2|\lambda|, \qquad \lambda \in \mathbb{R},$$

it is apparent that $\ell \in \mathcal{C}^1(\mathbb{R}) \setminus \mathcal{C}^2(\mathbb{R})$. Thus, $\mathfrak{L} \in \mathcal{C}^1(\mathbb{R}, \mathcal{L}(\mathbb{R}^2))$, but $\mathfrak{L}$ is not of class $\mathcal{C}^2$, and, hence, the regularity $r = 1$ is optimal. Moreover, for $\lambda \sim 0$ with $\lambda \neq 0$ we have that

$$\mathfrak{L}(\lambda)^{-1} = \begin{pmatrix} \dfrac{4 + \lambda^2 \operatorname{sign}\lambda - \lambda^3 \operatorname{sign}\lambda}{4\lambda^2(\lambda^2 - 2)\operatorname{sign}\lambda} & \dfrac{4 + 4\lambda - \lambda^2 \operatorname{sign}\lambda}{4\lambda^2(\lambda^2 - 2)\operatorname{sign}\lambda} \\ \dfrac{-1+\lambda}{4\lambda(\lambda^2 - 2)} & \dfrac{1}{4\lambda(\lambda^2 - 2)} \end{pmatrix}$$

and, therefore, the family

$$\lambda \mapsto \lambda^2 \mathfrak{L}(\lambda)^{-1} = \begin{pmatrix} \dfrac{4+\lambda^2 \operatorname{sign}\lambda - \lambda^3 \operatorname{sign}\lambda}{4(\lambda^2-2)\operatorname{sign}\lambda} & \dfrac{4+4\lambda-\lambda^2 \operatorname{sign}\lambda}{4(\lambda^2-2)\operatorname{sign}\lambda} \\ \dfrac{\lambda(-1+\lambda)}{4(\lambda^2-2)} & \dfrac{\lambda}{4(\lambda^2-2)} \end{pmatrix}$$

is bounded in a neighborhood of 0, but it is not continuous at 0. Consequently, $\lambda_0 := 0$ is an algebraic eigenvalue of order $\nu = 2$ of $\mathfrak{L}$, but, since $\mathfrak{L}$ is not of class $\mathcal{C}^2$, the multiplicity of $\mathfrak{L}$ at λ_0 is not defined. Incidentally, note that Theorem 5.0.1 cannot be applied because $r = 1 < \nu = 2$, and, actually, its thesis fails to be true in the present example. Therefore, its regularity requirements seem to be optimal.

By Theorem 7.8.1, $\mathfrak{L}$ cannot admit a local Smith form at 0. If we try to calculate the local Smith form, then we will arrive at a step of the construction from which one cannot go on. To illustrate this, we are going to try to perform the construction described in Section 7.4.

For each $i, j \in \{1, 2\}$, let ℓ_{ij} denote the (i, j)th entry of $\mathfrak{L}$. Note that

$$\operatorname{ord}_0 \ell_{12} = \operatorname{ord}_0 \ell_{22} = 1,$$

whereas the order of ℓ_{11} and of ℓ_{21} at zero is not defined. Naturally, the fact that the latter orders are not defined gives us a hint of the fact that $\mathfrak{L}$ does not admit a local Smith form at 0. In the first step, we use ℓ_{22} as a pivot element. To make a zero in the $(2, 1)$th entry, we replace, for each $\lambda \sim 0$, the column

$$\begin{pmatrix} \ell_{11}(\lambda) \\ \ell_{21}(\lambda) \end{pmatrix} \quad \text{with} \quad \begin{pmatrix} \ell_{11}(\lambda) \\ \ell_{21}(\lambda) \end{pmatrix} - \frac{(-1+\lambda)\lambda \operatorname{sign}\lambda}{-4+(-1+\lambda)\lambda^2 \operatorname{sign}\lambda} \begin{pmatrix} \ell_{12}(\lambda) \\ \ell_{22}(\lambda) \end{pmatrix},$$

which amounts to performing the matrix multiplication

$$\begin{aligned} &\begin{pmatrix} \dfrac{4\lambda^2(2-\lambda^2)}{\lambda^2(\lambda-1) - 4\operatorname{sign}\lambda} & 4\lambda(1+\lambda) - \lambda^3 \operatorname{sign}\lambda \\ 0 & \lambda[-4+(-1+\lambda)\lambda^2 \operatorname{sign}\lambda] \end{pmatrix} \\ &= \mathfrak{L}(\lambda) \begin{pmatrix} 1 & 0 \\ -\dfrac{(-1+\lambda)\lambda \operatorname{sign}\lambda}{-4+(-1+\lambda)\lambda^2 \operatorname{sign}\lambda} & 1 \end{pmatrix} =: \mathfrak{L}^1(\lambda). \end{aligned}$$

Now, for each $i, j \in \{1, 2\}$, let ℓ^1_{ij} denote the (i, j)th entry of $\mathfrak{L}^1$. To make a zero in the $(1, 2)$th entry of $\mathfrak{L}^1$, we replace, for each $\lambda \in \mathbb{R} \setminus \{0\}$, the row

$$\begin{pmatrix} \ell^1_{11}(\lambda) & \ell^1_{12}(\lambda) \end{pmatrix}$$

with

$$\begin{pmatrix} \ell^1_{11}(\lambda) & \ell^1_{12}(\lambda) \end{pmatrix} - \frac{4(1+\lambda) - \lambda^2 \operatorname{sign}\lambda}{-4+(-1+\lambda)\lambda^2 \operatorname{sign}\lambda} \begin{pmatrix} \ell^1_{21}(\lambda) & \ell^1_{22}(\lambda) \end{pmatrix}.$$

This amounts to performing the matrix multiplication

$$\begin{pmatrix} \dfrac{4\lambda^2(2-\lambda^2)}{\lambda^2(\lambda-1)-4\operatorname{sign}\lambda} & 0 \\ 0 & \lambda[-4+(-1+\lambda)\lambda^2\operatorname{sign}\lambda] \end{pmatrix} = \begin{pmatrix} 1 & -\dfrac{4(1+\lambda)-\lambda^2\operatorname{sign}\lambda}{-4+(-1+\lambda)\lambda^2\operatorname{sign}\lambda} \\ 0 & 1 \end{pmatrix} \mathfrak{L}^1(\lambda).$$

Finally, we notice that if we factorize

$$\begin{aligned} &\operatorname{diag}\left\{\frac{4\lambda^2(2-\lambda^2)}{\lambda^2(\lambda-1)-4\operatorname{sign}\lambda}, \lambda[-4+(-1+\lambda)\lambda^2\operatorname{sign}\lambda]\right\} \\ &= \operatorname{diag}\left\{\lambda^2,\lambda\right\} \times \operatorname{diag}\left\{\frac{4(2-\lambda^2)}{\lambda^2(\lambda-1)-4\operatorname{sign}\lambda}, -4+(-1+\lambda)\lambda^2\operatorname{sign}\lambda\right\}, \end{aligned}$$

then the function

$$\lambda \mapsto \frac{4(2-\lambda^2)}{\lambda^2(\lambda-1)-4\operatorname{sign}\lambda}$$

is not continuous at 0. If, instead of that, we make the factorization

$$\begin{aligned} &\operatorname{diag}\left\{\frac{4\lambda^2(2-\lambda^2)}{\lambda^2(\lambda-1)-4\operatorname{sign}\lambda}, \lambda[-4+(-1+\lambda)\lambda^2\operatorname{sign}\lambda]\right\} \\ &= \operatorname{diag}\left\{\lambda,\lambda\right\} \times \operatorname{diag}\left\{\frac{4\lambda(2-\lambda^2)}{\lambda^2(\lambda-1)-4\operatorname{sign}\lambda}, -4+(-1+\lambda)\lambda^2\operatorname{sign}\lambda\right\}, \end{aligned}$$

then, the matrix

$$\operatorname{diag}\left\{\frac{4\lambda(2-\lambda^2)}{\lambda^2(\lambda-1)-4\operatorname{sign}\lambda}, -4+(-1+\lambda)\lambda^2\operatorname{sign}\lambda\right\}$$

is not invertible at $\lambda = 0$. Consequently, no Smith form is available for $\mathfrak{L}$.

7.10 Local equivalence of operator families

This section uses Theorem 7.8.3 to classify the set of $\mathcal{C}^r$ families, $r \in \mathbb{N} \cup \{\infty, \omega\}$, through the concept of equivalence of families.

Throughout the rest of this book, given $\mathbb{K} \in \{\mathbb{R}, \mathbb{C}\}$, a point $\lambda_0 \in \mathbb{K}$, two non-zero $\mathbb{K}$-Banach spaces U and V, and $r \in \mathbb{N} \cup \{\infty, \omega\}$, we denote by $\mathcal{S}^r_{\lambda_0}(U,V)$ the set of all families $\mathfrak{L}$ of class $\mathcal{C}^r$ on a neighborhood of λ_0 with values in $\mathcal{L}(U,V)$ such that $\mathfrak{L}(\lambda_0) \in \mathrm{Fred}_0(U,V)$; the neighborhood may depend on $\mathfrak{L}$. We also set

$$\mathcal{S}^r_{\lambda_0}(U) := \mathcal{S}^r_{\lambda_0}(U,U),$$

and denote by $\mathcal{S}^r_{\lambda_0}$ the class of families $\mathfrak{L}$ for which there exist two Banach spaces U, V such that $\mathfrak{L} \in \mathcal{S}^r_{\lambda_0}(U,V)$.

Clearly, $\mathfrak{L}\mathfrak{M} \in \mathcal{S}^r_{\lambda_0}(U,V)$ if W is a Banach space, $\mathfrak{L} \in \mathcal{S}^r_{\lambda_0}(W,V)$ and $\mathfrak{M} \in \mathcal{S}^r_{\lambda_0}(U,W)$. Moreover, $\mathfrak{L}^{-1} \in \mathcal{S}^r_{\lambda_0}(V,U)$ if $\mathfrak{L} \in \mathcal{S}^r_{\lambda_0}(U,V)$ and $\mathfrak{L}(\lambda_0) \in \mathrm{Iso}(U,V)$.

The following definition introduces the concept of equivalence.

Definition 7.10.1 (Equivalence). Let U_1, U_2, V_1, V_2 be four Banach spaces, $\lambda_0 \in \mathbb{K}$ and $r, s \in \mathbb{N} \cup \{\infty, \omega\}$. The families $\mathfrak{L} \in \mathcal{S}^r_{\lambda_0}(U_1, V_1)$ and $\mathfrak{M} \in \mathcal{S}^r_{\lambda_0}(U_2, V_2)$ are said to be $\mathcal{C}^s$-equivalent at λ_0 when there exist an open neighborhood $\Omega \subset \mathbb{K}$ of λ_0 and two families $\mathfrak{E} \in \mathcal{S}^s_{\lambda_0}(V_2, V_1)$ and $\mathfrak{F} \in \mathcal{S}^s_{\lambda_0}(U_1, U_2)$ such that

$$\mathfrak{L}(\lambda) = \mathfrak{E}(\lambda)\, \mathfrak{M}(\lambda)\, \mathfrak{F}(\lambda), \qquad \lambda \in \Omega. \tag{7.43}$$

In such a case, we will write that $\mathfrak{L} \overset{s}{\equiv} \mathfrak{M}$.

Let $\mathfrak{J}$ be a set of isomorphic Banach spaces. It is easy to verify that the $\mathcal{C}^s$-equivalence indeed establishes a relation of equivalence in

$$\bigcup_{U,V \in \mathfrak{J}} \mathcal{S}^r_{\lambda_0}(U,V).$$

The following result shows the uniqueness of the Smith form at λ_0 when it exists, that is to say, when $\mathfrak{L}$ is in $\mathcal{S}^r_{\lambda_0}$ and $\lambda_0 \in \mathrm{Alg}_\nu(\mathfrak{L})$ for some $0 \leq \nu < r+1$.

Proposition 7.10.2. *Let U, V be two Banach spaces such that*

$$U = X \oplus Y, \qquad V = W \oplus Z, \tag{7.44}$$

for some closed subspaces $X, Y \subset U$, and $W, Z \subset V$, with

$$m = \dim X = \dim W < \infty.$$

For a given $r \in \mathbb{N} \cup \{\infty\}$ and

$$k_1, \dots, k_m \in \mathbb{N} \cap [0, r], \qquad \mathfrak{J} \in \mathcal{S}^r_{\lambda_0}(Y, Z)$$

such that $\mathfrak{J}_0 \in \mathrm{Iso}(Y,Z)$, let $\mathfrak{M} \in \mathcal{S}^r_{\lambda_0}(U,V)$ denote the operator family defined, for each $\lambda \sim \lambda_0$, by

$$\mathfrak{M}(\lambda) := \mathrm{diag}\left\{(\lambda - \lambda_0)^{k_1}, \dots, (\lambda - \lambda_0)^{k_m}\right\} \oplus \mathfrak{J}(\lambda)$$

with respect to certain bases of X and W, and the decompositions (7.44).

If a family $\mathfrak{L}$ in $\mathcal{S}^r_{\lambda_0}$ satisfies $\mathfrak{L} \overset{0}{\equiv} \mathfrak{M}$, then $\lambda_0 \in \mathrm{Alg}_\nu(\mathfrak{L})$ for some $0 \leq \nu < r+1$, the number $n := \dim N[\mathfrak{L}_0]$ satisfies $n \leq m$, and there exists a permutation τ of $\{1, \dots, m\}$ such that

$$k_{\tau(n+1)} = \cdots = k_{\tau(m)} = 0, \qquad k_{\tau(j)} = \kappa_j[\mathfrak{L}; \lambda_0], \quad 1 \leq j \leq n. \tag{7.45}$$

Proof. Obviously, there exist an open neighborhood Ω of λ_0 and a constant $C > 0$ such that

$$\|\mathfrak{M}(\lambda)^{-1}\| \leq \frac{C}{|\lambda - \lambda_0|^{\max\{k_1,\ldots,k_n\}}}, \qquad \lambda \in \Omega \setminus \{\lambda_0\}.$$

Thus, we find from (7.43) that there exist an open neighborhood $\tilde{\Omega} \subset \Omega$ of λ_0 and a constant $\tilde{C}$ such that

$$\|\mathfrak{L}(\lambda)^{-1}\| \leq \frac{\tilde{C}}{|\lambda - \lambda_0|^{\max\{k_1,\ldots,k_n\}}}, \qquad \lambda \in \tilde{\Omega} \setminus \{\lambda_0\}.$$

Moreover,

$$\max\{k_1, \ldots, k_n\} < r + 1.$$

Thus, $\lambda_0 \in \mathrm{Alg}_\nu(\mathfrak{L})$ for some integer $0 \leq \nu \leq r$. According to Theorem 7.6.3, $\kappa[\mathfrak{L}; \lambda_0]$ is well defined and finite. Moreover, thanks to Theorem 7.3.1, we find from (7.43) that

$$\kappa[\mathfrak{L}; \lambda_0] = \kappa[\mathfrak{M}; \lambda_0]$$

and, actually, when $\mathfrak{L}_0$ is not an isomorphism,

$$\kappa_j[\mathfrak{L}; \lambda_0] = \kappa_j[\mathfrak{M}; \lambda_0], \qquad 1 \leq j \leq n.$$

Now we set

$$\mathfrak{D}(\lambda) := \mathrm{diag}\left\{(\lambda - \lambda_0)^{k_1}, \ldots, (\lambda - \lambda_0)^{k_m}\right\}, \qquad \lambda \in \mathbb{K},$$

regarded as the restriction of $\mathfrak{M}(\lambda)$ from X to W, for $\lambda \sim \lambda_0$. As $\mathfrak{J}_0$ is an isomorphism, for any $s \geq 0$, every Jordan chain of $\mathfrak{M}$ at λ_0 of length $s+1$ must be of the form

$$\left(\begin{pmatrix} u_0 \\ 0 \end{pmatrix}, \ldots, \begin{pmatrix} u_s \\ 0 \end{pmatrix}\right),$$

for some Jordan chain $(u_0, \ldots, u_s)$ of $\mathfrak{D}$ at λ_0. Consequently, we actually have

$$\kappa[\mathfrak{L}; \lambda_0] = \kappa[\mathfrak{D}; \lambda_0]$$

and, in addition,

$$\kappa_j[\mathfrak{L}; \lambda_0] = \kappa_j[\mathfrak{D}; \lambda_0], \qquad 1 \leq j \leq n,$$

when $n \geq 1$. Suppose $n = 0$. Then,

$$\kappa[\mathfrak{D}, \lambda_0] = \kappa[\mathfrak{L}; \lambda_0] = 0$$

and, necessarily $k_j = 0$ for all $1 \leq j \leq m$, which concludes the proof.

Suppose $n \geq 1$, and let τ be a permutation of $\{1, \ldots, m\}$ such that

$$k_{\tau(1)} \geq \cdots \geq k_{\tau(m)}.$$

As $\mathfrak{L} \overset{0}{\equiv} \mathfrak{M}$ at λ_0, and $\mathfrak{J}_0$ is an isomorphism, then $n = \dim N[\mathfrak{D}_0]$. Therefore, owing to the result already established in the Example of Section 7.2, we conclude $n \leq m$ and (7.45), which ends the proof. $\square$

The following proposition will be used in the proof of the main result of this section. It is a generalization of a result already established in the proof of Theorem 7.8.1.

Proposition 7.10.3. *Let $r \in \mathbb{N} \cup \{\infty, \omega\}$ and $\mathfrak{L}, \mathfrak{M} \in \mathcal{S}^r_{\lambda_0}(U, V)$ be such that $\lambda_0 \in \mathrm{Alg}_\nu(\mathfrak{L})$, for some integer $0 \le \nu \le r$, and $\mathfrak{M}_n = \mathfrak{L}_n$ for all $0 \le n \le \nu$. Then there exists $\mathfrak{E} \in \mathcal{S}^{r-\nu}_{\lambda_0}(V)$ such that $\mathfrak{E}_0 = I_V$ and $\mathfrak{L} = \mathfrak{E}\mathfrak{M}$ locally at λ_0. In particular, $\mathfrak{L} \overset{r-\nu}{\equiv} \mathfrak{M}$ at λ_0, and $\lambda_0 \in \mathrm{Alg}_\nu(\mathfrak{M})$.*

Proof. Let Ω be an open neighborhood of λ_0 contained in the domains of definition of $\mathfrak{L}$ and $\mathfrak{M}$, and such that $\mathfrak{L}(\lambda) \in \mathrm{Iso}(U, V)$ for all $\lambda \in \Omega \backslash \{\lambda_0\}$. Let $\mathfrak{G} : \Omega \to \mathcal{L}(V)$ be the family defined by

$$\mathfrak{G}(\lambda) := \mathfrak{M}(\lambda)\mathfrak{L}(\lambda)^{-1}, \qquad \lambda \in \Omega \setminus \{\lambda_0\},$$

and

$$\mathfrak{G}(\lambda_0) := \lim_{\lambda \to \lambda_0} \mathfrak{G}(\lambda) = \lim_{\lambda \to \lambda_0} [\mathfrak{L}(\lambda) + o((\lambda - \lambda_0)^\nu)]\, \mathfrak{L}(\lambda)^{-1} = I_V.$$

In fact,

$$\mathfrak{G}(\lambda) = I_V + [\mathfrak{M}(\lambda) - \mathfrak{L}(\lambda)]\, \mathfrak{L}(\lambda)^{-1}, \qquad \lambda \in \Omega \setminus \{\lambda_0\}.$$

As $\mathfrak{M}_n = \mathfrak{L}_n$ for every $0 \le n \le \nu$, the function

$$\lambda \mapsto (\lambda - \lambda_0)^{-\nu} [\mathfrak{M}(\lambda) - \mathfrak{L}(\lambda)]$$

is of class $\mathcal{C}^{r-\nu}$ in an open neighborhood of λ_0. As $\lambda_0 \in \mathrm{Alg}_\nu(\mathfrak{L})$, by Theorem 5.0.1, the map

$$\lambda \mapsto (\lambda - \lambda_0)^\nu \mathfrak{L}(\lambda)^{-1}$$

is also of class $\mathcal{C}^{r-\nu}$. Consequently, $\mathfrak{G}$ is of class $\mathcal{C}^{r-\nu}$ as well. Finally, set

$$\mathfrak{E}(\lambda) := \mathfrak{G}(\lambda)^{-1}, \qquad \lambda \sim \lambda_0.$$

Then, $\mathfrak{E} \in \mathcal{S}^{r-\nu}_{\lambda_0}(V)$ and $\mathfrak{L} = \mathfrak{E}\mathfrak{M}$, which concludes the proof. □

The main result of this section provides a list of necessary and sufficient conditions for the local equivalence of two given families. It reads as follows.

Theorem 7.10.4. *Let $r \in \mathbb{N}^* \cup \{\infty, \omega\}$, an integer $1 \le \nu \le r$, four $\mathbb{K}$-Banach spaces U_1, U_2, V_1, V_2, families $\mathfrak{L} \in \mathcal{S}^r_{\lambda_0}(U_1, V_1)$ and $\mathfrak{M} \in \mathcal{S}^r_{\lambda_0}(U_2, V_2)$ be such that $\lambda_0 \in \mathrm{Alg}_{\nu_1}(\mathfrak{L}) \cap \mathrm{Alg}_{\nu_2}(\mathfrak{M})$ for some $\nu_1, \nu_2 \in \mathbb{N} \cap (0, \nu]$. Then the following conditions are equivalent:*

i) *$\mathfrak{L}$ is $\mathcal{C}^{r-\nu}$-equivalent to $\mathfrak{M}$ at λ_0.*
ii) *$\mathfrak{L}$ is $\mathcal{C}^0$-equivalent to $\mathfrak{M}$ at λ_0.*
iii) *$\kappa_j[\mathfrak{L}; \lambda_0] = \kappa_j[\mathfrak{M}; \lambda_0]$ for all $1 \le j \le n := \dim N[\mathfrak{L}_0] = \dim N[\mathfrak{M}_0]$.*
iv) *$\dim N[\mathfrak{L}_{[s]}] = \dim N[\mathfrak{M}_{[s]}]$ for all $s \in \mathbb{N} \cap [0, r]$.*

v) $\dim N[\mathfrak{L}_{[s]}] = \dim N[\mathfrak{M}_{[s]}]$ *for all* $0 \leq s \leq \nu - 1$.

vi) *For each* $0 \leq j \leq \nu - 1$ *there exist* $E_j \in \mathcal{L}(U_1, U_2)$ *and* $F_j \in \mathcal{L}(V_2, V_1)$ *such that* E_0 *and* F_0 *are isomorphisms, and*

$$\mathfrak{L}_{[\nu-1]} = \operatorname{trng}\{E_0, \ldots, E_{\nu-1}\} \cdot \mathfrak{M}_{[\nu-1]} \cdot \operatorname{trng}\{F_0, \ldots, F_{\nu-1}\}.$$

vii) *There exist* $G_0, \ldots, G_{\nu-1} \in \mathcal{L}(V_1, V_2)$ *such that* G_0 *is an isomorphism and*

$$N[\mathfrak{L}_{[\nu-1]}] = \operatorname{trng}\{G_0, \ldots, G_{\nu-1}\} N[\mathfrak{M}_{[\nu-1]}].$$

viii) $\chi_j[\mathfrak{L}; \lambda_0] = \chi_j[\mathfrak{M}; \lambda_0]$ *for all integers* $0 \leq j \leq \nu - 1$, *where the* χ_j*'s stand for the partial* χ*-multiplicities introduced in* (5.32).

ix) $\chi_j[\mathfrak{L}; \lambda_0] = \chi_j[\mathfrak{M}; \lambda_0]$ *for all integers* $0 \leq j \leq r$.

x) *There exist polynomials* $\Phi \in \mathbb{P}[\mathcal{L}(U_1)]$ *and* $\Psi \in \mathbb{P}[\mathcal{L}(U_2)]$ *such that* $\Phi_0 = I_{U_1}$ *and* $\Psi_0 = I_{U_2}$, *for which* λ_0 *is a transversal eigenvalue of the families* $\mathfrak{L}^\Phi := \mathfrak{L}\Phi$ *and* $\mathfrak{M}^\Psi := \mathfrak{M}\Psi$, *and, for each* $0 \leq j \leq \nu - 1$,

$$\dim \mathfrak{L}_j^\Phi \Big(\bigcap_{i=0}^{j-1} N[\mathfrak{L}_i^\Phi] \Big) = \dim \mathfrak{M}_j^\Psi \Big(\bigcap_{i=0}^{j-1} N[\mathfrak{M}_i^\Psi] \Big). \tag{7.46}$$

Therefore, $\nu_1 = \nu_2$.

xi) *For all* Φ, Ψ *in* $\mathcal{S}_{\lambda_0}^\nu$ *such that* Φ_0 *and* Ψ_0 *are isomorphisms and* λ_0 *is a transversal eigenvalue of the families* $\mathfrak{L}^\Phi := \mathfrak{L}\Phi$ *and* $\mathfrak{M}^\Psi := \mathfrak{M}\Psi$, *equality* (7.46) *occurs for every* $j \in \mathbb{N} \cap [0, \nu - 1]$. *Therefore,* $\nu_1 = \nu_2$.

Proof. Obviously, (i) implies (ii), (iv) implies (v), (vi) implies (vii), (viii) implies (ix), and (xi) implies (x).

By Theorem 7.3.1, (ii) implies (iii). By Theorem 7.6.2, (v) implies (iv). The equivalence between (v) and (x) follows from Theorem 4.3.2 and Propositions 7.6.1 and 4.2.2. The equivalence between (viii) and (x) follows from Theorems 4.3.2 and 5.4.1. By Theorem 7.5.3, (x) implies (iii). Due to (5.32), by Theorem 5.2.2, (ix) implies (viii). The equivalence between (x) and (xi) follows from Proposition 4.2.2 and Theorem 4.3.2.

It remains to prove, for example, that (iii) implies (i), that (vii) implies (v), and that (ii) implies (vi).

Suppose (iii). By Theorem 7.8.3, there is a topological direct sum decomposition $U_1 = X_0 \oplus X_1$, with $n = \dim X_0$, such that

$$\mathfrak{L} \overset{r-\nu}{\equiv} \mathfrak{S}_{\mathfrak{L}} \oplus I_{X_1},$$

where

$$\mathfrak{S}_{\mathfrak{L}}(\lambda) = \operatorname{diag}\left\{ (\lambda - \lambda_0)^{\kappa_1[\mathfrak{L};\lambda_0]}, \ldots, (\lambda - \lambda_0)^{\kappa_n[\mathfrak{L};\lambda_0]} \right\}, \quad \lambda \in \mathbb{K}.$$

Similarly, there is a topological direct sum decomposition $U_2 = Y_0 \oplus Y_1$, with $n = \dim Y_0$, such that

$$\mathfrak{M} \overset{r-\nu}{\equiv} \mathfrak{S}_{\mathfrak{M}} \oplus I_{Y_1},$$

where

$$\mathfrak{S}_{\mathfrak{M}}(\lambda) = \operatorname{diag}\left\{(\lambda-\lambda_0)^{\kappa_1[\mathfrak{M};\lambda_0]}, \dots, (\lambda-\lambda_0)^{\kappa_n[\mathfrak{M};\lambda_0]}\right\}, \quad \lambda \in \mathbb{K}.$$

By (iii), $\mathfrak{S}_{\mathfrak{L}} = \mathfrak{S}_{\mathfrak{M}}$. Moreover, X_1 and Y_1 are isomorphic, because they have the same finite codimension within two isomorphic spaces. Therefore,

$$\mathfrak{L} \overset{r-\nu}{\equiv} \mathfrak{M},$$

which ends the proof of (iii) implies (i).

Suppose (vii). Let $0 \leq s \leq \nu - 1$ and $x \in N[\mathfrak{L}_{[s]}]$. Then,

$$\operatorname{col}[\underbrace{0,\dots,0}_{\nu-s-1}, x] \in N[\mathfrak{L}_{[\nu-1]}]$$

and, according to (vii), there exists $z \in N[\mathfrak{M}_{[\nu-1]}]$ such that

$$\operatorname{col}[\underbrace{0,\dots,0}_{\nu-s-1}, x] = \operatorname{trng}\{G_0,\dots,G_{\nu-1}\}z.$$

Since G_0 is an isomorphism, necessarily $z = \operatorname{col}[0, y]$ for some $y \in N[\mathfrak{M}_{[s]}]$, and, therefore,

$$x = \operatorname{trng}\{G_0,\dots,G_s\}y.$$

As G_0 is an isomorphism, the previous reasoning can be reversed and, consequently,

$$N[\mathfrak{L}_{[s]}] = \operatorname{trng}\{G_0,\dots,G_s\}N[\mathfrak{M}_{[s]}],$$

which implies (v), and shows that (vii) implies (v).

Finally, we will show that (ii) implies (vi). Suppose (ii) and consider the Taylor polynomials $\hat{\mathfrak{L}} \in \mathbb{P}[\mathcal{L}(U_1,V_1)]$ and $\hat{\mathfrak{M}} \in \mathbb{P}[\mathcal{L}(U_2,V_2)]$ defined by

$$\hat{\mathfrak{L}}(\lambda) := \sum_{j=0}^{\nu} \mathfrak{L}_j(\lambda-\lambda_0)^j, \quad \hat{\mathfrak{M}}(\lambda) := \sum_{j=0}^{\nu} \mathfrak{M}_j(\lambda-\lambda_0)^j, \qquad \lambda \in \mathbb{K}.$$

By Proposition 7.10.3, $\mathfrak{L}$ and $\hat{\mathfrak{L}}$ are $\mathcal{C}^0$-equivalent at λ_0, as well as $\mathfrak{M}$ and $\hat{\mathfrak{M}}$. As we are assuming that $\mathfrak{L}$ and $\mathfrak{M}$ are $\mathcal{C}^0$-equivalent at λ_0, the analytic families $\hat{\mathfrak{L}}$ and $\hat{\mathfrak{M}}$ are $\mathcal{C}^0$-equivalent at λ_0. As we already know that (ii) and (i) are equivalent, $\hat{\mathfrak{L}}$ and $\hat{\mathfrak{M}}$ are actually $\mathcal{C}^\omega$-equivalent at λ_0. Therefore, there exist families $\mathfrak{E}, \mathfrak{F}$ in $\mathcal{S}^\omega_{\lambda_0}$ such that $\mathfrak{E}(\lambda_0)$ and $\mathfrak{F}(\lambda_0)$ are isomorphisms and $\hat{\mathfrak{L}} = \mathfrak{E}\hat{\mathfrak{M}}\mathfrak{F}$. By the Leibniz rule,

$$\mathfrak{L}_{[\nu-1]} = \mathfrak{E}_{[\nu-1]} \cdot \mathfrak{M}_{[\nu-1]} \cdot \mathfrak{F}_{[\nu-1]},$$

which implies (vi). This concludes the proof. □

7.11 Exercises

1. Let $s \in \mathbb{N}$, two $\mathbb{K}$-Banach spaces U, V, and, for each $j \in \{0, \dots, s\}$, an operator $T_j \in \mathcal{L}(U, V)$ be with $T_0 \in \mathrm{Fred}_0(U, V)$. It is requested to accomplish the following:

- Prove that, for each $\varepsilon \in \mathbb{K} \setminus \{0\}$,

$$\mathrm{trng}\,\{T_0, \dots, T_s\} = D(\varepsilon^{-1}, V; s)\,\mathrm{trng}\,\{T_0, \varepsilon T_1, \dots, \varepsilon^s T_s\}\,D(\varepsilon, U; s),$$

  where, given a Banach space W and $\varepsilon \in \mathbb{K}$, the operator $D(\varepsilon, W; s)$ stands for

$$D(\varepsilon, W; s) := \mathrm{diag}\,\{I_W, \varepsilon I_W, \dots, \varepsilon^s I_W\}.$$

- Show that

$$\mathrm{trng}\{T_0, \varepsilon T_1, \dots, \varepsilon^s T_s\} = \mathrm{diag}\{T_0, \dots, T_0\} + \mathrm{trng}\{0, \varepsilon T_1, \dots, \varepsilon^s T_s\}.$$

- Conclude that $\mathrm{trng}\{T_0, \dots, T_s\} \in \mathrm{Fred}_0(U^{s+1}, V^{s+1})$.

2. Let $\mathfrak{D} \in \mathbb{P}[\mathcal{L}(\mathbb{K}^m)]$ be the family defined by

$$\mathfrak{D}(\lambda) = \mathrm{diag}\,\{(\lambda - \lambda_0)^{k_1}, \dots, (\lambda - \lambda_0)^{k_n}, \underbrace{1, \dots, 1}_{m-n}\}, \qquad \lambda \in \mathbb{K},$$

for some integers $m \geq n \geq 1$ and $k_1 \geq \cdots \geq k_n \geq 1$. Prove, using the definition, that λ_0 is a k_1-transversal eigenvalue of $\mathfrak{D}$.

3. This exercise asks the following:

- Give an example of a family $\mathfrak{L} \in \mathcal{C}^\omega(\mathbb{K}, \mathcal{L}(\mathbb{K}^2))$ such that the matrix $\mathfrak{L}(\lambda)$ is triangular for all $\lambda \in \mathbb{K}$, and $\lambda_0 \in \mathbb{K}$ is an algebraic eigenvalue of $\mathfrak{L}$, but not a transversal eigenvalue of $\mathfrak{L}$.
- Give an example of a family $\mathfrak{L} \in \mathcal{C}^\omega(\mathbb{K}, \mathcal{L}(\mathbb{K}^2))$ such that $\lambda_0 \in \mathbb{K}$ is a transversal eigenvalue of $\mathfrak{L}$, but the matrix $\mathfrak{L}(\lambda_0)$ is not triangular.

4. Given $r \in \mathbb{N} \cup \{\infty\}$ and $\mathfrak{L} \in \mathcal{C}^r(\Omega, \mathcal{L}(U, V))$ with $\mathfrak{L}_0 \in \mathrm{Fred}_0(U, V)$, it is requested to accomplish the following:

- Prove the equivalence of the following assertions without invoking the multiplicity χ constructed in Chapter 4:
  i) $\kappa[\mathfrak{L}; \lambda_0] < \infty$.
  ii) $\lambda_0 \in \mathrm{Alg}_\nu(\mathfrak{L})$ for some integer $0 \leq \nu \leq r$.
  iii) The length of any Jordan chain of $\mathfrak{L}$ is bounded above by $\nu < r + 1$.
- Give an alternative proof of Theorem 7.8.1 without invoking the multiplicity χ.

5. This exercise proposes the construction of a local canonical form at any non-algebraic eigenvalue in the finite-dimensional setting (for simplicity). More precisely, given an integer $m \geq 1$, the space $U = \mathbb{K}^m$, the point $\lambda_0 \in \Omega$, and a family $\mathfrak{L} \in \mathcal{C}^\infty(\Omega, \mathcal{L}(U))$ with

$$\lambda_0 \in \mathrm{Eig}(\mathfrak{L}) \setminus \mathrm{Alg}(\mathfrak{L}),$$

it is requested to adapt the proof of Theorem 7.4.1 in order to show that there exist a decomposition

$$U = U_0 \oplus U_1 \tag{7.47}$$

with $1 \leq p := \dim U_0 \leq m$, and four operator families

$$\mathfrak{E}, \mathfrak{F} \in \mathcal{C}^\infty(\Omega, \mathcal{L}(U)), \quad \mathfrak{A} \in \mathcal{C}^\infty(\Omega, \mathcal{L}(U_0)), \quad \mathfrak{B} \in \mathcal{C}^\infty(\Omega, \mathcal{L}(U_1)),$$

such that $\mathfrak{E}(\lambda_0)$ and $\mathfrak{F}(\lambda_0)$ are isomorphisms, $\operatorname{ord}_{\lambda_0} \mathfrak{A} = \infty$,

$$\mathfrak{B}(\lambda) = \operatorname{diag}\{(\lambda - \lambda_0)^{k_{p+1}}, \dots, (\lambda - \lambda_0)^{k_m}\}, \qquad \lambda \sim \lambda_0,$$

for some integers $k_{p+1} \geq \cdots \geq k_m \geq 0$, if $p < m$, while $\mathfrak{B} = 0$ if $p = m$, and, with respect to (7.47), $\mathfrak{L}$ admits the factorization

$$\mathfrak{L} = \mathfrak{E}(\mathfrak{A} \oplus \mathfrak{B})\mathfrak{F}.$$

6. Let $\operatorname{Sub}(U)$ denote the set of non-zero linear subspaces of the Banach space U having finite dimension or finite codimension. Directly from the following definition of $\mathfrak{m}[\cdot; \lambda_0]$, prove that the function

$$\mathfrak{m}[\,\cdot\,; \lambda_0] \;:\; \bigcup_{V \in \operatorname{Sub}(U)} \mathcal{S}^\infty_{\lambda_0}(V) \to [0, \infty]$$

defined by

$$\mathfrak{m}[\mathfrak{L}; \lambda_0] := \sup_{s \geq 0} \dim N[\operatorname{trng}\{\mathfrak{L}_0, \dots, \mathfrak{L}_s\}], \qquad \mathfrak{L} \in \bigcup_{V \in \operatorname{Sub}(U)} \mathcal{S}^\infty_{\lambda_0}(V),$$

satisfies the following properties:

M1. For every $V \in \operatorname{Sub}(U)$ and $\mathfrak{E}, \mathfrak{L}, \mathfrak{F} \in \mathcal{S}^\infty_{\lambda_0}(V)$,

$$\mathfrak{m}[\mathfrak{E}\mathfrak{L}\mathfrak{F}; \lambda_0] \geq \mathfrak{m}[\mathfrak{L}; \lambda_0].$$

M2. If $V \in \operatorname{Sub}(U)$ has dimension 1, then

$$\mathfrak{m}[(\cdot - \lambda_0)^n I_V; \lambda_0] = n, \qquad n \in \mathbb{N}^*.$$

M3. There exists $\mathfrak{L} \in \mathcal{S}^\infty_{\lambda_0}(U)$ such that $\mathfrak{m}[\mathfrak{L}; \lambda_0] = 0$.

M4. If $V \in \operatorname{Sub}(U)$ is decomposed as $V = V_1 \oplus V_2$ for some $V_1, V_2 \in \operatorname{Sub}(V)$, and for each $i \in \{1, 2\}$ we have $\mathfrak{L}_i \in \mathcal{S}^\infty_{\lambda_0}(V_i)$, then

$$\mathfrak{m}[\mathfrak{L}_1 \oplus \mathfrak{L}_2; \lambda_0] = \mathfrak{m}[\mathfrak{L}_1; \lambda_0] + \mathfrak{m}[\mathfrak{L}_2; \lambda_0].$$

7. Let U be a Banach space and suppose that a mapping

$$m[\cdot; \lambda_0] : \bigcup_{V \in \operatorname{Sub}(U)} \mathcal{S}^\infty_{\lambda_0}(V) \to [0, \infty]$$

exists satisfying properties M1–M4 of Exercise 6. Use the result of Exercise 5 to prove that for every $V \in \operatorname{Sub}(U)$, and every $\mathfrak{L} \in \mathcal{S}^\infty_{\lambda_0}(V)$,

$$m[\mathfrak{L}; \lambda_0] = \chi[\mathfrak{L}; \lambda_0].$$

This is another uniqueness result, in the spirit of those of Chapter 6.

8. Use Exercise 7 to prove that, for each Banach space U, there exists a unique function

$$\mathfrak{m}_U[\,\cdot\,;\lambda_0] \;:\; \mathcal{S}^{\infty}_{\lambda_0}(U) \to [0,\infty]$$

satisfying the following properties:

M1. For each $\mathfrak{L}, \mathfrak{M} \in \mathcal{S}^{\infty}_{\lambda_0}(U)$,

$$\mathfrak{m}_U[\mathfrak{L}\mathfrak{M};\lambda_0] = \mathfrak{m}_U[\mathfrak{L};\lambda_0] + \mathfrak{m}_U[\mathfrak{M};\lambda_0].$$

M2. $\mathfrak{m}_U[(\cdot - \lambda_0)I_U;\lambda_0] = 1$ if $\dim U = 1$.

M3. There exists $\mathfrak{L} \in \mathcal{S}^{\infty}_{\lambda_0}(U)$ with $\mathfrak{m}_U[\mathfrak{L};\lambda_0] < \infty$ if $\dim U = \infty$.

M4. If $U = U_1 \oplus U_2$ with $U_i \in \mathrm{Sub}(U)$ and $\mathfrak{L}_i \in \mathcal{S}^{\infty}_{\lambda_0}(U_i)$ for each $i \in \{1,2\}$, then

$$\mathfrak{m}_U[\mathfrak{L}_1 \oplus \mathfrak{L}_2;\lambda_0] = \mathfrak{m}_{U_1}[\mathfrak{L}_1;\lambda_0] + \mathfrak{m}_{U_2}[\mathfrak{L}_2;\lambda_0].$$

9. Show that Theorem 6.0.1 can be obtained as an application of the uniqueness result of Exercise 8.

10. The main goal of this exercise is to show that none of the properties M1–M4 of Exercise 8 can be deduced from the remaining ones. More precisely, consider a non-zero Banach space U and the four maps

$$\mathfrak{m}^i_U[\,\cdot\,;\lambda_0] \;:\; \mathcal{S}^{\infty}_{\lambda_0}(U) \to \mathbb{N} \cup \{\infty\}, \qquad i \in \{1,2,3,4\},$$

defined, for each $\mathfrak{L} \in \mathcal{S}^{\infty}_{\lambda_0}(U)$, by

$$\mathfrak{m}^1_U[\mathfrak{L};\lambda_0] := \dim N[\mathfrak{L}(\lambda_0)], \qquad \mathfrak{m}^2_U[\mathfrak{L};\lambda_0] := 2\chi[\mathfrak{L};\lambda_0],$$

$$\mathfrak{m}^3_U[\mathfrak{L};\lambda_0] := \begin{cases} \chi[\mathfrak{L};\lambda_0] & \text{if } \dim U < \infty, \\ \infty & \text{if } \dim U = \infty, \end{cases}$$

$$\mathfrak{m}^4_U[\mathfrak{L};\lambda_0] := \begin{cases} \chi[\mathfrak{L};\lambda_0] & \text{if } \dim U = 1, \\ 2\chi[\mathfrak{L};\lambda_0] & \text{if } \dim U > 1, \end{cases}$$

and prove that, for each $i \in \{1,2,3,4\}$, the map $\mathfrak{m}^i_U$ satisfies property Mj of Exercise 8 for all $j \in \{1,2,3,4\} \setminus \{i\}$.

11. Let $J \in \mathcal{M}_n(\mathbb{K})$ be the matrix $J_{j,n}$ of (1.26). Calculate the local Smith form at λ_j of the family $\mathfrak{L}^J$ defined by

$$\mathfrak{L}^J(\lambda) := \lambda I_n - J, \qquad \lambda \in \mathbb{K}.$$

Use this calculation and the Jordan canonical form to calculate the local Smith form of the family $\mathfrak{L}^A$ defined by

$$\mathfrak{L}^A(\lambda) := \lambda I_n - A, \qquad \lambda \in \mathbb{K},$$

at every $\lambda_j \in \sigma(A)$, for each of the matrices A of Exercise 1 of Chapter 1.

12. Let U and V be two Banach spaces. Use the appropriate results of Section 7.6 to prove, directly from the definition, that the function

$$\begin{array}{rcl} \mathcal{S}^{\infty}_{\lambda_0}(U,V) & \to & \mathbb{N}\cup\{\infty\} \\ \mathfrak{L} & \mapsto & \sup_{s\geq 0} \dim N[\operatorname{trng}\{\mathfrak{L}_0,\dots,\mathfrak{L}_s\}] \end{array}$$

satisfies the product formula of Theorem 5.6.1.

13. Prove Theorem 7.3.1 without invoking Theorem 4.3.2. Indication: Adapt the proof of Theorem 7.3.1, and use Proposition 7.7.1 to choose a basis of $N[\mathfrak{L}_{[s]}]$ that simplifies the calculations.

14. This exercise proposes a dual construction for the Jordan chains described in this chapter; namely, the left Jordan chains. For the sake of simplicity, we restrict ourselves to considering the finite-dimensional setting. Fix an integer $m \geq 1$, and a family $\mathfrak{L} \in \mathcal{C}^r(\Omega, \mathcal{L}(\mathbb{K}^m))$ for some $r \in \mathbb{N}\cup\{\infty,\omega\}$. Let $0 \leq s \leq r$ be an integer. Every ordered set of row vectors

$$(u_0,\dots,u_s) \in (\mathbb{K}^m)^{s+1}$$

with $u_0 \neq 0$ and

$$\sum_{i=0}^{j} u_{j-i}\mathfrak{L}_i = 0, \qquad 0 \leq j \leq s,$$

is said to be a left Jordan chain of $\mathfrak{L}$ at λ_0 starting at u_0. You are asked to accomplish the following tasks:

- Define left rank in a natural way.
- Introduce the concept of canonical set of left Jordan chains, and give some algorithm to construct it.
- Introduce the associated concepts of left κ-multiplicity and left partial κ-multiplicities.
- Given a family $\mathfrak{L} \in \mathcal{C}^r(\Omega, \mathcal{L}(\mathbb{K}^m))$ with the length of all its left Jordan chains at λ_0 bounded above by some integer $k \leq r$, and $\mathfrak{E}, \mathfrak{F} \in \mathcal{C}^r(\Omega, \mathcal{L}(\mathbb{K}^m))$ such that $\mathfrak{E}(\lambda_0)$, $\mathfrak{F}(\lambda_0)$ are isomorphisms, prove that the left partial κ-multiplicities of $\mathfrak{L}$ at λ_0 equal those of $\mathfrak{E}\mathfrak{L}\mathfrak{F}$ if $\dim N[\mathfrak{L}_0] \geq 1$.
- State and prove a *left* counterpart of Theorem 7.3.1. Indication: Adapt the proof of Theorem 7.3.1 suggested in Exercise 13.
- State and prove a *left* counterpart of Theorem 7.4.1.
- Prove that the left partial κ-multiplicities actually equal the partial κ-multiplicities.

15. This exercise shows that Theorem 7.8.3 might fail when the condition $\mathfrak{L}_0 \in \operatorname{Fred}_0(U,V)$ is removed from its statement. Consider $p \in [1,\infty]$, and the Banach space ℓ_p of all sequences $x := (x(0), x(1), \dots) \in \mathbb{K}^{\mathbb{N}}$ such that

$$\|x\|_p := \left(\sum_{n=0}^{\infty} |x(n)|^p\right)^{1/p} < \infty,$$

if $p < \infty$, and

$$\|x\|_\infty := \sup_{n\geq 0}\{|x(n)|\} < \infty,$$

when $p = \infty$. For each $j \geq 0$, let $e_j \in \ell_p$ denote the vector defined by $e_j(i) = \delta_{ij}$ for all $i \geq 0$; here δ_{ij} denotes the Kronecker delta. For each $j \geq 0$, let $P_j \in \mathcal{L}(\ell_p)$ be the projection defined by

$$P_j x = x(j) e_j.$$

Clearly, P_j is not a Fredholm operator. Prove, step by step, the following:

- $\|P_j\| = 1$ for each $j \geq 0$.
- Set $\Omega := \{\lambda \in \mathbb{K} : |\lambda| < 1\}$. Then the family $\mathfrak{L} : \Omega \to \mathcal{L}(\ell_p)$ defined by

$$\mathfrak{L}(\lambda) := \sum_{n=0}^{\infty} \lambda^n P_n, \qquad \lambda \in \Omega,$$

  is analytic.
- The operator $\mathfrak{L}(\lambda)$ is injective for all $\lambda \in \Omega \setminus \{0\}$; however, $\mathfrak{L}(\lambda) \notin \mathrm{Iso}(\ell_p)$ for all $\lambda \in \Omega$.
- Extending Definitions 7.1.1, 7.2.1 and 7.2.2 to cover the case of non-Fredholm families, prove that no Jordan chain of $\mathfrak{L}$ at 0 can be continued indefinitely.
- $\operatorname{rank} x = \infty$ for all $x \in \ell_p \setminus \{0\}$.

16. Let $\mathfrak{L} \in \mathcal{C}^r(\Omega, \mathcal{L}(U,V))$ be such that $\lambda_0 \in \mathrm{Alg}_k(\mathfrak{L})$ for some $k \in \mathbb{N} \cap [0, r]$. Express the partial μ-multiplicities introduced by Definition 5.1.3 in terms of the partial κ-multiplicities of $\mathfrak{L}$ at λ_0.

17. Use Exercise 16 and the local Smith form of Theorem 7.8.3 to prove that the dual construction of Exercise 5 of Chapter 5 is actually equivalent to the construction of Section 5.1.

18. Prove that in Theorem 7.8.3 the family $\mathfrak{F}$ can be chosen to be analytic. Indication: Adapt the proof of Theorem 7.8.1, and apply Theorem 7.8.3 to the Taylor polynomial of $\mathfrak{L}$.

19. Adapt the proof of Theorem 7.8.3 and use Exercise 18 to show that the families $\mathfrak{E}$ and $\mathfrak{F}$ of (7.39) can be chosen so that $\mathfrak{E}$ is analytic and $\mathfrak{F}$ is of class $\mathcal{C}^{r-\nu}$. Combine this feature with Theorems 5.6.1, 7.3.1 and 7.8.3, and Exercise 18 to get the following general version of the product formula: For every $r, s \in \mathbb{N} \cup \{\infty\}$ and

$$\mathfrak{L} \in \mathcal{C}^r(\Omega, \mathcal{L}(U,V)), \qquad \mathfrak{M} \in \mathcal{C}^s(\Omega, \mathcal{L}(W,U))$$

such that

$$\mathfrak{L}_0 \in \mathrm{Fred}_0(U,V), \qquad \mathfrak{M}_0 \in \mathrm{Fred}_0(W,U)$$

and the multiplicities

$$\kappa[\mathfrak{L}; \lambda_0], \quad \kappa[\mathfrak{M}; \lambda_0], \quad \kappa[\mathfrak{L}\mathfrak{M}; \lambda_0]$$

are well defined, the following identity holds:

$$\kappa[\mathfrak{L}\mathfrak{M}; \lambda_0] = \kappa[\mathfrak{L}; \lambda_0] + \kappa[\mathfrak{M}; \lambda_0].$$

7.12 Comments on Chapter 7

The concept of Jordan chain is, of course, very well known for linear families of matrices (see, e.g., Gantmacher [39]). Jordan chains of families of operators (as discussed in this chapter) were introduced in the decade of 1940 to develop a spectral theory for polynomial, holomorphic and meromorphic families. One of the first uses of Jordan chains (in that case, in the meromorphic setting) can be found in Keldyš [66, 67]. The reader is referred to Gohberg [42] for an early exposition of these concepts and for a list of references to the original papers. Different expositions about Jordan chains, under closely related assumptions, can also be found in Markus & Sigal [95], Gohberg & Sigal [54], Gohberg & Rodman [53], and Gohberg, Lancaster & Rodman [50]. Jordan chains have been shown to be a powerful technical tool in a number of areas, as already discussed in the Preface of this book.

Later definitions and uses of Jordan chains for non-analytic families, in connection to bifurcation theory, are in Sarreither [116], Rabier [110], and López-Gómez & Mora-Corral [89]. The exposition of this chapter has followed the construction described in [89] and anticipated in [88]; however, in this chapter we have considerably polished and expanded the original material of [89].

The concept of local Smith form goes back to Smith [121], where its existence is proved in the ring of integers. A proof of the existence of the local Smith form in an abstract ring can be found in Jacobson [63]. For an exposition in the context of matrix polynomials see, for example, Gantmacher [39], or Gohberg, Lancaster & Rodman [50].

One of the first studies of equivalence of families can be found in Gohberg [42]. Lemma 7.8.2 has been borrowed from Gohberg, Goldberg & Kaashoek [43] (see also Gohberg & Sigal [54]), and our main source on the theory of equivalence of families has been Gohberg, Kaashoek & Lay [44]. The relationship between local and global equivalence is studied in Leiterer [78, 79], where the global Smith form for operator-valued analytic functions was introduced (see also Gohberg & Rodman [52]).

Many of the implications of Theorem 7.10.4 were already known in the context of holomorphic and meromorphic families; see Gohberg and Sigal [54] and Gohberg [42] for probably the first results, and Sigal [119] and Bart, Kaashoek and Lay [9] for extension to Fredholm families of an arbitrary index. Criteria for global equivalence of matrix polynomials are very well known (see, e.g., Gel'fand [40]).

Although in the context of polynomial operator families, the existence of a *local Smith form* is a very particular case of the available *global Smith form* described in Gohberg, Lancaster & Rodman [50], it should be noted that, not only the local results of this chapter are optimal for general operator families of class $\mathcal{C}^r$ regularity, for $r \in \mathbb{N}$, but also the Magnus factorizations introduced in Chapter 5 can actually be used to construct a global Smith form in the general context of this book. Indeed, let $\Omega \subset \mathbb{K}$ be a bounded open connected set, and suppose the family $\mathfrak{L} \in \mathcal{C}^\infty(\Omega, \mathcal{L}(U,V))$ satisfies $\mathfrak{L}(\Omega) \subset \mathrm{Fred}_0(U,V)$ and the set $\Sigma(\mathfrak{L})$ consists of a finite number of algebraic eigenvalues, say,

$$\Sigma(\mathfrak{L}) = \mathrm{Eig}(\mathfrak{L}) = \mathrm{Alg}(\mathfrak{L}) = \{\lambda_1, \ldots, \lambda_p\},$$

with

$$\lambda_i \neq \lambda_j, \qquad 1 \leq i < j \leq p,$$

for some integer $p \geq 1$. The theory developed in Chapter 8 shows that this occurs when $\mathfrak{L} \in \mathcal{C}^\omega(\Omega, \mathcal{L}(U,V))$ and the sets $\mathfrak{L}(\Omega) \cap \mathrm{Iso}(U,V)$ and $\Sigma(\mathfrak{L})$ are non-empty. As we are assuming $\Sigma(\mathfrak{L}) = \mathrm{Alg}(\mathfrak{L})$, for every $j \in \{1, \ldots, p\}$, there exists $\nu_j \in \mathbb{N}^*$ such that $\lambda_j \in \mathrm{Alg}_{\nu_j}(\mathfrak{L})$. According to Corollary 5.3.2, there are ν_1 finite-rank projections $\Pi_{h,1} \in \mathcal{L}(U) \setminus \{0\}$, for $1 \leq h \leq \nu_1$, and $\mathfrak{M} \in \mathcal{C}^\infty(\Omega, \mathcal{L}(U,V))$ such that

$$\mathfrak{M}(\lambda_1) \in \mathrm{Iso}(U,V) \tag{7.48}$$

and

$$\mathfrak{L}(\lambda) = \mathfrak{M}(\lambda)\mathfrak{P}_1(\lambda - \lambda_1), \qquad \lambda \in \Omega, \tag{7.49}$$

where we have denoted

$$\mathfrak{P}_1(t) := [t\Pi_{\nu_1,1} + I_U - \Pi_{\nu_1,1}] \cdots [t\Pi_{1,1} + I_U - \Pi_{1,1}], \qquad t \in \mathbb{K}.$$

If $p = 1$, then we are done. So, suppose $p \geq 2$. Then, as $\mathfrak{P}_1(t)$ is invertible for all $t \in \mathbb{K} \setminus \{0\}$, we find from (7.48) and (7.49) that

$$\Sigma(\mathfrak{M}) = \mathrm{Alg}(\mathfrak{M}) = \{\lambda_2, \ldots, \lambda_p\}.$$

Moreover, it is easily seen that

$$\lambda_j \in \mathrm{Alg}_{\nu_j}(\mathfrak{M}), \qquad 2 \leq j \leq p.$$

Thus, according to Corollary 5.3.2, there are ν_2 finite-rank projections $\Pi_{h,2} \in \mathcal{L}(U) \setminus \{0\}$, for $1 \leq h \leq \nu_2$, and $\mathfrak{N} \in \mathcal{C}^\infty(\Omega, \mathcal{L}(U,V))$ such that

$$\mathfrak{N}(\lambda_2) \in \mathrm{Iso}(U,V)$$

and

$$\mathfrak{M}(\lambda) = \mathfrak{N}(\lambda)\mathfrak{P}_2(\lambda - \lambda_2), \qquad \lambda \in \Omega, \tag{7.50}$$

where we have denoted

$$\mathfrak{P}_2(t) := [t\Pi_{\nu_2,2} + I_U - \Pi_{\nu_2,2}] \cdots [t\Pi_{1,2} + I_U - \Pi_{1,2}], \qquad t \in \mathbb{K}.$$

Consequently, substituting (7.50) into (7.49) shows that

$$\mathfrak{L}(\lambda) = \mathfrak{N}(\lambda)\mathfrak{P}_2(\lambda - \lambda_2)\mathfrak{P}_1(\lambda - \lambda_1), \qquad \lambda \in \Omega.$$

By construction, $\mathfrak{N}(\Omega) \subset \mathrm{Iso}(U,V)$ if $p=2$, and in the general case when $p \geq 3$, after finitely many steps, the previous argument shows the existence of $\sum_{j=1}^{p} \nu_j$ finite-rank projections

$$\Pi_{j,i} \in \mathcal{L}(U) \setminus \{0\}, \qquad 1 \leq i \leq p, \quad 1 \leq j \leq \nu_i,$$

and of a family $\mathfrak{E} \in \mathcal{C}^\infty(\Omega, \mathcal{L}(U))$ such that $\mathfrak{E}(\Omega) \subset \mathrm{Iso}(U)$ and

$$\mathfrak{L}(\lambda) = \mathfrak{E}(\lambda)\mathfrak{P}_p(\lambda - \lambda_p) \cdots \mathfrak{P}_1(\lambda - \lambda_1), \qquad \lambda \in \Omega,$$

where, for every $1 \leq i \leq p$,

$$\mathfrak{P}_i(t) := [t\Pi_{\nu_i,i} + I_U - \Pi_{\nu_i,i}] \cdots [t\Pi_{1,i} + I_U - \Pi_{1,i}], \qquad t \in \mathbb{K}.$$

Clearly, the product polynomial $\mathfrak{P}$ defined by

$$\mathfrak{P}(\lambda) := \mathfrak{P}_p(\lambda - \lambda_p) \cdots \mathfrak{P}_1(\lambda - \lambda_1), \qquad \lambda \in \mathbb{K},$$

satisfies that $\mathfrak{P}(\lambda)$ is a finite-rank perturbation of the identity I_U for all $\lambda \in \mathbb{K}$. Hence, $\mathfrak{P}(\mathbb{K}) \subset \mathrm{Fred}_0(U)$. Moreover, if

$$\dim U = \dim V < \infty, \tag{7.51}$$

then $\mathfrak{P}$ is a matrix polynomial, as discussed in Chapter 10 and, therefore, according to the global results of Gohberg, Lancaster & Rodman [50], $\mathfrak{P}$ possesses a global Smith form. Obviously, such form provides us with a global Smith form for $\mathfrak{L}$ in Ω. As $\mathfrak{L}$ is a general family, this result indeed provides a substantial extension of the previous finite-dimensional global results of Gohberg, Lancaster & Rodman [50]. Further, it is rather obvious that the regularity of $\mathfrak{L}$ can be relaxed to assume $\mathfrak{L} \in \mathcal{C}^r(\Omega, \mathcal{L}(U,V))$ with

$$r = \sum_{i=1}^{p} \nu_j,$$

or even

$$r = \max\{\nu_1, \dots, \nu_p\}.$$

In such a case, the existence of $\mathfrak{E} \in \mathcal{C}(\Omega, \mathcal{L}(U,V))$ satisfying $\mathfrak{E}(\Omega) \subset \mathrm{Iso}(U,V)$ and

$$\mathfrak{L}(\lambda) = \mathfrak{E}(\lambda)\mathfrak{P}(\lambda), \qquad \lambda \in \Omega,$$

is still guaranteed by the theory developed in Chapter 5. At a further stage, condition (7.51) can be overcome by using the fact that $\mathfrak{P}$ is a finite-rank perturbation of I_U. Therefore, the theory developed in this book does not inherit exclusively a local nature, but essentially a global nature in the sense that it provides a general scheme to get global results through a canonical procedure.

As already anticipated in Section 3.6, the condition that

$$\lambda_0 \in \bigcup_{0 \leq \nu < r+1} \mathrm{Alg}_\nu(\mathfrak{L}),$$

has been shown to be pivotal for characterizing the existence of the local Smith form of $\mathfrak{L}$ at λ_0. As a consequence of the results of Chapter 8, any analytic family must satisfy that condition at every isolated eigenvalue. This feature might explain why most of the work on Jordan chains has been developed for holomorphic and meromorphic pencils.

The proof of the result of Exercise 12 can be found in Sarreither [117], where a number of different results about the multiplicity of a product can be found.

The proof of the result of Exercise 5 can be found in Mora-Corral [99]. Similarly, Exercises 7, 8 and 10 are based on the results of [99].

Chapter 8

Analytic and Classical Families. Stability

This chapter focuses attention on the analytic operator families. After studying some universal spectral properties of these families, this chapter deals with the classical families $\mathfrak{L}^A$ of the form

$$\mathfrak{L}^A(\lambda) := \lambda I_U - A, \qquad \lambda \in \mathbb{K}, \tag{8.1}$$

for a given $A \in \mathcal{L}(U)$, in order to show that, in this particular case, the classic concepts of algebraic ascent and multiplicity equal the generalized concepts introduced in the previous four chapters. Consequently, the algebraic multiplicity analyzed in this book, from a series of different perspectives, is indeed a generalization of the classic concept of algebraic multiplicity.

The plan of this chapter is the following. Section 8.1 characterizes the isolated eigenvalues of a general analytic family. Section 8.2 shows that the discrete spectrum of an analytic family consists of isolated eigenvalues. Section 8.3 shows the coincidence between the classic concept of multiplicity and the generalized concepts introduced in this book. Finally, Section 8.4 proves the global invariance by homotopy of the complex global multiplicity of a holomorphic family in a bounded domain.

Throughout this chapter we consider $\mathbb{K} \in \{\mathbb{R}, \mathbb{C}\}$, two $\mathbb{K}$-Banach spaces U and V, an open subset $\Omega \subset \mathbb{K}$, a point $\lambda_0 \in \Omega$, and a general family $\mathfrak{L} \in \mathcal{C}^\omega(\Omega, \mathcal{L}(U, V))$ with $\mathfrak{L}_0 \in \mathrm{Fred}_0(U, V)$. Under these assumptions, we already know that $\mathfrak{L}(\lambda) \in \mathrm{Fred}_0(U, V)$ for $\lambda \sim \lambda_0$. As in Chapter 7, we will also adopt the notation (7.1).

8.1 Isolated eigenvalues

The following result characterizes the isolated eigenvalues of an analytic operator family. It might be regarded as an improvement of Theorem 7.8.3.

Theorem 8.1.1. *Let $\lambda_0 \in \Omega$ and $\mathfrak{L} \in \mathcal{C}^\omega(\Omega, \mathcal{L}(U,V))$ be with*

$$\mathfrak{L}_0 := \mathfrak{L}(\lambda_0) \in \mathrm{Fred}_0(U,V).$$

Then, the following conditions are equivalent:

i) *There exists an open neighborhood $\Omega' \subset \Omega$ of λ_0 such that*

$$\mathfrak{L}(\lambda) \in \mathrm{Iso}(U,V) \quad \textit{for all} \quad \lambda \in \Omega' \setminus \{\lambda_0\}.$$

ii) $\lambda_0 \in \mathrm{Alg}_\nu(\mathfrak{L})$ *for some $\nu \in \mathbb{N}$.*

iii) *$\mathfrak{L}$ possesses a local Smith form at λ_0, in the sense of Definition 7.4.3.*

iv) *The set of lengths of the Jordan chains of $\mathfrak{L}$ at λ_0 is bounded above.*

v) *No Jordan chain of $\mathfrak{L}$ at λ_0 can be continued indefinitely.*

vi) *$\mathfrak{m}[\mathfrak{L};\lambda_0] < \infty$, where $\mathfrak{m}$ is the unique multiplicity of Theorem 6.0.1.*

Proof. By definition of algebraic eigenvalue, (ii) implies (i). According to Theorems 7.8.3, 7.6.3 and 6.0.1, conditions (ii), (iii), (iv), (v) and (vi) are equivalent.

Suppose (i). By Lemma 7.8.2, there exist an open neighborhood $\tilde{\Omega} \subset \Omega'$ of λ_0, a topological direct sum decomposition $U = U_0 \oplus U_1$, with

$$\dim U_0 = \dim N[\mathfrak{L}_0] < \infty,$$

and three operator families

$$\mathfrak{E} \in \mathcal{C}^\omega(\tilde{\Omega}, \mathcal{L}(U,V)), \quad \mathfrak{F} \in \mathcal{C}^\omega(\tilde{\Omega}, \mathcal{L}(U)), \quad \mathfrak{M} \in \mathcal{C}^\omega(\tilde{\Omega}, \mathcal{L}(U_0)),$$

with $\mathfrak{E}_0 \in \mathrm{Iso}(U,V)$ and $\mathfrak{F}_0 \in \mathrm{Iso}(U)$, such that

$$\mathfrak{L}(\lambda) = \mathfrak{E}(\lambda)\,[\mathfrak{M}(\lambda) \oplus I_{U_1}]\,\mathfrak{F}(\lambda), \qquad \lambda \in \tilde{\Omega}. \tag{8.2}$$

As we are assuming that $\mathfrak{L}(\lambda) \in \mathrm{Iso}(U,V)$ for all $\lambda \in \tilde{\Omega} \setminus \{\lambda_0\}$, choosing a basis in U_0 shows that

$$\det \mathfrak{M}(\lambda) \neq 0, \qquad \lambda \in \tilde{\Omega} \setminus \{\lambda_0\}.$$

As $\det \mathfrak{M} \in \mathcal{C}^\omega(\tilde{\Omega})$, necessarily

$$\mathrm{ord}_{\lambda_0} \mathfrak{M} < \infty.$$

Therefore, thanks to Proposition 3.1.1, λ_0 must be an algebraic eigenvalue of $\mathfrak{M}$. Obviously, according to (8.2), this implies (ii) and concludes the proof. □

The family $\mathfrak{L}$ defined through (4.20) is not analytic, despite being of class $\mathcal{C}^\infty$; the value 0 is an isolated eigenvalue of $\mathfrak{L}$, but $0 \notin \mathrm{Alg}(\mathfrak{L})$. Consequently, the analyticity is crucial in Theorem 8.1.1, as well as in the remaining results of this chapter.

8.2 The structure of the spectrum

The next result ascertains the structure of the spectrum of an analytic family of Fredholm operators with index zero.

Theorem 8.2.1. *Suppose the following:*

i) Ω *is an open connected subset of* $\mathbb{K}$.

ii) $\mathfrak{L} \in \mathcal{C}^\omega(\Omega, \mathcal{L}(U,V))$ *satisfies* $\mathfrak{L}(\Omega) \subset \mathrm{Fred}(U,V)$.

iii) $\mathfrak{L}(\Omega) \cap \mathrm{Iso}(U,V) \neq \varnothing$.

Then, $\mathfrak{L}(\Omega) \subset \mathrm{Fred}_0(U,V)$, *and, hence, by the open mapping theorem,* $\Sigma(\mathfrak{L}) = \mathrm{Eig}(\mathfrak{L})$. *Moreover,* $\Sigma(\mathfrak{L})$ *is discrete in* Ω *and, hence, due to Theorem* 8.1.1,

$$\Sigma(\mathfrak{L}) = \mathrm{Eig}(\mathfrak{L}) \subset \mathrm{Alg}(\mathfrak{L}).$$

Proof. By (i) and (ii), $\mathfrak{L}(\Omega)$ is a connected set contained in $\mathrm{Fred}(U,V)$. Due to (iii), it satisfies

$$\mathfrak{L}(\Omega) \cap \mathrm{Fred}_0(U,V) \neq \varnothing.$$

Therefore, $\mathfrak{L}(\Omega) \subset \mathrm{Fred}_0(U,V)$, because the Fredholm index is continuous. Consequently, $\Sigma(\mathfrak{L})$ consists of eigenvalues and, necessarily, must be closed. In addition, according to (iii),

$$\Sigma(\mathfrak{L}) \subsetneq \Omega.$$

Let S be the set of accumulation points of $\Sigma(\mathfrak{L})$. Then, S is closed in Ω and $S \subset \Sigma(\mathfrak{L})$. Assume that $\Sigma(\mathfrak{L})$ is not discrete and let $\lambda_0 \in S$. By Lemma 7.8.2, there exist an open neighborhood $\tilde{\Omega} \subset \Omega$ of λ_0, a topological direct sum decomposition $U = U_0 \oplus U_1$, with

$$\dim U_0 = \dim N[\mathfrak{L}_0] < \infty,$$

and three operator families

$$\mathfrak{E} \in \mathcal{C}^\omega(\tilde{\Omega}, \mathcal{L}(U,V)), \quad \mathfrak{F} \in \mathcal{C}^\omega(\tilde{\Omega}, \mathcal{L}(U)), \quad \mathfrak{M} \in \mathcal{C}^\omega(\tilde{\Omega}, \mathcal{L}(U_0)),$$

with $\mathfrak{E}_0 \in \mathrm{Iso}(U,V)$ and $\mathfrak{F}_0 \in \mathrm{Iso}(U)$, satisfying (8.2). Clearly,

$$\Sigma(\mathfrak{M}) = \Sigma(\mathfrak{L}) \cap \tilde{\Omega}$$

and, consequently, choosing a basis in U_0, the point λ_0 must be an accumulation point of zeros of $\det \mathfrak{M} \in \mathcal{C}^\omega(\tilde{\Omega})$. Thus, by the identity theorem for analytic functions,

$$\det \mathfrak{M} = 0 \quad \text{in} \quad \tilde{\Omega}.$$

Therefore, $\tilde{\Omega} \subset S$, and, hence, S is an open subset of Ω. As it is also closed, and Ω is connected, necessarily $S = \Omega$ and $\Sigma(\mathfrak{L}) = \Omega$, which contradicts (iii). Consequently, $\Sigma(\mathfrak{L})$ must be discrete in Ω.

Finally, Theorem 8.1.1 concludes the proof, because any isolated eigenvalue of an analytic family is algebraic. □

8.3 Classic algebraic multiplicity

Throughout this section, the attention will be focused on the classical case when

$$\mathfrak{L}^A(\lambda) = \lambda I_U - A, \qquad \lambda \in \mathbb{K}, \tag{8.3}$$

for a given $A \in \mathcal{L}(U)$. Our main goal is to show that the classic algebraic multiplicity of A equals the *generalized* algebraic multiplicity $\chi[\mathfrak{L}^A;\cdot]$ analyzed in this book.

According to Corollary 6.5.4, in the finite-dimensional case when $U = \mathbb{K}^m$ for some integer $m \geq 1$, it is already known that

$$\chi[\mathfrak{L}^A;\lambda_0] = m_a(\lambda_0), \qquad \lambda_0 \in \sigma(A) = \Sigma(\mathfrak{L}^A) = \mathrm{Eig}(\mathfrak{L}^A),$$

where $m_a(\lambda_0)$ denotes the algebraic multiplicity of λ_0 as an eigenvalue of the matrix A. Obtaining a general version of this result becomes slightly more delicate.

Now we consider a $\mathbb{K}$-Banach space U and an operator $A \in \mathcal{L}(U)$. Clearly,

$$\mathfrak{L}^A \in \mathcal{C}^\omega(\mathbb{K}, \mathcal{L}(U)).$$

Moreover, if the modulus of $\lambda \in \mathbb{K}$ is sufficiently large, one has that

$$\|\lambda^{-1}A\| < 1$$

and, therefore, for such range of values of λ,

$$\mathfrak{L}^A(\lambda) = \lambda I_U - A = \lambda\left(I_U - \lambda^{-1}A\right) \in \mathrm{Iso}(U).$$

If one assumes that, for some open set $\Omega \subset \mathbb{K}$,

$$\mathfrak{L}^A(\Omega) \subset \mathrm{Fred}(U), \qquad \mathfrak{L}^A(\Omega) \cap \mathrm{Iso}(U) \neq \varnothing,$$

then, according to Theorem 8.2.1, we find that

$$\sigma(A) \cap \Omega = \Sigma(\mathfrak{L}^A) \cap \Omega = \mathrm{Eig}(\mathfrak{L}^A) \cap \Omega = \mathrm{Eig}(A) \cap \Omega.$$

Here and throughout the rest of the book, given an operator $B \in \mathcal{L}(U)$, the set $\mathrm{Eig}(B)$ denotes the set of eigenvalues of the operator B.

Fix any $\lambda_0 \in \sigma(A) \cap \Omega$. Then, the *classic algebraic multiplicity* of A at λ_0, denoted by $m_a(\lambda_0)$, is defined through

$$m_a(\lambda_0) := \dim \bigcup_{j=1}^{\infty} N[(\lambda_0 I_U - A)^j].$$

By Theorem 1.2.1, this quantity indeed provides us with the algebraic multiplicity of λ_0 as an eigenvalue of A, if $U = \mathbb{K}^m$. In the general case when U is an arbitrary Banach space, either

$$N[(\lambda_0 I_U - A)^j] \subsetneq N[(\lambda_0 I_U - A)^{j+1}] \quad \text{for all} \quad j \geq 0, \tag{8.4}$$

or, alternatively, there exists an integer $\nu(\lambda_0) \geq 1$ such that

$$N[(\lambda_0 I_U - A)^i] \subsetneq N[(\lambda_0 I_U - A)^j] \subset N[(\lambda_0 I_U - A)^{\nu(\lambda_0)}] = N[(\lambda_0 I_U - A)^\ell]$$

for all $0 \leq i < j \leq \nu(\lambda_0) \leq \ell$, much like in the finite-dimensional setting. When the second option occurs, $\nu(\lambda_0)$ is referred to the *algebraic ascent* of λ_0 as an eigenvalue of A, and $N[(\lambda_0 I_U - A)^{\nu(\lambda_0)}]$ is named the *ascent generalized eigenspace* associated with λ_0. Consequently, in all these circumstances, we are led to the identity

$$m_a(\lambda_0) = \sup_{j \geq 0} \dim N[(\lambda_0 I_U - A)^j].$$

Obviously, alternative (8.4) cannot occur if $\dim U < \infty$.

The key result revealing why

$$m_a(\lambda_0) = \chi[\mathfrak{L}^A; \lambda_0]$$

is the following one.

Proposition 8.3.1. *Given $A \in \mathcal{L}(U)$, let $\mathfrak{L}^A$ be the family defined by (8.3). Then, for every integer $s \geq 1$ and $\lambda_0 \in \mathbb{K}$, the operator*

$$\begin{array}{rcl} N[\mathrm{trng}\{\mathfrak{L}_0^A, \dots, \mathfrak{L}_{s-1}^A\}] & \to & N[(\lambda_0 I_U - A)^s] \\ \mathrm{col}[x_0, \dots, x_{s-1}] & \mapsto & x_{s-1} \end{array} \tag{8.5}$$

is a linear topological isomorphism whose inverse is given by

$$\begin{array}{rcl} N[(\lambda_0 I_U - A)^s] & \to & N[\mathrm{trng}\{\mathfrak{L}_0^A, \dots, \mathfrak{L}_{s-1}^A\}] \\ y & \mapsto & \mathrm{col}[x_0, \dots, x_{s-1}] \end{array} \tag{8.6}$$

where we have denoted

$$x_{s-i} := (A - \lambda_0 I_U)^{i-1} y, \qquad 1 \leq i \leq s.$$

Consequently,

$$\dim N[\mathfrak{L}_{[s-1]}^A] = \dim N[(\lambda_0 I - A)^s], \qquad s \geq 1.$$

Proof. Note that

$$\mathfrak{L}_0^A = \lambda_0 I_U - A, \qquad \mathfrak{L}_1^A = I_U, \qquad \mathfrak{L}_j^A = 0, \quad j \geq 2.$$

Fix $s \geq 1$ and let $(x_0, \dots, x_{s-1}) \in U^s$. If

$$\mathrm{col}[x_0, \dots, x_{s-1}] \in N[\mathrm{trng}\{\mathfrak{L}_0^A, \dots, \mathfrak{L}_{s-1}^A\}],$$

by an elementary induction, it is apparent that

$$x_i \in N[(\lambda_0 I_U - A)^{i+1}], \qquad 0 \leq i \leq s-1.$$

Therefore, the mapping (8.5) is well defined, linear and continuous.

Conversely, for any $y \in N[(\lambda_0 I_U - A)^s]$, it is easy to verify that, when one sets

$$x_{s-i} := (A - \lambda_0 I_U)^{i-1} y, \qquad 1 \le i \le s,$$

it follows that

$$\mathrm{col}[x_0, \dots, x_{s-1}] \in N[\mathrm{trng}\{\mathfrak{L}_0^A, \dots, \mathfrak{L}_{s-1}^A\}].$$

Consequently, (8.6) also is well defined, linear and continuous.

Finally, it can be checked straightforwardly that these maps are mutually inverses. This concludes the proof. □

Owing to Proposition 8.3.1 and Theorem 7.6.2, the following fundamental result holds.

Theorem 8.3.2. *Let $\lambda_0 \in \mathbb{K}$ and $A \in \mathcal{L}(U)$ be such that λ_0 is an isolated eigenvalue of A, and*

$$\lambda_0 I_U - A \in \mathrm{Fred}_0(U).$$

Then

$$m_a(\lambda_0) = \chi[\mathfrak{L}^A; \lambda_0] < \infty \quad \textit{and} \quad \nu(\lambda_0) = \alpha[\mathfrak{L}^A; \lambda_0],$$

where

$$\mathfrak{L}^A(\lambda) := \lambda I_U - A, \qquad \lambda \in \mathbb{K}.$$

Moreover, $\nu(\lambda_0)$ equals the order of λ_0 as a pole of the (analytic) resolvent operator

$$\lambda \mapsto \mathcal{R}(\lambda; A) := (\lambda I_U - A)^{-1}, \qquad \lambda \in \mathbb{K} \setminus \sigma(A).$$

Proof. Clearly, $\mathfrak{L}^A \in \mathcal{C}^\omega(\mathbb{K}, \mathcal{L}(U))$. Moreover, by the assumptions, $\mathfrak{L}^A$ satisfies

$$\mathfrak{L}^A(\lambda_0) \in \mathrm{Fred}_0(U)$$

and condition (i) of Theorem 8.1.1. Thus, by Theorem 8.1.1, there exists an integer $\nu \ge 1$ such that $\lambda_0 \in \mathrm{Alg}_\nu(\mathfrak{L}^A)$. By Definition 4.3.1, ν equals the order of the pole of $\mathcal{R}(\cdot; A)$ at λ_0. Thanks to Theorem 5.3.1,

$$\nu = \alpha[\mathfrak{L}^A; \lambda_0].$$

Moreover, according to Theorem 7.6.2, we have that

$$\dim N[\mathfrak{L}^A_{[j-1]}] < \dim N[\mathfrak{L}^A_{[j]}] \le \dim N[\mathfrak{L}^A_{[\nu-1]}] = \dim N[\mathfrak{L}^A_{[\ell]}],$$

whenever $1 \le j \le \nu - 1 \le \ell$, and that

$$\dim N[\mathfrak{L}^A_{[\nu-1]}] = \chi[\mathfrak{L}^A; \lambda_0].$$

Therefore, due to Proposition 8.3.1, it becomes apparent that $\nu(\lambda_0) = \nu$ and that

$$m_a(\lambda_0) = \dim N[(\lambda_0 I_U - A)^\nu] = \dim N[\mathfrak{L}^A_{[\nu-1]}] = \chi[\mathfrak{L}^A; \lambda_0].$$

This concludes the proof. □

In the special, but important, case when $A \in \mathcal{K}(U)$ is a compact operator, the general theory provides us with the following classic result.

Corollary 8.3.3. *Suppose $A \in \mathcal{K}(U)$ and let $\sigma(A)$ denote the spectrum of A. Then, $\sigma(A) \setminus \{0\}$ consists of isolated eigenvalues with finite algebraic multiplicity.*

Proof. For each $\lambda \in \mathbb{K} \setminus \{0\}$, the operator

$$\mathfrak{L}^A(\lambda) := \lambda I_U - A = \lambda(I_U - \lambda^{-1}A)$$

is a non-zero multiple of a compact perturbation of the identity. Thus,

$$\mathfrak{L}^A(\mathbb{K} \setminus \{0\}) \subset \mathrm{Fred}_0(U).$$

Moreover, $\mathfrak{L}^A(\lambda) \in \mathrm{Iso}(U)$ if $|\lambda| > \|A\|$. Therefore, according to Theorem 8.2.1 (applied to $\Omega := \mathbb{K} \setminus \{0\}$), we find that $\sigma(A) \cap (\mathbb{K} \setminus \{0\})$ is a discrete set contained in $\mathrm{Eig}(A)$. Theorem 8.3.2 concludes the proof. □

8.4 Stability of the complex algebraic multiplicity

When $\mathbb{K} = \mathbb{C}$, the fact $\mathfrak{L} \in \mathcal{C}^1(\Omega, \mathcal{L}(U,V))$ entails $\mathfrak{L} \in \mathcal{H}(\Omega, \mathcal{L}(U,V))$. So, for complex operator families, we have actually been dealing with holomorphic families, except in the least interesting case when the family is merely continuous, where none of the algebraic invariants introduced in this book can be defined, unless the family consists of isomorphisms.

Although at first glance one might be tempted to think that the theories of complex and real multiplicities are essentially the same, except perhaps for some incidental simplification of the proofs in the complex setting, there is, however, a crucial difference between both theories. Namely, whereas the complex multiplicity is stable, in the sense that small perturbations of the family do not affect the value of the multiplicity, it turns out that the real multiplicity is unstable. Indeed, in the simplest case when $U = V = \mathbb{R}$, the multiplicity of $\mathfrak{L}$ at λ_0 is given through $\mathrm{ord}_{\lambda_0} \mathfrak{L}$, and it is well known that the order does not enjoy any reasonable stability property.

This section shows the stability of the complex multiplicity for holomorphic operator families.

Throughout this section we suppose that $\mathbb{K} = \mathbb{C}$ and consider a non-empty bounded connected open set $\Omega \subset \mathbb{C}$ and a family $\mathfrak{L}$ such that

$$\mathfrak{L} \in \mathcal{C}(\bar{\Omega}, \mathcal{L}(U,V)), \quad \mathfrak{L}|_\Omega \in \mathcal{H}(\Omega, \mathcal{L}(U,V)) \tag{8.7}$$

and

$$\mathfrak{L}(\Omega) \subset \mathrm{Fred}(U,V) \quad \text{and} \quad \mathfrak{L}(\partial\Omega) \subset \mathrm{Iso}(U,V), \tag{8.8}$$

where, as usual, we have denoted by $\bar{\Omega}$ and $\partial\Omega$, the closure and the boundary of Ω, respectively. Then, as the Fredholm index is an integer-valued continuous function and $\bar{\Omega}$ is connected, we have that

$$\mathfrak{L}(\bar{\Omega}) \subset \text{Fred}_0(U,V).$$

Moreover, by Theorem 8.2.1, the set $\Sigma(\mathfrak{L})$ is contained in $\text{Alg}(\mathfrak{L})$ and consists of a finite number of algebraic eigenvalues. It should be noted that, thanks to (8.8), $\mathfrak{L}(\lambda) \in \text{Iso}(U,V)$ for all $\lambda \in \Omega$ sufficiently close to $\partial\Omega$. From now on, we will denote by $\mathfrak{m}$ the multiplicity whose existence and uniqueness was established by Theorem 6.0.1. By the theory developed in the previous chapters, we already know that it equals the multiplicities χ, μ and κ introduced in Chapters 4, 5 and 7, respectively. The following concept makes sense.

Definition 8.4.1. The multiplicity of $\mathfrak{L}$ in Ω, denoted by $\mathfrak{m}[\mathfrak{L};\Omega]$, is defined through

$$\mathfrak{m}[\mathfrak{L};\Omega] := \sum_{\lambda\in\Omega} \mathfrak{m}[\mathfrak{L};\lambda]. \tag{8.9}$$

When

$$\Sigma(\mathfrak{L}) = \{\lambda_1,\dots,\lambda_n\}$$

for some integer $n \geq 1$, and $\lambda_i \neq \lambda_j$ for $1 \leq i < j \leq n$, then (8.9) becomes

$$\mathfrak{m}[\mathfrak{L};\Omega] := \sum_{i=1}^{n} \mathfrak{m}[\mathfrak{L};\lambda_i],$$

while $\mathfrak{m}[\mathfrak{L};\Omega] = 0$ if $\mathfrak{L}(\Omega) \subset \text{Iso}(U,V)$. Consequently, under hypotheses (8.7) and (8.8), $\mathfrak{m}[\mathfrak{L};\Omega] \in \mathbb{N}$.

8.4.1 Local stability

The next result establishes the local stability of the multiplicity $\mathfrak{m}[\mathfrak{L};\Omega]$.

Proposition 8.4.2. *Suppose $\mathfrak{L}, \mathfrak{M} \in \mathcal{H}(\Omega, \mathcal{L}(U,V))$ satisfy $\lambda_0 \in \text{Alg}(\mathfrak{L})$, and $\mathfrak{L}(\lambda_0) \in \text{Fred}_0(U,V)$. Then, there exist $r, \delta > 0$ such that the estimate*

$$\sup_{\lambda\in C_r(\lambda_0)} \|\mathfrak{L}(\lambda) - \mathfrak{M}(\lambda)\| \leq \delta \tag{8.10}$$

implies

$$\mathfrak{m}[\mathfrak{M}; D_r(\lambda_0)] = \mathfrak{m}[\mathfrak{L}; D_r(\lambda_0)] = \mathfrak{m}[\mathfrak{L};\lambda_0].$$

Proof. It is based on the simultaneous application to $\mathfrak{L}$ and $\mathfrak{M}$ of the finite-dimensional reduction described in the proof of Lemma 7.8.2.

Since $\mathfrak{L}_0 \in \text{Fred}_0(U,V)$, there exists a finite-rank operator $F \in \mathcal{L}(U,V)$ for which

$$\mathfrak{L}_0 + F \in \text{Iso}(U,V).$$

For sufficiently small $\delta, r > 0$, by (8.10) and the Cauchy integral formula, $\mathfrak{M}_0$ is close enough to $\mathfrak{L}_0$ so that

$$\mathfrak{M}_0 + F \in \mathrm{Iso}(U, V)$$

as well.

Let $\mathfrak{E}_{\mathfrak{L}}, \mathfrak{E}_{\mathfrak{M}} \in \mathcal{H}(\Omega, \mathcal{L}(U, V))$ be the families defined by

$$\mathfrak{E}_{\mathfrak{L}}(\lambda) := \mathfrak{L}(\lambda) + F, \quad \mathfrak{E}_{\mathfrak{M}}(\lambda) := \mathfrak{M}(\lambda) + F, \qquad \lambda \in \Omega.$$

Obviously, $\mathfrak{E}_{\mathfrak{L}}(\lambda)$ and $\mathfrak{E}_{\mathfrak{M}}(\lambda)$ are isomorphisms for λ sufficiently close to λ_0. As F has finite rank, $N[F]$ has a finite-dimensional topological complement U_0 in U. Let $P \in \mathcal{L}(U)$ be the projection satisfying

$$R[P] = U_0 \quad \text{and} \quad N[P] = N[F].$$

Then, P has finite rank and $FP = F$. Thus, for every $\lambda \sim \lambda_0$ and $\mathfrak{N} \in \{\mathfrak{L}, \mathfrak{M}\}$, we have that

$$I_U - \mathfrak{E}_{\mathfrak{N}}(\lambda)^{-1}F = \big[I_U - P\mathfrak{E}_{\mathfrak{N}}(\lambda)^{-1}FP\big]\big[I_U - (I_U - P)\mathfrak{E}_{\mathfrak{N}}(\lambda)^{-1}FP\big],$$

and, setting

$$\mathfrak{F}_{\mathfrak{N}}(\lambda) := I_U - (I_U - P)\mathfrak{E}_{\mathfrak{N}}(\lambda)^{-1}FP, \qquad \lambda \sim \lambda_0,$$

one gets that the families $\mathfrak{F}_{\mathfrak{L}}$ and $\mathfrak{F}_{\mathfrak{M}}$ are holomorphic in a neighborhood of λ_0. These families take invertible values for $\lambda \sim \lambda_0$ and actually

$$\mathfrak{F}_{\mathfrak{N}}(\lambda)^{-1} = I_U + (I_U - P)\mathfrak{E}_{\mathfrak{N}}(\lambda)^{-1}FP, \qquad \mathfrak{N} \in \{\mathfrak{L}, \mathfrak{M}\}, \quad \lambda \sim \lambda_0.$$

Consequently, for each $\lambda \sim \lambda_0$ and $\mathfrak{N} \in \{\mathfrak{L}, \mathfrak{M}\}$, we are led to

$$\mathfrak{N}(\lambda) = \mathfrak{E}_{\mathfrak{N}}(\lambda)\big[I_U - P\mathfrak{E}_{\mathfrak{N}}(\lambda)^{-1}FP\big]\mathfrak{F}_{\mathfrak{N}}(\lambda).$$

Moreover, setting

$$\mathfrak{D}_{\mathfrak{N}}(\lambda) := I_U - P\mathfrak{E}_{\mathfrak{N}}(\lambda)^{-1}FP, \qquad \mathfrak{N} \in \{\mathfrak{L}, \mathfrak{M}\}, \quad \lambda \sim \lambda_0,$$

we have that, for each $\mathfrak{N} \in \{\mathfrak{L}, \mathfrak{M}\}$ and $\lambda \sim \lambda_0$,

$$\mathfrak{N}(\lambda) = \mathfrak{E}_{\mathfrak{N}}(\lambda)\ \mathfrak{D}_{\mathfrak{N}}(\lambda)\ \mathfrak{F}_{\mathfrak{N}}(\lambda)$$

and

$$\mathfrak{D}_{\mathfrak{N}}(\lambda)(U_0) \subset U_0, \qquad \mathfrak{D}_{\mathfrak{N}}(\lambda)|_{N[F]} = I|_{N[F]}.$$

We now define, for each $\mathfrak{N} \in \{\mathfrak{L}, \mathfrak{M}\}$ and $\lambda \sim \lambda_0$,

$$\mathfrak{G}_{\mathfrak{N}}(\lambda) := \mathfrak{D}_{\mathfrak{N}}(\lambda)|_{U_0} \in \mathcal{L}(U_0).$$

Then, with respect to the decomposition

$$U = R[P] \oplus N[P],$$

the family $\mathfrak{D}_{\mathfrak{N}}$ can be expressed as

$$\mathfrak{D}_{\mathfrak{N}}(\lambda) = \operatorname{diag}\{\mathfrak{G}_{\mathfrak{N}}(\lambda), I_{N[F]}\} = \mathfrak{G}_{\mathfrak{N}}(\lambda) \oplus I_{N[F]}, \qquad \lambda \sim \lambda_0.$$

According to the proof of (ii) implies (i) in Theorem 7.8.3, we find that

$$\mathfrak{m}[\mathfrak{L}; \lambda_0] = \mathfrak{m}[\mathfrak{G}_{\mathfrak{L}}; \lambda_0] \quad \text{and} \quad \mathfrak{m}[\mathfrak{M}; \lambda_0] = \mathfrak{m}[\mathfrak{G}_{\mathfrak{M}}; \lambda_0].$$

Thus, by Corollary 6.5.1,

$$\mathfrak{m}[\mathfrak{L}; \lambda_0] = \operatorname{ord}_{\lambda_0} \det \mathfrak{G}_{\mathfrak{L}} \quad \text{and} \quad \mathfrak{m}[\mathfrak{M}; \lambda_0] = \operatorname{ord}_{\lambda_0} \det \mathfrak{G}_{\mathfrak{M}}.$$

By standard results in one-dimensional complex function theory, there exist $\delta', r' > 0$ such that if

$$\sup_{\lambda \in C_{r'}(\lambda_0)} \| \det \mathfrak{G}_{\mathfrak{L}}(\lambda) - \det \mathfrak{G}_{\mathfrak{M}}(\lambda) \| \leq \delta', \tag{8.11}$$

then

$$\operatorname{ord}_{\lambda_0} \det \mathfrak{G}_{\mathfrak{L}} = \operatorname{ord}_{\lambda_0} \det \mathfrak{G}_{\mathfrak{M}}$$

and, hence,

$$\mathfrak{m}[\mathfrak{L}; \lambda_0] = \mathfrak{m}[\mathfrak{M}; \lambda_0].$$

Finally, by the construction of $\mathfrak{G}_{\mathfrak{L}}$ and $\mathfrak{G}_{\mathfrak{M}}$, it is apparent that there exist $\delta, r > 0$ such that condition (8.10) implies (8.11), because the determinant is continuous and there exists a constant $C > 0$ such that

$$\|\mathfrak{G}_{\mathfrak{L}}(\lambda) - \mathfrak{G}_{\mathfrak{M}}(\lambda)\| \leq C \|\mathfrak{L}(\lambda) - \mathfrak{M}(\lambda)\|$$

for every λ sufficiently close to λ_0. This concludes the proof. □

8.4.2 Global stability

Once we have established the local stability of the multiplicity, we can proceed to obtain its global stability.

Theorem 8.4.3 (of stability). *Let $\Omega \neq \varnothing$ be a bounded connected open subset of $\mathbb{C}$, and $\mathfrak{L}$ an operator family satisfying* (8.7) *and* (8.8). *Then, there exists $\delta > 0$ such that, for every operator family $\mathfrak{M}$ satisfying*

$$\mathfrak{M} \in \mathcal{C}(\bar{\Omega}, \mathcal{L}(U, V)), \quad \mathfrak{M}|_{\Omega} \in \mathcal{H}(\Omega, \mathcal{L}(U, V))$$

and

$$\sup_{\lambda \in \partial\Omega} \|\mathfrak{L}(\lambda) - \mathfrak{M}(\lambda)\| \leq \delta, \tag{8.12}$$

one has

$$\mathfrak{M}(\Omega) \subset \operatorname{Fred}_0(U, V), \qquad \mathfrak{M}(\partial\Omega) \subset \operatorname{Iso}(U, V), \tag{8.13}$$

and

$$\mathfrak{m}[\mathfrak{L}; \Omega] = \mathfrak{m}[\mathfrak{M}; \Omega]. \tag{8.14}$$

Proof. By the discussion carried out before Definition 8.4.1, we know that

$$\mathfrak{L}(\bar{\Omega}) \subset \mathrm{Fred}_0(U, V).$$

Moreover, $\Sigma(\mathfrak{L})$ is finite, possibly empty, and

$$\mathfrak{m}[\mathfrak{L}; \Omega] = \sum_{\lambda \in \Sigma(\mathfrak{L})} \mathfrak{m}[\mathfrak{L}; \lambda] \in \mathbb{N}.$$

If $\delta > 0$ is sufficiently small, then condition (8.12) implies $\mathfrak{M}(\lambda) \in \mathrm{Iso}(U, V)$ for all $\lambda \in \Omega$ sufficiently close to $\partial\Omega$. In addition, by the Cauchy integral formula, for each $\varepsilon > 0$ there exists $\delta > 0$ such that inequality (8.12) guarantees

$$\sup_{\lambda \in \bar{\Omega}} \|\mathfrak{L}(\lambda) - \mathfrak{M}(\lambda)\| \leq \varepsilon. \tag{8.15}$$

It suffices to prove that, under condition (8.15), all the conclusions of the theorem hold true for sufficiently small $\varepsilon > 0$. Clearly, as $\bar{\Omega}$ is compact, there exists $\varepsilon_0 > 0$ such that (8.13) is satisfied for every $\varepsilon \in (0, \varepsilon_0]$ and $\mathfrak{M}$ satisfying (8.15). Moreover, if $\Sigma(\mathfrak{L}) = \varnothing$, then, by shortening ε, if necessary, one gets $\Sigma(\mathfrak{M}) = \varnothing$ as long as $\mathfrak{M}$ satisfies (8.15). Therefore, in such circumstances, we find that

$$\mathfrak{m}[\mathfrak{L}; \Omega] = \mathfrak{m}[\mathfrak{M}; \Omega] = 0,$$

which ends the proof of the theorem. Assume that, for some integer $n \geq 1$,

$$\Sigma(\mathfrak{L}) = \{\lambda_1, \dots, \lambda_n\}.$$

As $\Sigma(\mathfrak{L})$ is finite, due to Proposition 8.4.2, there exist $r, \delta > 0$ with

$$\bigcup_{i=1}^{n} \bar{D}_r(\lambda_i) \subset \Omega, \qquad \bar{D}_r(\lambda_i) \cap \bar{D}_r(\lambda_j) = \varnothing \ \text{ if } \ 1 \leq i < j \leq n,$$

for which inequality (8.15) implies

$$\mathfrak{m}[\mathfrak{L}; \lambda_i] = \mathfrak{m}[\mathfrak{M}; D_r(\lambda_i)], \qquad 1 \leq i \leq n,$$

and, hence,

$$\mathfrak{m}[\mathfrak{L}; \Omega] = \sum_{i=1}^{n} \mathfrak{m}[\mathfrak{M}; D_r(\lambda_i)].$$

As

$$K := \bar{\Omega} \setminus \bigcup_{i=1}^{n} D_r(\lambda_i)$$

is a compact subset of $\bar{\Omega}$ such that $\mathfrak{L}(K) \subset \mathrm{Iso}(U, V)$, the number $\varepsilon > 0$ can be shortened, if necessary, so that $\mathfrak{M}(K) \subset \mathrm{Iso}(U, V)$. For such a choice of $\varepsilon > 0$, identity (8.14) is satisfied, and the proof is completed. □

8.4.3 Homotopy invariance

Based on Theorem 8.4.3, the invariance by homotopy of $\mathfrak{m}[\mathfrak{L};\Omega]$ readily follows. Precisely, the following result holds true.

Corollary 8.4.4 (Invariance by homotopy). *Let $\Omega \neq \varnothing$ be a bounded connected open subset of $\mathbb{C}$, and X a connected topological space. Then, for every $\mathfrak{H} \in \mathcal{C}(X \times \bar{\Omega}, \mathcal{L}(U,V))$ satisfying*

- $\mathfrak{H}(x,\cdot)$ *is holomorphic in Ω for all $x \in X$,*
- $\mathfrak{H}(X \times \Omega) \subset \mathrm{Fred}(U,V)$,
- $\mathfrak{H}(X \times \partial\Omega) \subset \mathrm{Iso}(U,V)$,

and any $x_0 \in X$, one has that

$$\mathfrak{m}[\mathfrak{H}(x,\cdot);\Omega] = \mathfrak{m}[\mathfrak{H}(x_0,\cdot);\Omega] \quad \textit{for all} \quad x \in X.$$

Proof. By Theorem 8.4.3, the map $h : X \to \mathbb{N}$ defined by

$$h(x) := \mathfrak{m}[\mathfrak{H}(x,\cdot);\Omega], \qquad x \in X,$$

is continuous. As X is connected and $\mathbb{N}$ has the discrete topology, h must be constant, which concludes the proof. □

In most applications, the space X is the interval $[0,1]$.

8.5 Exercises

1. For a given $A \in \mathcal{L}(U)$, consider the classical family

$$\mathfrak{L}^A(\lambda) := \lambda I_U - A, \qquad \lambda \in \mathbb{K},$$

and suppose $\lambda_0 \in \mathbb{K}$ is an isolated point of $\sigma(A)$. According to Theorem 8.1.1, $\lambda_0 \in \mathrm{Alg}_\nu(\mathfrak{L}^A)$ for some integer $\nu \geq 1$. Let $\ell_1,\dots,\ell_\nu \geq 0$ be ν integers so that the partial κ-multiplicities of $\mathfrak{L}^A$ at λ_0 are given through (7.23). Prove that

$$\dim N[(\lambda_0 I_U - A)^s] = \sum_{j=1}^{s-1} j\ell_j + s\sum_{j=s}^{\nu} \ell_j, \qquad 0 \leq s \leq \nu.$$

2. Given $p \in [1,\infty]$, let ℓ_p be the space of sequences introduced in Exercise 15 of Chapter 7. Define $S \in \mathcal{L}(\ell_p)$ as

$$(Sx)(n) = x(n+1), \qquad x \in \ell_p, \quad n \geq 0.$$

Prove that

$$\dim N[S^\alpha] = \alpha, \qquad \alpha \in \mathbb{N}.$$

3. Let $\Omega \subset \mathbb{K}$ be open, $\lambda_0 \in \Omega$ and $\mathfrak{L} \in \mathcal{C}^\infty(\Omega, \mathcal{L}(U))$ such that $\mathfrak{L}_0 \in \mathrm{Fred}_0(U)$,

$$N[\mathfrak{L}_0] \cap N[\mathfrak{L}_1] = \{0\}.$$

and

$$\mathfrak{L}(\lambda)\mathfrak{L}_0 = \mathfrak{L}_0\mathfrak{L}(\lambda), \qquad \lambda \sim \lambda_0.$$

Let $\{\mathfrak{M}_{\{j\}}\}_{j=0}^{\infty}$ be a sequence of derived families from $\mathfrak{L}$ at λ_0, as discussed in Chapter 5. Prove, step by step, the following assertions:

- For every integer $j \geq 0$ and $\lambda \sim \lambda_0$,

$$\mathfrak{L}_0^j \mathfrak{M}_{\{j\}}(\lambda) = \mathfrak{L}(\lambda)\mathfrak{L}_0^j.$$

- For every integer $j \geq 0$,

$$\mathfrak{L}_0^j \mathfrak{M}'_{\{j\}}(\lambda_0) = \mathfrak{L}_1\mathfrak{L}_0^j.$$

- For every integer $j \geq 0$,

$$N[\mathfrak{L}_0^{j+1}] = N[\mathfrak{M}_{\{j\}}(\lambda_0)] \oplus N[\mathfrak{L}_0^j].$$

- For every integer $j \geq 1$,

$$N[\mathfrak{L}_0^j] = N[\mathfrak{M}_{\{0\}}(\lambda_0)] \oplus \cdots \oplus N[\mathfrak{M}_{\{j-1\}}(\lambda_0)].$$

4. Use the result of Exercise 3 to give an alternative proof of Theorem 8.3.2 directly from the results of Chapter 5.

5. Given a linear compact operator $K \in \mathcal{K}(U)$, let $\mathfrak{L}^K$ be the family defined as

$$\mathfrak{L}^K(\lambda) := \lambda I_U - K, \qquad \lambda \in \mathbb{K}.$$

Prove that, for every open subset $\Omega \subset \mathbb{K} \setminus \{0\}$, the (finite or infinite) number $\mathfrak{m}[\mathfrak{L}^K; \Omega]$ equals the total number of eigenvalues of K in Ω, counted according to their classical algebraic multiplicities.

6. Let U be a real Banach space, set

$$\mathcal{O} := \{K \in \mathcal{K}(U) \;:\; I_U - K \in \mathrm{Iso}(U)\},$$

and consider the map, denoted by Ind (often referred to as the *topological index map*), defined by

$$\begin{array}{rcl} \mathrm{Ind} : \mathcal{O} & \to & \mathbb{N} \\ K & \mapsto & (-1)^{\mathfrak{n}[K]} \end{array}$$

where $\mathfrak{n}(K)$ stands for the number of real eigenvalues of K greater than 1, counted according to their algebraic multiplicities. Then, define the map, called *topological degree* and denoted by Deg,

$$\mathrm{Deg}(I_U - K) := \mathrm{Ind}(K), \qquad K \in \mathcal{O}.$$

Prove the following:

- $\text{Deg}(I_U) = 1$.
- If $U = \mathbb{R}^m$ for some integer $m \geq 1$, then

$$\text{Deg}(I_U - K) = \text{sign}\det(I_U - K), \qquad K \in \mathcal{O}.$$

- For every $K_1, K_2 \in \mathcal{O}$,

$$\text{Deg}\left([I_U - K_1][I_U - K_2]\right) = \text{Deg}(I_U - K_1) \cdot \text{Deg}(I_U - K_2).$$

- Let $U = \bigoplus_{j=1}^{n} U_j$ be a topological direct sum decomposition of U for some integer $n \geq 1$, and $K \in \mathcal{O}$ such that $K(U_j) \subset U_j$ for all $j \in \{1, \ldots, n\}$. For each $j \in \{1, \ldots, n\}$, define $K_j \in \mathcal{L}(U_j)$ as the restriction of K to U_j Then,

$$\text{Deg}(I - K) = \prod_{j=1}^{n} \text{Deg}\left(I_{U_j} - K_j\right).$$

7. In this exercise we suppose $\mathbb{K} = \mathbb{R}$, and, given $\mathfrak{K} \in \mathcal{C}^{\omega}(\Omega, \mathcal{L}(U))$ such that $\mathfrak{K}(\Omega) \subset \mathcal{K}(U)$, we consider the family $\mathfrak{L}$ defined by

$$\mathfrak{L}(\lambda) := I_U - \mathfrak{K}(\lambda), \qquad \lambda \in \Omega.$$

Let $\lambda_0 \in \Omega$ be an isolated point of $\Sigma(\mathfrak{L})$. Use a sequence of derived families from $\mathfrak{L}$ at λ_0, as discussed in Chapter 5, and the properties of Exercise 6 to show the existence of a constant $\eta \in \{-1, 1\}$ such that

$$\text{Deg}(\mathfrak{L}(\lambda)) = \text{Ind}(\mathfrak{K}(\lambda)) = \eta \,\text{sign}(\lambda - \lambda_0)^{\mathfrak{m}[\mathfrak{L};\lambda_0]}, \qquad \lambda \sim \lambda_0.$$

Consequently, $\text{Ind}(\mathfrak{K}(\lambda))$ changes as λ crosses λ_0 if and only if $\mathfrak{m}[\mathfrak{L}; \lambda_0]$ is odd.

8. Let $\Omega \neq \varnothing$ be a bounded open connected subset of $\mathbb{C}$, and $\mathfrak{L}$ an operator family satisfying (8.7) and (8.8). Suppose that there exists $\lambda_0 \in \mathbb{C} \setminus \partial\Omega$ such that

$$\{(\lambda - \lambda_0)x : -\infty < x < 0\} \cap \sigma(\mathfrak{L}(\lambda)) = \varnothing \quad \text{for all} \quad \lambda \in \partial\Omega.$$

Prove the following:

- If $\lambda_0 \in \Omega$, then $\mathfrak{m}[\mathfrak{L}; \Omega] = 1$. Consequently, there exists $\lambda^* \in \Omega$ such that $\mathfrak{L}(\lambda) \in \text{Iso}(U)$ for each $\lambda \in \Omega \setminus \{\lambda^*\}$, while $\lambda^* \in \Sigma(\mathfrak{L})$ and $\mathfrak{m}[\mathfrak{L}; \lambda^*] = 1$.
- If $\lambda_0 \notin \Omega$, then $\mathfrak{m}[\mathfrak{L}; \Omega] = 0$. Therefore, $\mathfrak{L}(\Omega) \subset \text{Iso}(U)$.

8.6 Comments on Chapter 8

One of the most pioneering studies of Fredholm-operator-valued holomorphic functions can be found in Gohberg [41]. The paper of Gohberg [42] presents a description of related results with references to the original sources. Stability results started in Gohberg & Kreĭn [45]. The stability of the multiplicity is due to Eni [28]; see also Markus & Sigal [95] and Gohberg & Sigal [54]. Later results appeared in Wolf [128] and Boer & Thijsse [15].

A number of different proofs of the results of Sections 8.1, 8.2 and 8.3 can be found in, for example, Gohberg, Goldberg & Kaashoek [43], Ramm [113], and López-Gómez [82], though Theorem 8.1.1 is a substantial improvement of all those results, and we believe that the proof of Theorem 8.3.2 given here is simpler than the previous ones.

Exercises 3 and 4 propose an alternative proof of Theorem 8.3.2. In this occasion, instead of invoking the results of Chapter 7, we are simply using the results of Chapter 5. Although the proof is based on some rather elementary facts and basic definitions, it runs slightly more lengthily than the one given in Section 8.3. It goes back to Magnus [92].

The relevance of Corollary 8.3.3 is based upon its huge number of applications to the theory of partial differential equations and integral equations. It has been a milestone in the development of modern functional analysis.

Theorem 8.4.3 can be regarded as an infinite-dimensional version of the celebrated Rouché's theorem. Actually, the theorem of invariance by homotopy is a global version of the Rouché theorem.

Exercises 6 and 7 summarize some of the most important properties of the topological index for compact perturbations of the identity from the point of view of nonlinear analysis. They are pivotal results in bifurcation theory, as it will become apparent in Chapter 12 (see López-Gómez [82] for further details).

Chapter 9

Algebraic Multiplicity Through Logarithmic Residues

The stability results of Section 8.4 can be regarded as infinite-dimensional versions of the classic Rouché theorem. A closely related topic in complex function theory is the so-called *argument principle*, otherwise known as the *logarithmic residue theorem*, which has been established by Theorem 3.4.1 (for classical families) and Corollary 6.5.2 in a finite-dimensional setting.

The main goal of this chapter is to obtain a number of infinite-dimensional generalizations of all these results. As a straightforward consequence of the main theorem of this chapter, it follows that if λ_0 is an isolated eigenvalue of a holomorphic family $\mathfrak{L} \in \mathcal{H}(\Omega, \mathcal{L}(U,V))$ with $\mathfrak{L}_0 \in \mathrm{Fred}_0(U,V)$, then, for every sufficiently small $r > 0$, the operator

$$\frac{1}{2\pi i}\int_{C_r(\lambda_0)} \mathfrak{L}'(\lambda)\mathfrak{L}(\lambda)^{-1}d\lambda$$

has finite rank, and

$$\chi[\mathfrak{L};\lambda_0] = \mathrm{tr}\,\frac{1}{2\pi i}\int_{C_r(\lambda_0)} \mathfrak{L}'(\lambda)\mathfrak{L}(\lambda)^{-1}d\lambda. \tag{9.1}$$

Strikingly, this result has a real non-analytic counterpart, which constitutes the main theorem of this chapter. More precisely, given two real Banach spaces U, V, an open interval $\Omega \subset \mathbb{R}$, a family $\mathfrak{L} \in \mathcal{C}^r(\Omega, \mathcal{L}(U,V))$ for some $r \in \mathbb{N} \cup \{\infty\}$ such that $\mathfrak{L}_0 \in \mathrm{Fred}_0(U,V)$ and $\lambda_0 \in \mathrm{Alg}_\nu(\mathfrak{L})$ for a certain integer $\nu \geq 1$ with $2\nu \leq r$, the main result of this chapter establishes the existence of (unique) finite-rank operators

$$\left(\mathfrak{L}'\mathfrak{L}^{-1}\right)_n \in \mathcal{L}(V), \qquad n \in \{-\nu, \dots, -1\},$$

such that, for $\lambda \neq \lambda_0$ sufficiently close to λ_0, the following asymptotic expansion at λ_0 holds true,

$$\mathfrak{L}'(\lambda)\mathfrak{L}(\lambda) = \sum_{n=-\nu}^{-1} \left(\mathfrak{L}'\mathfrak{L}^{-1}\right)_n (\lambda - \lambda_0)^n + o\left((\lambda - \lambda_0)^{-1}\right),$$

and, crucially,

$$\chi[\mathfrak{L}; \lambda_0] = \operatorname{tr}\left(\mathfrak{L}'\mathfrak{L}^{-1}\right)_{-1}. \tag{9.2}$$

Identity (9.2) provides us with a sharp generalization of (9.1) to the real non-analytic setting.

The plan of this chapter is as follows. Section 9.1 shows the existence of a finite Laurent development (or asymptotic expansion) of $\mathfrak{L}^{-1}$ at any $\lambda_0 \in \mathrm{Alg}_\nu(\mathfrak{L})$ with $\nu \leq r$ and analyzes some of its properties. Section 9.2 introduces the concept of trace of a finite-rank operator and studies its most basic properties, focusing special attention on those necessary to prove the main theorem of this chapter. Section 9.3 establishes the main theorem, Theorem 9.3.2, which has been already discussed above. Section 9.4 applies Theorem 9.3.2 for deriving (9.1) in the holomorphic setting. Finally, Section 9.5 constructs the underlying spectral projector, thus generalizing Theorem 3.4.1(d) substantially.

9.1 Finite Laurent developments of $\mathfrak{L}^{-1}$

The following result introduces the concept of finite Laurent development for the inverse $\mathfrak{L}^{-1}$ of a real non-analytic operator family $\mathfrak{L}$. It establishes the existence of a unique asymptotic expansion of $\mathfrak{L}^{-1}$ at any algebraic eigenvalue.

Theorem 9.1.1. *Let $\lambda_0 \in \Omega$ and $\mathfrak{L} \in \mathcal{C}^r(\Omega, \mathcal{L}(U,V))$ be such that $\mathfrak{L}_0 \in \mathrm{Fred}_0(U,V)$ for some $r \in \mathbb{N}$, and $\lambda_0 \in \mathrm{Alg}_\nu(\mathfrak{L})$ for some $\nu \in \{0, \dots, r\}$. Then, there exist $r - \nu + 1$ (unique) operators*

$$(\mathfrak{L}^{-1})_i \in \mathcal{L}(V,U), \qquad i \in \{-\nu, \dots, r - 2\nu\},$$

such that

$$\mathfrak{L}(\lambda)^{-1} = \sum_{i=-\nu}^{r-2\nu} (\lambda - \lambda_0)^i (\mathfrak{L}^{-1})_i + o\big((\lambda - \lambda_0)^{r-2\nu}\big) \quad \textit{as } \lambda \to \lambda_0. \tag{9.3}$$

Moreover, $(\mathfrak{L}^{-1})_{-\nu} \neq 0$,

$$\dim R[(\mathfrak{L}^{-1})_i] < \infty \qquad \textit{for all} \quad -\nu \leq i \leq \min\{-1, r - 2\nu\},$$

and if $r \geq 2\nu$, then $(\mathfrak{L}^{-1})_0 \in \mathrm{Fred}_0(V,U)$.

According to Theorem 9.1.1, we can introduce the following concept.

Definition 9.1.2. The finite asymptotic expansion (9.3) will be referred to as the finite Laurent development of $\mathfrak{L}^{-1}$ at λ_0. Accordingly, the operators $(\mathfrak{L}^{-1})_i$, for $-\nu \leq i \leq r-2\nu$, will be referred to as the Laurent coefficients of $\mathfrak{L}^{-1}$ at λ_0.

Proof of Theorem 9.1.1. The uniqueness of the operators

$$(\mathfrak{L}^{-1})_{-\nu}, \dots, (\mathfrak{L}^{-1})_{r-2\nu}$$

can be obtained very easily. Indeed, multiplying (9.3) by $(\lambda-\lambda_0)^\nu$ we obtain a Taylor expansion, whose operator coefficients are known to be unique.

In the case $\nu = 0$, we have $\mathfrak{L}_0 \in \mathrm{Iso}(U,V)$ and, hence, $\mathfrak{L}^{-1}$ is of class $\mathcal{C}^r$ in an open neighborhood of λ_0. In such a case, (9.3) is the Taylor expansion of $\mathfrak{L}^{-1}$ at λ_0 and we are done.

Suppose $1 \leq \nu \leq r$. Then, the existence of the asymptotic expansion follows from the theory developed in Chapter 5. By Corollary 5.3.2, there exist ν finite-rank projections $\Pi_j \in \mathcal{L}(U)$, for $0 \leq j \leq \nu-1$, an open neighborhood $D \subset \Omega$ of λ_0, and a family $\mathfrak{T} \in \mathcal{C}^{r-\nu}(D, \mathcal{L}(V,U))$ such that $\mathfrak{T}(D) \subset \mathrm{Iso}(V,U)$ and

$$\mathfrak{L}(\lambda)^{-1} = \mathfrak{C}_{\Pi_0}(\lambda-\lambda_0)^{-1} \cdots \mathfrak{C}_{\Pi_{\nu-1}}(\lambda-\lambda_0)^{-1}\mathfrak{T}(\lambda), \qquad \lambda \in D \setminus \{\lambda_0\}, \tag{9.4}$$

where, for any finite-rank projection $P \in \mathcal{L}(U)$ and $t \in \mathbb{K} \setminus \{0\}$, it should be remembered that

$$\mathfrak{C}_P(t)^{-1} = t^{-1}P + I_U - P.$$

Clearly, the product of the first ν operator factors on the right-hand side of (9.4) can be expressed in the form

$$\mathfrak{C}_{\Pi_0}(\lambda-\lambda_0)^{-1} \cdots \mathfrak{C}_{\Pi_{\nu-1}}(\lambda-\lambda_0)^{-1} = \sum_{i=-\nu}^{0} M_i(\lambda-\lambda_0)^i, \tag{9.5}$$

for certain operators $M_j \in \mathcal{L}(U)$, for $-\nu \leq j \leq 0$, whose explicit knowledge is not necessary here, though it is easily realized that

$$M_0 = [I_U - \Pi_0] \cdots [I_U - \Pi_{\nu-1}] \tag{9.6}$$

and that, for every $j \in \{-\nu, \dots, -1\}$, the operator M_j has finite rank, because Π_i has finite rank for all $0 \leq i \leq \nu-1$. As this entails $I_U - \Pi_i \in \mathrm{Fred}_0(U)$, equality (9.6) implies that $M_0 \in \mathrm{Fred}_0(U)$.

On the other hand, since $\mathfrak{T} \in \mathcal{C}^{r-\nu}(D, \mathcal{L}(V,U))$, we have that

$$\mathfrak{T}(\lambda) = \sum_{i=0}^{r-\nu} (\lambda-\lambda_0)^i \mathfrak{T}_i + o((\lambda-\lambda_0)^{r-\nu}) \quad \text{as } \lambda \to \lambda_0.$$

Thus, according to (9.5) and (9.4), we find that

$$\mathfrak{L}(\lambda)^{-1} = \sum_{i=-\nu}^{r-2\nu} \sum_{j=\max\{0,i\}}^{\min\{r-\nu,i+\nu\}} M_{i-j}\mathfrak{T}_j\,(\lambda-\lambda_0)^i + o\big((\lambda-\lambda_0)^{r-2\nu}\big)$$

as $\lambda \to \lambda_0$. Consequently, setting

$$(\mathfrak{L}^{-1})_i := \sum_{j=\max\{0,i\}}^{\min\{r-\nu,i+\nu\}} M_{i-j}\mathfrak{T}_j \in \mathcal{L}(V,U), \quad -\nu \leq i \leq r-2\nu, \tag{9.7}$$

yields (9.3). As, for every $j \in \{-\nu, \dots, -1\}$, the operator M_j has finite rank, it is apparent that $(\mathfrak{L}^{-1})_i$ has finite rank for all $-\nu \leq i \leq \min\{-1, r-2\nu\}$.

Moreover, as we are assuming that $\lambda_0 \in \mathrm{Alg}_\nu(\mathfrak{L})$, then ν is the minimum integer satisfying (4.18) and, therefore, $(\mathfrak{L}^{-1})_{-\nu} \neq 0$.

Finally, suppose $r \geq 2\nu$. Then, according to (9.7),

$$(\mathfrak{L}^{-1})_0 = \sum_{j=0}^{\nu} M_{-j}\mathfrak{T}_j.$$

As $\mathfrak{T}_0 \in \mathrm{Iso}(V,U)$ and $M_0 \in \mathrm{Fred}_0(U)$, then $M_0\mathfrak{T}_0 \in \mathrm{Fred}_0(V,U)$. Moreover,

$$\sum_{j=1}^{\nu} M_{-j}\mathfrak{T}_j$$

is a finite-rank operator. Therefore, $(\mathfrak{L}^{-1})_0 \in \mathrm{Fred}_0(V,U)$, because it is a finite-rank perturbation of a Fredholm operator of index zero. This completes the proof. □

In the analytic case, combining Theorem 8.2.1 with Theorem 9.1.1 provides us with the following fundamental result. It establishes that any $\lambda_0 \in \Sigma(\mathfrak{L})$ must be an algebraic eigenvalue of $\mathfrak{L}$ and that, therefore, is a pole of $\mathfrak{L}^{-1}$, which admits a (unique) finite Laurent development at λ_0.

Corollary 9.1.3. *Suppose:*

i) *Ω is an open connected subset of $\mathbb{K}$.*
ii) *$\mathfrak{L} \in \mathcal{C}^\omega(\Omega, \mathcal{L}(U,V))$ satisfies $\mathfrak{L}(\Omega) \subset \mathrm{Fred}(U,V)$.*
iii) *$\mathfrak{L}(\Omega) \cap \mathrm{Iso}(U,V) \neq \varnothing$.*

Then, $\mathfrak{L}(\Omega) \subset \mathrm{Fred}_0(U,V)$, the set $\Sigma(\mathfrak{L}) = \mathrm{Eig}(\mathfrak{L})$ is contained $\mathrm{Alg}(\mathfrak{L})$ and is discrete in Ω, the inverse family $\mathfrak{L}^{-1}$ defined as

$$\mathfrak{L}^{-1}(\lambda) := \mathfrak{L}(\lambda)^{-1}, \qquad \lambda \in \Omega \setminus \Sigma(\mathfrak{L}),$$

is analytic in $\Omega \setminus \Sigma(\mathfrak{L})$, and every $\lambda_0 \in \Sigma(\mathfrak{L})$ is a pole of $\mathfrak{L}^{-1}$.

Moreover, for each $\nu \in \mathbb{N}$ and $\lambda_0 \in \mathrm{Alg}_\nu(\mathfrak{L})$, there exist an open neighborhood $D \subset \Omega$ of λ_0 and a (unique) sequence of operators $\{(\mathfrak{L}^{-1})_n\}_{n=-\nu}^{\infty}$ in $\mathcal{L}(V,U)$ such that $(\mathfrak{L}^{-1})_{-\nu} \neq 0$, and

$$\mathfrak{L}^{-1}(\lambda) = \sum_{n=-\nu}^{\infty} (\mathfrak{L}^{-1})_n (\lambda-\lambda_0)^n, \qquad \lambda \in D \setminus \{\lambda_0\}, \tag{9.8}$$

uniformly on compact subsets of $D \setminus \{\lambda_0\}$. *In addition,*

$$(\mathfrak{L}^{-1})_0 \in \mathrm{Fred}_0(V,U), \qquad \dim R[(\mathfrak{L}^{-1})_n] < \infty, \quad n \in \{-\nu, \dots, -1\}.$$

Proof. The first part of the corollary is a direct consequence of Theorem 8.2.1. Let $\lambda_0 \in \Sigma(\mathfrak{L})$. Then, $\lambda_0 \in \mathrm{Alg}_\nu(\mathfrak{L})$ for some integer $\nu \geq 1$. According to Theorem 9.1.1 applied to $r = 2\nu$, there exist $\nu + 1$ (unique) operators $(\mathfrak{L}^{-1})_i \in \mathcal{L}(V,U)$, for $i \in \{-\nu, \dots, 0\}$, an open neighborhood $D \subset \Omega$ of λ_0, and $\mathfrak{R} \in \mathcal{C}^\omega(D \setminus \{\lambda_0\}, \mathcal{L}(V,U))$, with $\mathfrak{R}(\lambda) = o(1)$ as $\lambda \to \lambda_0$, such that

$$\mathfrak{L}^{-1}(\lambda) = \sum_{i=-\nu}^{0} (\mathfrak{L}^{-1})_i (\lambda - \lambda_0)^i + \mathfrak{R}(\lambda), \qquad \lambda \in D \setminus \{\lambda_0\}.$$

Moreover, the operator $(\mathfrak{L}^{-1})_{-\nu}$ is non-zero, $(\mathfrak{L}^{-1})_0 \in \mathrm{Fred}_0(V,U)$, and

$$\dim R[(\mathfrak{L}^{-1})_n] < \infty \qquad \text{for all} \quad n \in \{-\nu, \dots, -1\}.$$

By the theorem of Riemann on removable singularities, setting $\mathfrak{R}(\lambda_0) = 0$, we have that $\mathfrak{R} \in \mathcal{C}^\omega(D, \mathcal{L}(V,U))$. This concludes the proof very easily. The existence, uniqueness and convergence of the whole series (9.8) follows from standard facts in the theory of analytic functions. □

Definition 9.1.4. The infinite development (9.8) is called the Laurent series of $\mathfrak{L}^{-1}$ at λ_0. Accordingly, the operators $(\mathfrak{L}^{-1})_n$, for $n \in \mathbb{Z} \cap [-\nu, \infty)$, are referred to as the Laurent coefficients of $\mathfrak{L}^{-1}$ at λ_0.

More generally, suppose $p \leq q$ are integers, W is a $\mathbb{K}$-Banach space, and a map $\mathfrak{R} : \Omega \setminus \{\lambda_0\} \to W$ satisfies

$$\mathfrak{R}(\lambda) = \sum_{n=p}^{q} (\lambda - \lambda_0)^n \mathfrak{R}_n + o((\lambda - \lambda_0)^q), \qquad \lambda \in \Omega \setminus \{\lambda_0\}.$$

Then, the coefficient operators $\mathfrak{R}_n \in W$ must be unique for $p \leq n \leq q$. We will sometimes denote them by

$$c(\mathfrak{R}, \lambda_0, n) := \mathfrak{R}_n, \qquad p \leq n \leq q.$$

According to this notation, (9.3) can be equivalently written in the form

$$\mathfrak{L}(\lambda)^{-1} = \sum_{i=-\nu}^{r-2\nu} c(\mathfrak{L}^{-1}, \lambda_0, i)(\lambda - \lambda_0)^i + o\big((\lambda - \lambda_0)^{r-2\nu}\big).$$

9.2 The trace operator

This section extends the finite-dimensional concept of trace to the infinite-dimensional setting, and obtains a series of important properties concerning the traces of the product of two finite Laurent series at a $\lambda_0 \in \Omega$.

9.2.1 Concept of trace and basic properties

The following definition extends the classic finite-dimensional concept of *trace* to an infinite-dimensional setting.

Definition 9.2.1. Let U be a $\mathbb{K}$-Banach space and $A \in \mathcal{L}(U)$ a finite-rank operator. Then, the trace of A, denoted by $\operatorname{tr} A$, is defined through

$$\operatorname{tr} A := \operatorname{tr} A_R,$$

where A_R stands for the operator defined by restriction of A,

$$A|_{R[A]} : R[A] \to R[A],$$

which is finite dimensional.

The next result collects the basic properties of this concept. They are easily obtained from the corresponding properties of its finite-dimensional counterpart.

Proposition 9.2.2. *Let $A \in \mathcal{L}(U)$ be a finite-rank operator. Then,*

i) *For every finite-dimensional subspace $M \subset U$ with $R[A] \subset M$,*

$$\operatorname{tr} A = \operatorname{tr} A_M,$$

where A_M stands for the restriction $A|_M : M \to M$.

ii) *For every finite-rank operator $B \in \mathcal{L}(U)$ and $\beta \in \mathbb{K}$,*

$$\operatorname{tr}(\beta A + B) = \beta \operatorname{tr} A + \operatorname{tr} B.$$

iii) *If A is a projection then*

$$\operatorname{tr} A = \operatorname{rank} A.$$

iv) *For every Banach space V, every finite-rank operator $B \in \mathcal{L}(U, V)$, and $C \in \mathcal{L}(V, U)$,*

$$\operatorname{tr}(BC) = \operatorname{tr}(CB).$$

Proof. Let M be a finite-dimensional subspace of U such that $R[A] \subset M$. Let $A_R \in \mathcal{L}(R[A])$ denote the restriction of A from $R[A]$ to itself, and $A_M \in \mathcal{L}(M)$ the restriction of A from M to itself. By definition,

$$\operatorname{tr} A = \operatorname{tr} A_R.$$

Let U_1 be a topological complement of $R[A]$ in M, and let A_1 denote the restriction of A from U_1 to $R[A]$. With respect to the decomposition $M = R[A] \oplus U_1$, the operator A_M can be expressed in the form

$$A_M = \begin{pmatrix} A_R & A_1 \\ 0 & 0 \end{pmatrix}$$

and, therefore, $\operatorname{tr} A_M = \operatorname{tr} A_R$, which concludes the proof of property (i).

Let $B \in \mathcal{L}(U)$ be a finite-rank operator, and $\beta \in \mathbb{K}$. Let U_1 be a finite-dimensional subspace of U such that

$$R[A] + R[B] \subset U_1.$$

Let $A_1 \in \mathcal{L}(U_1)$ and $B_1 \in \mathcal{L}(U_1)$ denote the restrictions of A and B from U_1 to itself, respectively. Then, by the properties of the finite-dimensional trace operator, we find from property (i) that

$$\operatorname{tr}(\beta A + B) = \operatorname{tr}(\beta A_1 + B_1) = \beta \operatorname{tr} A_1 + \operatorname{tr} B_1 = \beta \operatorname{tr} A + \operatorname{tr} B,$$

which proves property (ii).

Suppose A is a finite-rank projection. Then, the restriction A_R of A from $R[A]$ to itself equals the identity operator in $R[A]$, whence

$$\operatorname{tr} A = \operatorname{tr} A_R = \operatorname{tr} I_{R[A]} = \dim R[A] = \operatorname{rank} A,$$

which ends the proof of property (iii).

Finally, assume that, for another $\mathbb{K}$-Banach space V, the operator $B \in \mathcal{L}(U, V)$ has finite rank and $C \in \mathcal{L}(V, U)$. Since $R[BC] \subset R[B]$, we obtain that BC is a finite-rank operator. Moreover, $R[CB] = C(R[B])$ and, hence,

$$\operatorname{rank}(CB) \leq \operatorname{rank} B.$$

Let B_1 denote the restriction of B from $R[CB]$ to $R[B]$, and C_1 the restriction of C from $R[B]$ to $R[CB]$. By (i) and the properties of the finite-dimensional trace we obtain that

$$\operatorname{tr}(BC) = \operatorname{tr}(B_1 C_1) = \operatorname{tr}(C_1 B_1) = \operatorname{tr}(CB),$$

which proves property (iv). □

9.2.2 Traces of the coefficients of the product of two families

This subsection gives some useful properties concerning the traces of the Laurent operator coefficients of a product of two operator families.

Proposition 9.2.3. *Let $p \leq q$ be integers, $\mathfrak{L} : \Omega \setminus \{\lambda_0\} \to \mathcal{L}(U, V)$ a map of the form*

$$\mathfrak{L}(\lambda) = \sum_{n=p}^{q} \mathfrak{L}_n (\lambda - \lambda_0)^n + o((\lambda - \lambda_0)^q),$$

with

$$\dim R[\mathfrak{L}_n] < \infty \quad \text{for all} \quad p \leq n \leq \min\{q, -1\}, \tag{9.9}$$

and $\mathfrak{M} \in \mathcal{C}^\infty(\Omega, \mathcal{L}(V, U))$. Then, for every $p \leq n \leq \min\{q, -1\}$, the operators

$$c(\mathfrak{L}\mathfrak{M}, \lambda_0, n) \in \mathcal{L}(V) \quad \text{and} \quad c(\mathfrak{M}\mathfrak{L}, \lambda_0, n) \in \mathcal{L}(U)$$

have finite rank and

$$\operatorname{tr} c(\mathfrak{L}\mathfrak{M}, \lambda_0, n) = \operatorname{tr} c(\mathfrak{M}\mathfrak{L}, \lambda_0, n).$$

Proof. For λ in a perforated neighborhood of λ_0, we have that

$$\mathfrak{L}(\lambda)\mathfrak{M}(\lambda) = \sum_{n=p}^{q}\sum_{j=p}^{n} \mathfrak{L}_j \mathfrak{M}_{n-j}(\lambda-\lambda_0)^n + o((\lambda-\lambda_0)^q),$$

$$\mathfrak{M}(\lambda)\mathfrak{L}(\lambda) = \sum_{n=p}^{q}\sum_{j=p}^{n} \mathfrak{M}_{n-j}\mathfrak{L}_j(\lambda-\lambda_0)^n + o((\lambda-\lambda_0)^q).$$

Therefore, by the uniqueness of the coefficients in these asymptotic developments, we find that, for every $p \leq n \leq \min\{q,-1\}$,

$$c(\mathfrak{L}\mathfrak{M},\lambda_0,n) = \sum_{j=p}^{n} \mathfrak{L}_j\mathfrak{M}_{n-j}, \qquad c(\mathfrak{M}\mathfrak{L},\lambda_0,n) = \sum_{j=p}^{n} \mathfrak{M}_{n-j}\mathfrak{L}_j.$$

According to (9.9), each of these coefficient operators must have finite rank. Therefore, by properties (ii), (iv) of Proposition 9.2.2, we find that

$$\operatorname{tr} c(\mathfrak{L}\mathfrak{M},\lambda_0,n) = \sum_{j=p}^{n} \operatorname{tr}(\mathfrak{L}_j\mathfrak{M}_{n-j}) = \sum_{j=p}^{n} \operatorname{tr}(\mathfrak{M}_{n-j}\mathfrak{L}_j) = \operatorname{tr} c(\mathfrak{M}\mathfrak{L},\lambda_0,n),$$

which completes the proof. □

Proposition 9.2.4. *Consider $r \in \mathbb{N}\cup\{\infty\}$. Let*

$$\mathfrak{L} \in \mathcal{C}^r(\Omega,\mathcal{L}(U,V)) \quad \text{and} \quad \mathfrak{F} \in \mathcal{C}^\infty(\Omega,\mathcal{L}(U))$$

be such that $\mathfrak{L}_0 \in \mathrm{Fred}_0(U,V)$ and $\mathfrak{F}_0 \in \mathrm{Iso}(U)$. Assume that $\lambda_0 \in \mathrm{Alg}_\nu(\mathfrak{L})$ for some integer $\nu \geq 1$ with $2\nu \leq r+1$. Then, for each $i \in \{-\nu,\dots,-1\}$, the operators $c((\mathfrak{L}\mathfrak{F})'(\mathfrak{L}\mathfrak{F})^{-1},\lambda_0,i)$ and $c(\mathfrak{L}'\mathfrak{L}^{-1},\lambda_0,i)$ have finite rank and

$$\operatorname{tr} c((\mathfrak{L}\mathfrak{F})'(\mathfrak{L}\mathfrak{F})^{-1},\lambda_0,i) = \operatorname{tr} c(\mathfrak{L}'\mathfrak{L}^{-1},\lambda_0,i).$$

Proof. In some perforated neighborhood of λ_0, we can express

$$(\mathfrak{L}\mathfrak{F})'(\mathfrak{L}\mathfrak{F})^{-1} = \mathfrak{L}'\mathfrak{L}^{-1} + \mathfrak{L}\mathfrak{F}'\mathfrak{F}^{-1}\mathfrak{L}^{-1}. \tag{9.10}$$

The family $\mathfrak{L}\mathfrak{F}'\mathfrak{F}^{-1}$ is of class $\mathcal{C}^r$ in a neighborhood of λ_0. By Theorem 9.1.1, there exists an open neighborhood $\tilde{\Omega} \subset \Omega$ of λ_0 such that, for each $\lambda \in \tilde{\Omega}\setminus\{\lambda_0\}$, the operator $\mathfrak{L}(\lambda)^{-1}$ exists and

$$\mathfrak{L}(\lambda)^{-1} = \sum_{i=-\nu}^{-1} c(\mathfrak{L}^{-1},\lambda_0,i)(\lambda-\lambda_0)^i + o\big((\lambda-\lambda_0)^{-1}\big).$$

Moreover,

$$\dim R[c(\mathfrak{L}^{-1},\lambda_0,i)] < \infty \qquad \text{for all} \quad i \in \{-\nu,\dots,-1\}. \tag{9.11}$$

Thus, due to Proposition 9.2.3, for every $i \in \{-\nu, \dots, -1\}$, the operators

$$c(\mathfrak{L}\mathfrak{F}'\mathfrak{F}^{-1}\mathfrak{L}^{-1}, \lambda_0, i) \quad \text{and} \quad c(\mathfrak{F}'\mathfrak{F}^{-1}, \lambda_0, i)$$

have finite rank and

$$\operatorname{tr} c(\mathfrak{L}\mathfrak{F}'\mathfrak{F}^{-1}\mathfrak{L}^{-1}, \lambda_0, i) = \operatorname{tr} c(\mathfrak{F}'\mathfrak{F}^{-1}, \lambda_0, i).$$

Actually, since $\mathfrak{F}'\mathfrak{F}^{-1}$ is of class $\mathcal{C}^\infty$ in a neighborhood of λ_0,

$$c(\mathfrak{F}'\mathfrak{F}^{-1}, \lambda_0, i) = 0, \qquad i \in \{-\nu, \dots, -1\},$$

and, therefore,

$$\operatorname{tr} c(\mathfrak{L}\mathfrak{F}'\mathfrak{F}^{-1}\mathfrak{L}^{-1}, \lambda_0, i) = 0, \qquad i \in \{-\nu, \dots, -1\}. \tag{9.12}$$

On the other hand, as $\mathfrak{L}'$ is of class $\mathcal{C}^{r-1}$, it follows from (9.11) that

$$\dim R[c(\mathfrak{L}'\mathfrak{L}^{-1}, \lambda_0, i)] < \infty \qquad \text{for all} \quad i \in \{-\nu, \dots, -1\}.$$

Consequently, by (9.10),

$$\dim R[c((\mathfrak{L}\mathfrak{F})'(\mathfrak{L}\mathfrak{F})^{-1}, \lambda_0, i)] < \infty \qquad \text{for all} \quad i \in \{-\nu, \dots, -1\},$$

and, thanks to Proposition 9.2.2(ii), we conclude from (9.10) and (9.12) that

$$\operatorname{tr} c((\mathfrak{L}\mathfrak{F})'(\mathfrak{L}\mathfrak{F})^{-1}, \lambda_0, i) = \operatorname{tr} c(\mathfrak{L}'\mathfrak{L}^{-1}, \lambda_0, i), \qquad i \in \{-\nu, \dots, -1\}.$$

This ends the proof. □

9.3 The multiplicity through a logarithmic residue

The next result shows that transversalizing a family $\mathfrak{L}$ at $\lambda_0 \in \Omega$ extraordinarily simplifies the finite Laurent expansion of $\mathfrak{L}'\mathfrak{L}^{-1}$ at any $\lambda_0 \in \mathrm{Alg}_k(\mathfrak{L})$ for some integer $k \geq 1$. More precisely, it characterizes the ranges of

$$c(\mathfrak{L}, \lambda_0, i) \cdot c(\mathfrak{L}^{-1}, \lambda_0, i), \qquad -k \leq i \leq 0,$$

through the subspaces in which the direct sum (4.7) decomposes. This property will enable us to obtain the main result of this chapter, which asserts

$$\operatorname{tr} c(\mathfrak{L}'\mathfrak{L}^{-1}, \lambda_0, -1) = \chi[\mathfrak{L}; \lambda_0].$$

Lemma 9.3.1. *Let $r \in \mathbb{N} \cup \{\infty\}$ and $\lambda_0 \in \Omega$. Suppose $\mathfrak{L} \in \mathcal{C}^r(\Omega, \mathcal{L}(U,V))$ satisfies $\mathfrak{L}_0 \in \mathrm{Fred}_0(U,V)$, and λ_0 is a k-transversal eigenvalue of $\mathfrak{L}$ for some integer $k \geq 1$ with $2k \leq r$. Then,*

$$R[\mathfrak{L}_0(\mathfrak{L}^{-1})_0] = R[\mathfrak{L}_0], \qquad R[\mathfrak{L}_i(\mathfrak{L}^{-1})_{-i}] = \mathfrak{L}_i(\bigcap_{j=0}^{i-1} N[\mathfrak{L}_j]), \quad 1 \leq i \leq k, \tag{9.13}$$

and

$$\begin{aligned} &\sum_{i=0}^{k} \mathfrak{L}_i(\mathfrak{L}^{-1})_{-i} = I_V, \\ &\mathfrak{L}_i(\mathfrak{L}^{-1})_{-i}\mathfrak{L}_j(\mathfrak{L}^{-1})_{-j} = \delta_{ij}\mathfrak{L}_i(\mathfrak{L}^{-1})_{-i}, \qquad 0 \leq i,j \leq k. \end{aligned} \tag{9.14}$$

Proof. Thanks to Theorem 9.1.1, the inverse operator $\mathfrak{L}^{-1}$ admits, for λ in a perforated neighborhood of λ_0, a unique finite Laurent expansion of the form

$$\mathfrak{L}(\lambda)^{-1} = \sum_{n=-k}^{0} (\mathfrak{L}^{-1})_n(\lambda-\lambda_0)^n + o(1), \tag{9.15}$$

for some operators

$$(\mathfrak{L}^{-1})_{-k}, \dots, (\mathfrak{L}^{-1})_0 \in \mathcal{L}(V,U)$$

such that $(\mathfrak{L}^{-1})_{-k} \neq 0$ and

$$\dim R[(\mathfrak{L}^{-1})_n] < \infty, \qquad n \in \{-k, \dots, -1\}.$$

A direct calculation yields

$$\begin{aligned} I_V &= \mathfrak{L}(\lambda)\mathfrak{L}(\lambda)^{-1} \\ &= \left[\sum_{n=0}^{k} \mathfrak{L}_n(\lambda-\lambda_0)^n + o((\lambda-\lambda_0)^k)\right]\left[\sum_{n=-k}^{0} (\mathfrak{L}^{-1})_n(\lambda-\lambda_0)^n + o(1)\right] \\ &= \sum_{n=-k}^{0}\sum_{i=0}^{n+k} \mathfrak{L}_i(\mathfrak{L}^{-1})_{n-i}(\lambda-\lambda_0)^n + o(1), \end{aligned}$$

for λ in a certain perforated neighborhood of λ_0. Consequently, owing to the uniqueness of the Laurent coefficients, guaranteed by Theorem 9.1.1, we find that

$$\sum_{i=0}^{k} \mathfrak{L}_i(\mathfrak{L}^{-1})_{-i} = I_V \tag{9.16}$$

and

$$\sum_{i=0}^{n+k} \mathfrak{L}_i(\mathfrak{L}^{-1})_{n-i} = 0, \qquad n \in \{-k, \dots, -1\}. \tag{9.17}$$

In particular, the first identity of (9.14) holds. Moreover, according to (9.17), we have that, for each $v_{-k}, \dots, v_{-1} \in V$,

$$\sum_{i=0}^{n+k} \mathfrak{L}_i(\mathfrak{L}^{-1})_{n-i} v_{n-i} = 0, \qquad n \in \{-k, \dots, -1\},$$

and, consequently, as λ_0 is a k-transversal eigenvalue of $\mathfrak{L}$, Lemma 4.1.1 implies that

$$\mathfrak{L}_i(\mathfrak{L}^{-1})_{n-i} v_{n-i} = 0, \qquad -k \leq n \leq -1, \quad 0 \leq i \leq n+k.$$

In other words,

$$R[(\mathfrak{L}^{-1})_{-i}] \subset \bigcap_{j=0}^{i-1} N[\mathfrak{L}_j], \qquad 1 \leq i \leq k, \tag{9.18}$$

and, hence,

$$R[\mathfrak{L}_0(\mathfrak{L}^{-1})_0] \subset R[\mathfrak{L}_0], \qquad R[\mathfrak{L}_i(\mathfrak{L}^{-1})_{-i}] \subset \mathfrak{L}_i(\bigcap_{j=0}^{i-1} N[\mathfrak{L}_j]), \quad 1 \leq i \leq k.$$

Therefore, according to (9.16), by the uniqueness of the projections associated with every factor of a direct sum, the operators

$$\mathfrak{L}_0(\mathfrak{L}^{-1})_0, \quad \mathfrak{L}_1(\mathfrak{L}^{-1})_{-1}, \quad \dots, \quad \mathfrak{L}_k(\mathfrak{L}^{-1})_{-k}, \tag{9.19}$$

must be the projections associated with the direct sum decomposition (4.7), which concludes the proof. □

Note that (9.13) and (9.14) indeed establish that the operators (9.19) are the associated projections to each of the factors of the direct sum (4.7).

The main result of this chapter can be stated as follows.

Theorem 9.3.2. *Let $r \in \mathbb{N} \cup \{\infty\}$ and $\mathfrak{L} \in \mathcal{C}^r(\Omega, \mathcal{L}(U, V))$ be such that $\mathfrak{L}_0 \in \mathrm{Fred}_0(U, V)$ and $\lambda_0 \in \mathrm{Alg}_\nu(\mathfrak{L})$ for some integer $\nu \geq 1$ with $2\nu \leq r$. Then, for every $n \in \{-\nu, \dots, -1\}$, the operator $c(\mathfrak{L}'\mathfrak{L}^{-1}, \lambda_0, n) \in \mathcal{L}(V)$ has finite rank and*

$$\operatorname{tr} c(\mathfrak{L}'\mathfrak{L}^{-1}, \lambda_0, n) = \begin{cases} 0, & n \in \{-\nu, \dots, -2\}, \\ \chi[\mathfrak{L}; \lambda_0], & n = -1. \end{cases} \tag{9.20}$$

Proof. Assume that λ_0 is a ν-transversal eigenvalue of $\mathfrak{L}$. Then, by (9.15) and

$$\mathfrak{L}'(\lambda) = \sum_{n=0}^{\nu} (n+1)\mathfrak{L}_{n+1}(\lambda - \lambda_0)^n + o((\lambda - \lambda_0)^\nu),$$

we obtain that, for $\lambda \sim \lambda_0$ and $\lambda \neq \lambda_0$,

$$\mathfrak{L}'(\lambda)\mathfrak{L}(\lambda)^{-1} = \sum_{n=-\nu}^{0} \sum_{i=1}^{n+\nu+1} i\mathfrak{L}_i(\mathfrak{L}^{-1})_{n+1-i}(\lambda - \lambda_0)^n + o(1).$$

Thus, from (9.18) it is apparent that, whenever $\nu \geq 2$,

$$c(\mathfrak{L}'\mathfrak{L}^{-1}, \lambda_0, n) = 0, \qquad n \in \{-\nu, \dots, -2\}.$$

Moreover,

$$c(\mathfrak{L}'\mathfrak{L}^{-1}, \lambda_0, -1) = \sum_{i=1}^{\nu} i\mathfrak{L}_i c(\mathfrak{L}^{-1}, \lambda_0, -i) = \sum_{i=1}^{\nu} i\mathfrak{L}_i (\mathfrak{L}^{-1})_{-i}.$$

Consequently, due to Lemma 9.3.1, $c(\mathfrak{L}'\mathfrak{L}^{-1}, \lambda_0, -1)$ is a linear combination of finite-rank projections. In particular, it has finite rank. Therefore, since $\mathfrak{L}_i(\mathfrak{L}^{-1})_{-i}$ is a projection for every $1 \leq i \leq \nu$, we find from Proposition 9.2.2 and Definition 4.3.3 that

$$\operatorname{tr} c(\mathfrak{L}'\mathfrak{L}^{-1}, \lambda_0, -1) = \sum_{i=1}^{\nu} i \dim \mathfrak{L}_i\Big(\bigcap_{j=0}^{i-1} N[\mathfrak{L}_j]\Big) = \chi[\mathfrak{L}; \lambda_0],$$

which concludes the proof of the theorem when λ_0 is a ν-transversal eigenvalue of $\mathfrak{L}$.

Suppose now that $\lambda_0 \in \mathrm{Alg}_\nu(\mathfrak{L})$. Then, according to Theorem 5.3.1, there exists $\Phi \in \mathbb{P}[\mathcal{L}(U)]$ such that $\Phi(\lambda_0) = I_U$ and λ_0 is a ν-transversal eigenvalue of the transversalized family

$$\mathfrak{L}^\Phi := \mathfrak{L}\Phi.$$

Thanks to Proposition 9.2.4, we have that

$$\operatorname{tr} c((\mathfrak{L}^\Phi)'(\mathfrak{L}^\Phi)^{-1}, \lambda_0, n) = \operatorname{tr} c(\mathfrak{L}'\mathfrak{L}^{-1}, \lambda_0, n), \qquad n \in \{-\nu, \dots, -1\}.$$

Moreover, since λ_0 is a ν-transversal eigenvalue of $\mathfrak{L}^\Phi$, we already know that

$$\operatorname{tr} c((\mathfrak{L}^\Phi)'(\mathfrak{L}^\Phi)^{-1}, \lambda_0, n) = \begin{cases} 0, & n \in \{-\nu, \dots, -2\}, \\ \chi[\mathfrak{L}^\Phi; \lambda_0], & n = -1. \end{cases}$$

Finally, by Definition 4.3.3,

$$\chi[\mathfrak{L}; \lambda_0] = \chi[\mathfrak{L}^\Phi; \lambda_0].$$

Therefore, (9.20) holds true, which concludes the proof. □

Theorem 9.3.2 is satisfied for $\nu = 0$ as well. Indeed, in such a case, $\mathfrak{L}_0 \in \mathrm{Iso}(U, V)$ and, hence, $\mathfrak{L}'\mathfrak{L}^{-1}$ is of class $\mathcal{C}^{r-1}$ in a neighborhood of λ_0. Consequently,

$$c(\mathfrak{L}'\mathfrak{L}^{-1}, \lambda_0, -1) = 0$$

and, therefore,

$$\operatorname{tr} c(\mathfrak{L}'\mathfrak{L}^{-1}, \lambda_0, -1) = \operatorname{tr} 0 = 0 = \chi[\mathfrak{L}; \lambda_0].$$

9.4 Holomorphic and classical families

When $\mathbb{K} = \mathbb{C}$ and $\mathfrak{L} \in \mathcal{H}(\Omega; \mathcal{L}(U, V))$, Theorem 9.3.2 provides us with the following classic result, which extends to arbitrary complex Banach spaces the finite-dimensional result established by Corollary 6.5.2.

Theorem 9.4.1. *Suppose:*

i) Ω *is an open connected subset of* $\mathbb{C}$.
ii) $\mathfrak{L} \in \mathcal{H}(\Omega, \mathcal{L}(U, V))$ *satisfies* $\mathfrak{L}(\Omega) \subset \mathrm{Fred}(U, V)$.
iii) $\mathfrak{L}(\Omega) \cap \mathrm{Iso}(U, V) \neq \varnothing$.

Then, for every $\lambda_0 \in \Sigma(\mathfrak{L})$,

$$\chi[\mathfrak{L}; \lambda_0] = \mathrm{tr}\, \frac{1}{2\pi i} \int_\gamma \mathfrak{L}'(\lambda)\mathfrak{L}(\lambda)^{-1}\, d\lambda,$$

where γ *is any Cauchy contour whose inner domain* D_γ *satisfies*

$$\bar{D}_\gamma \subset \Omega, \qquad \bar{D}_\gamma \cap \Sigma(\mathfrak{L}) = \{\lambda_0\}.$$

Proof. Thanks to Theorem 8.2.1, $\mathfrak{L}(\Omega) \subset \mathrm{Fred}_0(U, V)$, and

$$\Sigma(\mathfrak{L}) = \mathrm{Eig}(\mathfrak{L}) \subset \mathrm{Alg}(\mathfrak{L})$$

is a discrete set in Ω. Fix $\lambda_0 \in \Sigma(\mathfrak{L})$ and let γ be a Cauchy contour whose inner domain D_γ satisfies the requirements of the statement. Also, let $\nu \geq 1$ be an integer such that $\lambda_0 \in \mathrm{Alg}_\nu(\mathfrak{L})$. Then, $\lambda_0 \in \mathrm{Alg}_k(\mathfrak{L}'\mathfrak{L}^{-1})$ for some $k \in \{1, \dots, \nu\}$ and, therefore, the Laurent series

$$\mathfrak{L}'(\lambda)\mathfrak{L}^{-1}(\lambda) = \sum_{n=-k}^{\infty} c(\mathfrak{L}'\mathfrak{L}^{-1}, \lambda_0, n)(\lambda - \lambda_0)^n \tag{9.21}$$

converges uniformly in compact subsets of

$$\big(\bar{D}_\gamma + D_\delta(0)\big) \setminus \{\lambda_0\},$$

for sufficiently small $\delta > 0$. Therefore, integrating (9.21) along γ, shows that

$$c(\mathfrak{L}'\mathfrak{L}^{-1}, \lambda_0, -1) = \frac{1}{2\pi i} \int_\gamma \mathfrak{L}'(\lambda)\mathfrak{L}^{-1}(\lambda)\, d\lambda.$$

Consequently, by Theorem 9.3.2,

$$\chi[\mathfrak{L}; \lambda_0] = \mathrm{tr}\, c(\mathfrak{L}'\mathfrak{L}^{-1}, \lambda_0, -1) = \mathrm{tr}\, \frac{1}{2\pi i} \int_\gamma \mathfrak{L}'(\lambda)\mathfrak{L}^{-1}(\lambda)\, d\lambda,$$

which concludes the proof. □

According to Theorem 9.4.1, when $A \in \mathcal{L}(U)$ and the classical family $\mathfrak{L}^A$ defined as

$$\mathfrak{L}^A(\lambda) := \lambda I_U - A, \qquad \lambda \in \mathbb{C},$$

satisfies $\mathfrak{L}^A(\Omega) \subset \text{Fred}(U)$ for some open connected subset $\Omega \subset \mathbb{C}$, then

$$\chi[\mathfrak{L}^A; \lambda_0] = \operatorname{tr} \frac{1}{2\pi i} \int_\gamma (\lambda I_U - A)^{-1}\, d\lambda,$$

for all $\lambda_0 \in \text{Eig}(A) \cap \Omega$ and any Cauchy contour γ whose inner domain D_γ satisfies $\bar{D}_\gamma \subset \Omega$ and $\sigma(A) \cap \bar{D}_\gamma = \{\lambda_0\}$. Moreover, this is always the case if $A \in \mathcal{K}(U)$ and $\lambda_0 \neq 0$. Further, by Theorem 8.3.2, under these circumstances, we actually have

$$\chi[\mathfrak{L}^A; \lambda_0] = m_a(\lambda_0) = \dim N[(\lambda_0 I_U - A)^{\nu(\lambda_0)}]$$

and $\lambda_0 \in \text{Alg}_{\nu(\lambda_0)}(\mathfrak{L})$. Therefore,

$$m_a(\lambda_0) = \operatorname{tr} \frac{1}{2\pi i} \int_\gamma (\lambda I_U - A)^{-1}\, d\lambda.$$

Actually, in the next section we will show that

$$((\mathfrak{L}^A)^{-1})_{-1} := c((\mathfrak{L}^A)^{-1}, \lambda_0, -1) = \frac{1}{2\pi i} \int_\gamma (\lambda I_U - A)^{-1}\, d\lambda \in \mathcal{L}(U)$$

is a projection onto $N[(\lambda I_U - A)^{\nu(\lambda_0)}]$; thus generalizing the corresponding finite-dimensional result of Theorem 3.4.1.

9.5 Spectral projection

According to Theorem 7.6.2, if $r \in \mathbb{N}^* \cup \{\infty\}$ and $\mathfrak{L} \in \mathcal{C}^r(\Omega, \mathcal{L}(U,V))$ satisfies $\mathfrak{L}_0 \in \text{Fred}_0(U,V)$, then $\lambda_0 \in \text{Alg}_\nu(\mathfrak{L})$ for some integer $1 \leq \nu \leq r$ if and only if

$$\begin{aligned} &\dim N[\mathfrak{L}_{[j-1]}] < \dim N[\mathfrak{L}_{[j]}], \qquad \text{for all } 1 \leq j \leq \nu - 1, \\ &\dim N[\mathfrak{L}_{[\nu-1]}] = \dim N[\mathfrak{L}_{[\nu]}], \end{aligned}$$

and, in such a case,

$$\chi[\mathfrak{L}; \lambda_0] = \dim N[\mathfrak{L}_{[\nu-1]}].$$

The following result provides us with a projection of U^ν onto $N[\mathfrak{L}_{[\nu-1]}]$.

Theorem 9.5.1. *Let $r \in \mathbb{N} \cup \{\infty\}$ and $\mathfrak{L} \in \mathcal{C}^r(\Omega, \mathcal{L}(U,V))$ be such that $\mathfrak{L}_0 \in \text{Fred}_0(U,V)$ and $\lambda_0 \in \text{Alg}_\nu(\mathfrak{L})$ for some integer $\nu \geq 1$ with $2\nu \leq r+1$. Then, the operator*

$$P := \begin{pmatrix} (\mathfrak{L}^{-1})_{-\nu} & & \\ \vdots & \ddots & \\ (\mathfrak{L}^{-1})_{-1} & \cdots & (\mathfrak{L}^{-1})_{-\nu} \end{pmatrix} \begin{pmatrix} \mathfrak{L}_\nu & \cdots & \mathfrak{L}_1 \\ \vdots & \ddots & \vdots \\ \mathfrak{L}_{2\nu-1} & \cdots & \mathfrak{L}_\nu \end{pmatrix} \in \mathcal{L}(U^\nu)$$

is a projection onto $N[\mathfrak{L}_{[\nu-1]}]$. *Moreover,*

$$R[\mathrm{trng}\{(\mathfrak{L}^{-1})_{-\nu},\dots,(\mathfrak{L}^{-1})_{-1}\}] = N[\mathfrak{L}_{[\nu-1]}] \tag{9.22}$$

and, therefore, due to Theorem 7.6.2,

$$\chi[\mathfrak{L};\lambda_0] = \dim R[P] = \dim R[\mathrm{trng}\{(\mathfrak{L}^{-1})_{-\nu},\dots,(\mathfrak{L}^{-1})_{-1}\}].$$

Proof. As $\mathfrak{L}\mathfrak{L}^{-1} = I_V$, identifying terms with the same order in the asymptotic expansion of $\mathfrak{L}\mathfrak{L}^{-1}$, one easily arrives at

$$\mathrm{trng}\{\mathfrak{L}_0,\dots,\mathfrak{L}_{\nu-1}\}\cdot\mathrm{trng}\{(\mathfrak{L}^{-1})_{-\nu},\dots,(\mathfrak{L}^{-1})_{-1}\} = 0. \tag{9.23}$$

Similarly, it follows from

$$\mathfrak{L}^{-1}(\lambda)\mathfrak{L}(\lambda) = I_U, \qquad \lambda\sim\lambda_0, \quad \lambda\neq\lambda_0,$$

that

$$\begin{pmatrix} (\mathfrak{L}^{-1})_0 & \cdots & (\mathfrak{L}^{-1})_{-\nu+1} \\ \vdots & \ddots & \vdots \\ (\mathfrak{L}^{-1})_{\nu-1} & \cdots & (\mathfrak{L}^{-1})_0 \end{pmatrix}\begin{pmatrix} \mathfrak{L}_0 & & \\ \vdots & \ddots & \\ \mathfrak{L}_{\nu-1} & \cdots & \mathfrak{L}_0 \end{pmatrix} = I_{U^\nu} - P. \tag{9.24}$$

By (9.23),

$$R[P]\subset N[\mathrm{trng}\{\mathfrak{L}_0,\dots,\mathfrak{L}_{\nu-1}\}] = N[\mathfrak{L}_{[\nu-1]}].$$

Let $x\in N[\mathfrak{L}_{[\nu-1]}]$. Then, thanks to (9.24), $Px = x$. Consequently, P is a projection onto $N[\mathfrak{L}_{[\nu-1]}]$. Moreover, it follows from the definition of P that

$$R[P]\subset R\left[\mathrm{trng}\{(\mathfrak{L}^{-1})_{-\nu},\dots,(\mathfrak{L}^{-1})_{-1}\}\right],$$

and, according to (9.23),

$$R\left[\mathrm{trng}\{(\mathfrak{L}^{-1})_{-\nu},\dots,(\mathfrak{L}^{-1})_{-1}\}\right]\subset N\left[\mathfrak{L}_{[\nu-1]}\right].$$

Therefore, (9.22) holds. The remaining assertions follow from Theorem 7.6.2. This concludes the proof. □

As a consequence of Theorem 9.5.1, one can easily obtain the following classic result, which generalizes Theorem 3.4.1(d).

Theorem 9.5.2. *Given* $A\in\mathcal{L}(U)$, *let* $\mathfrak{L}^A\in\mathbb{P}[\mathcal{L}(U)]$ *be defined by*

$$\mathfrak{L}^A(\lambda) := \lambda I_U - A, \qquad \lambda\in\mathbb{K},$$

and assume that $\lambda_0\in\mathbb{K}$ *is an isolated eigenvalue of* $\mathfrak{L}^A$ *such that* $\mathfrak{L}^A_0\in\mathrm{Fred}_0(U)$. *Then,*

$$\left((\mathfrak{L}^A)^{-1}\right)_{-1}\in\mathcal{L}(U)$$

is a projection onto $N[(\lambda I_U - A)^{\nu(\lambda_0)}]$.

If, in addition, $\mathbb{K} = \mathbb{C}$ *then for any Cauchy contour* γ *whose inner domain* D_γ *satisfies* $\bar{D}_\gamma \subset \Omega$ *and* $\sigma(A) \cap \bar{D}_\gamma = \{\lambda_0\}$, *one has that*

$$c\left(\left(\mathfrak{L}^A\right)^{-1}, \lambda_0, -1\right) = \frac{1}{2\pi i}\int_\gamma (\lambda I_U - A)^{-1}\, d\lambda. \tag{9.25}$$

Proof. By Theorem 8.3.2,

$$\lambda_0 \in \mathrm{Alg}_{\nu(\lambda_0)}(\mathfrak{L}^A).$$

As usual, we set

$$\left((\mathfrak{L}^A)^{-1}\right)_{-j} := c\left(\left(\mathfrak{L}^A\right)^{-1}, \lambda_0, j\right), \qquad j \in \{-\nu(\lambda_0), \dots, -1\}.$$

Since

$$\mathfrak{L}_1^A = I_U, \qquad \mathfrak{L}_j^A = 0, \quad j \geq 2,$$

we find from Theorem 9.5.1 that the operator

$$P := \begin{pmatrix} 0 & \cdots & 0 & \left((\mathfrak{L}^A)^{-1}\right)_{-\nu(\lambda_0)} \\ 0 & \cdots & 0 & \left((\mathfrak{L}^A)^{-1}\right)_{-\nu(\lambda_0)+1} \\ \vdots & \ddots & \vdots & \vdots \\ 0 & \cdots & 0 & \left((\mathfrak{L}^A)^{-1}\right)_{-1} \end{pmatrix} \in \mathcal{L}(U^{\nu(\lambda_0)})$$

is a projection onto $N[\mathfrak{L}^A_{[\nu(\lambda_0)-1]}]$.

Now, we denote by $\mathcal{P}_{\nu(\lambda_0)-1} \in \mathcal{L}(U^{\nu(\lambda_0)}, U)$ the operator defined through

$$\mathcal{P}_{\nu(\lambda_0)-1}\,\mathrm{col}[u_0, \dots, u_{\nu(\lambda_0)-1}] = u_{\nu(\lambda_0)-1}, \qquad u_0, \dots, u_{\nu(\lambda_0)-1} \in U.$$

As P is a projection onto $N[\mathfrak{L}^A_{[\nu(\lambda_0)-1]}]$, we have that

$$R[\left((\mathfrak{L}^A)^{-1}\right)_{-1}] \subset \mathcal{P}_{\nu(\lambda_0)-1}\left(N[\mathfrak{L}^A_{[\nu(\lambda_0)-1]}]\right).$$

Conversely, for any

$$u_{\nu(\lambda_0)-1} \in \mathcal{P}_{\nu(\lambda_0)-1}\left(N[\mathfrak{L}^A_{[\nu(\lambda_0)-1]}]\right),$$

there exists $u_0, \dots, u_{\nu(\lambda_0)-2} \in U$ such that

$$u := \mathrm{col}[u_0, \dots, u_{\nu(\lambda_0)-1}] \in N[\mathfrak{L}^A_{[\nu(\lambda_0)-1]}].$$

Moreover, since $Pu = u$, we have that

$$\left((\mathfrak{L}^A)^{-1}\right)_{-1} u_{\nu(\lambda_0)-1} = u_{\nu(\lambda_0)-1}.$$

Consequently, $\left((\mathfrak{L}^A)^{-1}\right)_{-1}$ is a projection onto $\mathcal{P}_{\nu(\lambda_0)-1}(N[\mathfrak{L}^A_{[\nu(\lambda_0)-1]}])$. On the other hand, by Proposition 8.3.1, we have that

$$\mathcal{P}_{\nu(\lambda_0)-1}(N[\mathfrak{L}^A_{[\nu(\lambda_0)-1]}]) = N[(\lambda_0 I_U - A)^{\nu(\lambda_0)}].$$

Therefore, $\left((\mathfrak{L}^A)^{-1}\right)_{-1}$ is a projection onto $N[(\lambda_0 I_U - A)^{\nu(\lambda_0)}]$.

If $\mathbb{K} = \mathbb{C}$, then, by Theorem 9.4.1, it is apparent that (9.25) is satisfied. This concludes the proof. □

9.6 Exercises

1. The statement of Theorem 9.3.2 holds true substituting $\mathfrak{L}^{-1}\mathfrak{L}'$ for $\mathfrak{L}'\mathfrak{L}^{-1}$. State this formally and prove it.

2. The proof of Theorem 9.3.2 can be easily adapted to show that, for every scalar-valued function f of class $\mathcal{C}^{k-1}$ in a neighborhood of λ_0,

$$\operatorname{tr} c(f\mathfrak{L}'\mathfrak{L}^{-1}, \lambda_0, -1) = \operatorname{tr} c(f\mathfrak{L}^{-1}\mathfrak{L}', \lambda_0, -1) = f(\lambda_0)\chi[\mathfrak{L}; \lambda_0].$$

State this formally and prove it.

3. This exercise provides an example of an infinite-dimensional operator family

$$\mathfrak{L} \in \mathcal{C}^r(\Omega, \mathcal{L}(U, V)) \quad \text{with} \quad \mathfrak{L}(\Omega) \subset \operatorname{Fred}_0(U, V)$$

for which the Laurent series of $\mathfrak{L}^{-1}$ can be determined explicitly. Precisely, consider integers $n, r \geq 1$, an open interval $D \subset \mathbb{R}$ with $0 \in D$, a function $g \in \mathcal{C}^r(D, \mathbb{R})$ satisfying $g(0) = 0$, and define

$$f(\lambda) := (n\pi)^2 + g(\lambda), \qquad \lambda \in D.$$

Let U denote the Banach space of functions $u \in \mathcal{C}^2[0,1]$ such that $u(0) = u(1) = 0$, and $V := \mathcal{C}[0,1]$. The following tasks are requested to be accomplished:

- Prove that the operator

$$\begin{array}{rccc} \frac{d^2}{dx^2} : & U & \to & V \\ & u & \mapsto & u'' \end{array}$$

  is an isomorphism.
- Use the Ascoli–Arzelà theorem to prove that the linear operator

$$\left(\frac{d^2}{dx^2}\right)^{-1} : V \to V$$

  is compact.
- Let $\mathfrak{L}$ denote the family defined by

$$\mathfrak{L}(\lambda)u := u + f(\lambda)\left(\frac{d^2}{dx^2}\right)^{-1} u, \qquad \lambda \in D, \quad u \in V.$$

  Prove that $\mathfrak{L} \in \mathcal{C}^r(D, \mathcal{L}(V))$ satisfies $\mathfrak{L}(D) \subset \operatorname{Fred}_0(V)$.

- Prove that, for each $\lambda \in D$ such that

$$\sin \sqrt{f(\lambda)} \neq 0,$$

the operator $\mathfrak{L}(\lambda)$ is invertible and actually, for every $v \in V$ and $x \in [0,1]$,

$$\mathfrak{L}(\lambda)^{-1}v(x) = v(x) + \sqrt{f(\lambda)}\cos(\sqrt{f(\lambda)}x)\int_0^x \sin(\sqrt{f(\lambda)}t)v(t)dt + \sin(\sqrt{f(\lambda)}x)$$
$$\times \left[-\frac{\sqrt{f(\lambda)}\cos\sqrt{f(\lambda)}}{\sin\sqrt{f(\lambda)}}\int_0^1 \sin(\sqrt{f(\lambda)}t)v(t)dt + \sqrt{f(\lambda)}\int_0^1 \cos(\sqrt{f(\lambda)}t)v(t)dt\right].$$

- Prove that

$$N[\mathfrak{L}(\lambda_0)] = \operatorname{span}[\sin(n\pi\cdot)],$$

and that 0 is an isolated eigenvalue of $\mathfrak{L}$ if and only if it is an isolated zero of g.
- From now on, suppose $\operatorname{ord}_0 g = k$ for some $k \in \mathbb{N}^*$ with $r \geq 2k-1$ to make sure that

$$\operatorname{ord}_0 \sin\sqrt{f} = k \quad \text{and} \quad 0 \in \operatorname{Alg}_k(\mathfrak{L}).$$

- Use the identity

$$\lim_{\lambda\to 0} \lambda^k \left(\sin\sqrt{f(\lambda)}\right)^{-1} = (-1)^n 2n\pi g_k^{-1}$$

to infer that, for every $v \in V$ and $x \in [0,1]$,

$$c(\mathfrak{L}^{-1},0,-k)v(x) = -2n^2\pi^2 g_k^{-1}\int_0^1 \sin(n\pi t)v(t)dt\,\sin(n\pi x).$$

- Show

$$\mathfrak{L}'(\lambda)\mathfrak{L}(\lambda)^{-1} = kg_k\left(\frac{d^2}{dx^2}\right)^{-1} c(\mathfrak{L}^{-1},0,-k)\lambda^{-1} + o(\lambda^{-1})$$

and

$$\left(\frac{d^2}{dx^2}\right)^{-1}\sin(n\pi\cdot) = -(n\pi)^{-2}\sin(n\pi\cdot).$$

- Conclude that, for every $v \in V$ and $x \in [0,1]$,

$$c(\mathfrak{L}'\mathfrak{L}^{-1},0,-1)v(x) = 2k\int_0^1 \sin(n\pi t)v(t)dt\,\sin(n\pi x)$$

and that $c(\mathfrak{L}'\mathfrak{L}^{-1},0,-1)$ is a rank-one operator whose trace is k. Therefore, $\chi[\mathfrak{L};0] = k$.

4. Use the local Smith form of Theorem 7.8.3 to give an alternative proof of Theorems 9.1.1 and 9.3.2.

5. Imposing the necessary regularity requirements on the families involved, prove the results of Exercise 9 of Chapter 4 and Exercise 7 of Chapter 5 by using the representation of the multiplicity as a logarithmic residue.

6. Imposing the necessary regularity requirements on the families involved, prove the results of Exercise 8 of Chapter 5 by using the representation of the multiplicity as a logarithmic residue.

7. Prove the following generalization of Proposition 9.2.3: Let

$$\mathfrak{L} : \Omega \setminus \{\lambda_0\} \to \mathcal{L}(V, U), \qquad \mathfrak{M} : \Omega \setminus \{\lambda_0\} \to \mathcal{L}(U, V),$$

be two maps satisfying

$$\mathfrak{L}(\lambda) = \sum_{n=p_1}^{p_2} \mathfrak{L}_n (\lambda - \lambda_0)^n + o((\lambda - \lambda_0)^{p_2}),$$
$$\mathfrak{M}(\lambda) = \sum_{n=q_1}^{q_2} \mathfrak{M}_n (\lambda - \lambda_0)^n + o((\lambda - \lambda_0)^{q_2}),$$

for some integers p_1, p_2, q_1, q_2, with $p_1 \leq p_2$ and $q_1 \leq q_2$, and certain finite-rank operators $\mathfrak{L}_i$ and $\mathfrak{M}_j$ for $p_1 \leq i \leq \min\{p_2, -1\}$ and $q_1 \leq j \leq \min\{q_2, -1\}$. Then, for every

$$p_1 + q_1 \leq n \leq \min\{p_1 + q_2, p_2 + q_1, -1\},$$

the operators

$$c(\mathfrak{L}\mathfrak{M}, \lambda_0, n) \in \mathcal{L}(V) \quad \text{and} \quad c(\mathfrak{M}\mathfrak{L}, \lambda_0, n) \in \mathcal{L}(U)$$

have finite rank and

$$\operatorname{tr} c(\mathfrak{L}\mathfrak{M}, \lambda_0, n) = \operatorname{tr} c(\mathfrak{M}\mathfrak{L}, \lambda_0, n).$$

8. Let U, V, W be three $\mathbb{K}$-Banach spaces, $r, s \in \mathbb{N} \cup \{\infty\}$, and

$$\mathfrak{L} \in \mathcal{C}^r(\Omega, \mathcal{L}(U, V)), \qquad \mathfrak{M} \in \mathcal{C}^s(\Omega, \mathcal{L}(W, V)),$$

be such that $\mathfrak{L}_0 \in \mathrm{Fred}_0(U, V)$ and $\mathfrak{M}_0 \in \mathrm{Fred}_0(W, V)$. Suppose

$$\lambda_0 \in \mathrm{Alg}_k(\mathfrak{L}) \cap \mathrm{Alg}_\ell(\mathfrak{M})$$

for some integers $k, \ell \geq 1$ satisfying $2k \leq r + 1$ and $2\ell \leq s + 1$. Prove the following:

- $\lambda_0 \in \mathrm{Alg}_p(\mathfrak{L}\mathfrak{M})$ for some integer

$$\max\{k, \ell\} \leq p \leq k + \ell$$

  and $\mathfrak{L}\mathfrak{M} \in \mathcal{C}^t$ for some $t \geq \min\{r, s\}$.
- Suppose $2p \leq t + 1$ in order to prove that, for each $-k - \ell \leq n \leq -1$, the operators $c((\mathfrak{L}\mathfrak{M})'(\mathfrak{L}\mathfrak{M})^{-1}, \lambda_0, n)$, $c(\mathfrak{L}'\mathfrak{L}^{-1}, \lambda_0, n)$ and $c(\mathfrak{M}'\mathfrak{M}^{-1}, \lambda_0, n)$ have finite rank and

$$\operatorname{tr} c((\mathfrak{L}\mathfrak{M})'(\mathfrak{L}\mathfrak{M})^{-1}, \lambda_0, n) = \operatorname{tr} c(\mathfrak{L}'\mathfrak{L}^{-1}, \lambda_0, n) + \operatorname{tr} c(\mathfrak{M}'\mathfrak{M}^{-1}, \lambda_0, n).$$

- Show that, for $n = -1$, the previous identity provides us with the product formula.

Note that this is a generalization of Proposition 9.2.4.

9. Prove that Theorem 9.3.2 also holds for families of class $\mathcal{C}^{2\nu-1}$ regularity. State this formally and prove it.

10. Let $r \in \mathbb{N} \cup \{\infty\}$ and $\mathfrak{L}, \mathfrak{M} \in \mathcal{C}^r(\Omega, \mathcal{L}(U, V))$ be such that $\mathfrak{L}_0 \in \mathrm{Fred}_0(U, V)$ and $\lambda_0 \in \mathrm{Alg}_k(\mathfrak{L})$ for some integer $k \geq 1$ with $r \geq 2k - 1$. Adapt the proof of Proposition 7.10.3 in order to show that if

$$\mathfrak{L}_n = \mathfrak{M}_n, \qquad 0 \leq n \leq 2k - 1,$$

then

$$\mathfrak{L}(\lambda)^{-1} - \mathfrak{M}(\lambda)^{-1} = o((\lambda - \lambda_0)^{-1}).$$

11. Let $r \in \mathbb{N} \cup \{\infty\}$ and $\mathfrak{L} \in \mathcal{C}^r(\Omega, \mathcal{L}(U,V))$ be such that $\mathfrak{L}_0 \in \mathrm{Fred}_0(U,V)$ and $\lambda_0 \in \mathrm{Alg}_\nu(\mathfrak{L})$ for some integer $\nu \geq 1$ with $2\nu \leq r$, and, for every $n \in \{-\nu, \dots, -1\}$, set

$$(\mathfrak{L}^{-1})_n := c(\mathfrak{L}^{-1}, \lambda_0, n)$$

and

$$S_n := \mathrm{trng}\{(\mathfrak{L}^{-1})_{-\nu}, (\mathfrak{L}^{-1})_{-\nu+1}, \dots, (\mathfrak{L}^{-1})_n\} \in \mathcal{L}(V^{n+\nu+1}, U^{n+\nu+1}).$$

Prove the following assertions:

- $N[\mathfrak{L}_0] = \{u \in U : \mathrm{col}[0, \dots, 0, u] \in R[S_{-1}]\}$.
- For every $u \in N[\mathfrak{L}_0] \setminus \{0\}$,

$$\mathrm{rank}\, u = \max\{s \in \{1, \dots, \nu\} : \mathrm{col}[0, \dots, 0, u] \in R[S_{-s}]\}.$$

- Let $u \in N[\mathfrak{L}_0] \setminus \{0\}$ satisfy $\mathrm{rank}\, u = s$ and

$$S_{-s}\, \mathrm{col}[x_s, \dots, x_k] = \mathrm{col}[0, \dots, 0, u].$$

  Set

$$u_0 = u, \qquad u_j = \sum_{n=s}^{\nu} (\mathfrak{L}^{-1})_{j-n} x_n, \quad 1 \leq j \leq s-1.$$

  Show that $(u_0, \dots, u_{s-1})$ is a Jordan chain of $\mathfrak{L}$ at λ_0.

12. Let $r \in \mathbb{N} \cup \{\infty\}$ and $\mathfrak{L} \in \mathcal{C}^r(\Omega, \mathcal{L}(U,V))$ be such that $\mathfrak{L}_0 \in \Phi_0(U,V)$ and $\lambda_0 \in \mathrm{Alg}_\nu(\mathfrak{L})$ for some integer $\nu \geq 1$ with $2\nu \leq r+1$. Prove the following assertions:

- The operator Q defined by

$$Q := \begin{pmatrix} \mathfrak{L}_\nu & \cdots & \mathfrak{L}_1 \\ \vdots & \ddots & \vdots \\ \mathfrak{L}_{2\nu-1} & \cdots & \mathfrak{L}_\nu \end{pmatrix} \begin{pmatrix} (\mathfrak{L}^{-1})_{-\nu} & & \\ \vdots & \ddots & \\ (\mathfrak{L}^{-1})_{-1} & \cdots & (\mathfrak{L}^{-1})_{-\nu} \end{pmatrix} \in \mathcal{L}(V^\nu)$$

  is a projection.
- The mapping

$$\begin{array}{rcl} R[\mathrm{trng}\{(\mathfrak{L}^{-1})_{-\nu}, \dots, (\mathfrak{L}^{-1})_{-1}\}] & \to & R[Q] \\ x & \mapsto & \begin{pmatrix} \mathfrak{L}_\nu & \cdots & \mathfrak{L}_1 \\ \vdots & \ddots & \vdots \\ \mathfrak{L}_{2\nu-1} & \cdots & \mathfrak{L}_\nu \end{pmatrix} x \end{array}$$

  is an isomorphism whose inverse is given by the restriction of

$$\mathrm{trng}\{(\mathfrak{L}^{-1})_{-\nu}, \dots, (\mathfrak{L}^{-1})_{-1}\}$$

  from $R[Q]$ to $R[\mathrm{trng}\{(\mathfrak{L}^{-1})_{-\nu}, \dots, (\mathfrak{L}^{-1})_{-1}\}]$.

13. Let $r \in \mathbb{N}^* \cup \{\infty\}$ and $\mathfrak{L} \in \mathcal{C}^r(\Omega, \mathcal{L}(U,V))$ be such that $\mathfrak{L}_0 \in \mathrm{Fred}_0(U,V)$ and $\lambda_0 \in \mathrm{Alg}_\nu(\mathfrak{L})$ for some integer $\nu \geq 1$ with $2\nu \leq r+1$, and set

$$(\mathfrak{L}^{-1})_i := c(\mathfrak{L}^{-1}, \lambda_0, i), \qquad i \in \{-\nu, \dots, -1\},$$

$$A := \mathrm{trng}\{\mathfrak{L}_0, \dots, \mathfrak{L}_{\nu-1}\}, \qquad B := \mathrm{trng}\{(\mathfrak{L}^{-1})_{-\nu}, \dots, (\mathfrak{L}^{-1})_{-1}\},$$

and

$$C := \begin{pmatrix} \mathfrak{L}_\nu & \cdots & \mathfrak{L}_1 \\ \vdots & \ddots & \vdots \\ \mathfrak{L}_{2\nu-1} & \cdots & \mathfrak{L}_\nu \end{pmatrix}, \qquad D := \begin{pmatrix} (\mathfrak{L}^{-1})_0 & \cdots & (\mathfrak{L}^{-1})_{-\nu+1} \\ \vdots & \ddots & \vdots \\ (\mathfrak{L}^{-1})_{\nu-1} & \cdots & (\mathfrak{L}^{-1})_0 \end{pmatrix}.$$

Use Exercise 12 to prove the following:

- The next identities are satisfied:

$$\begin{aligned} \chi[\mathfrak{L}; \lambda_0] &= \dim N[A] = \dim N[DA] = \dim N[AD] \\ &= \mathrm{rank}\, B = \mathrm{rank}(BC) = \mathrm{rank}(CB). \end{aligned}$$

- The next identities are satisfied:

$$\mathrm{tr}\left(\sum_{n=1}^{\nu} n\mathfrak{L}_n (\mathfrak{L}^{-1})_{-n}\right) = \mathrm{tr}\, c(\mathfrak{L}'\mathfrak{L}^{-1}, \lambda_0, -1),$$

$$\mathrm{tr}\left(\sum_{n=1}^{\nu} n(\mathfrak{L}^{-1})_{-n}\mathfrak{L}_n\right) = \mathrm{tr}\, c(\mathfrak{L}^{-1}\mathfrak{L}', \lambda_0, -1).$$

- Infer, from the previous items, the validity of Theorem 9.3.2.

14. Let $r \in \mathbb{N}^* \cup \{\infty\}$ and $\mathfrak{L} \in \mathcal{C}^r(\Omega, \mathcal{L}(U,V))$ be such that $\mathfrak{L}_0 \in \mathrm{Fred}_0(U,V)$ and $\lambda_0 \in \mathrm{Alg}_\nu(\mathfrak{L})$ for some integer $\nu \geq 1$ with $2\nu \leq r+1$. Let $\ell_1, \dots, \ell_\nu \geq 0$ be a series of integer numbers such that the partial κ-multiplicities of $\mathfrak{L}$ at λ_0 are given by (7.23). Set

$$(\mathfrak{L}^{-1})_n := c(\mathfrak{L}^{-1}, \lambda_0, n), \qquad -\nu \leq n \leq -1,$$

and

$$r_s := \dim R[\mathrm{trng}\{(\mathfrak{L}^{-1})_{-\nu}, \dots, (\mathfrak{L}^{-1})_s\}], \qquad -\nu \leq s \leq -1.$$

Do the following:

- Prove that each of the spaces

$$R[\mathrm{trng}\{(\mathfrak{L}^{-1})_{-\nu}, \dots, (\mathfrak{L}^{-1})_s\}], \qquad -\nu \leq s \leq -1,$$

  is finite dimensional.
- Use the Smith form at λ_0 to get the identity

$$\mathrm{col}[r_{-1}, \dots, r_{-\nu}] = \mathrm{trng}\{1, \dots, \nu\}\, \mathrm{col}[\ell_\nu, \dots, \ell_1].$$

15. This exercise completes Theorem 7.10.4. Consider $r \in \mathbb{N} \cup \{\infty, \omega\}$, four $\mathbb{K}$-Banach spaces U_1, U_2, V_1 and V_2, two families, $\mathfrak{L} \in \mathcal{S}^r_{\lambda_0}(U_1, V_1)$ and $\mathfrak{M} \in \mathcal{S}^r_{\lambda_0}(U_2, V_2)$, such that λ_0 is an algebraic eigenvalue of both $\mathfrak{L}$ and $\mathfrak{M}$ of (possibly different) order less than or equal to k, for some integer $k \geq 1$ with $2k \leq r+1$. Set

$$(\mathfrak{L}^{-1})_i := c(\mathfrak{L}^{-1}, \lambda_0, i), \quad (\mathfrak{M}^{-1})_i := c(\mathfrak{M}^{-1}, \lambda_0, i), \qquad -k \leq i \leq -1.$$

Prove the equivalence of the following assertions:

- $\mathfrak{L}$ is $\mathcal{C}^{r-k}$-equivalent to $\mathfrak{M}$ at λ_0.
- For all $-k \leq s \leq -1$,

$$\dim R[\operatorname{trng}\{(\mathfrak{L}^{-1})_{-k}, \dots, (\mathfrak{L}^{-1})_s\}] = \dim R[\operatorname{trng}\{(\mathfrak{M}^{-1})_{-k}, \dots, (\mathfrak{M}^{-1})_s\}].$$

- For each $1 \leq j \leq k$ there exist $E_j \in \mathcal{L}(U_2, U_1)$ and $F_j \in \mathcal{L}(V_1, V_2)$ such that E_1 and F_1 are isomorphisms and

$$\begin{aligned}\operatorname{trng}\{&(\mathfrak{L}^{-1})_{-k}, \dots, (\mathfrak{L}^{-1})_{-1}\} \\ &= \operatorname{trng}\{E_1, \dots, E_k\} \cdot \operatorname{trng}\{(\mathfrak{M}^{-1})_{-k}, \dots, (\mathfrak{M}^{-1})_{-1}\} \cdot \operatorname{trng}\{F_1, \dots, F_k\}.\end{aligned}$$

- There exist $E_1, \dots, E_k \in \mathcal{L}(U_2, U_1)$ such that E_1 is an isomorphism and

$$\begin{aligned}R[\operatorname{trng}\{&(\mathfrak{L}^{-1})_{-k}, \dots, (\mathfrak{L}^{-1})_{-1}\}] \\ &= \operatorname{trng}\{E_1, \dots, E_k\} R[\operatorname{trng}\{(\mathfrak{M}^{-1})_{-k}, \dots, (\mathfrak{M}^{-1})_{-1}\}].\end{aligned}$$

9.7 Comments on Chapter 9

The logarithmic residue theorem giving rise to (9.1), which has been formally established by Theorem 9.4.1, is a classic result attributable to Gohberg & Sigal [54], although in the most pioneering versions of this result the multiplicity $\chi[\mathfrak{L}; \lambda_0]$ could not appear, of course.

The main result of the chapter, established by Theorem 9.3.2, comes from López-Gómez & Mora-Corral [90], as well as most of the results of Sections 9.1, 9.2 and 9.3. The corollaries of Section 9.4 are due to Markus & Sigal [95] and Gohberg & Sigal [54], and the construction of the spectral projector in Section 9.5 was carried out by Magnus & Mora-Corral [94]. Nevertheless, Corollary 9.1.3 might be considered to be a classic result attributable to Markus & Sigal [95] (see Gohberg, Goldberg & Kaashoek [43], and Ramm [113], for a number of different approaches).

Far beyond the concept of trace introduced in Section 9.2, there are fairly general concepts of trace (see, for example, the review paper of Gohberg & Kreĭn [45] and the books of Kato [65], and Gohberg, Goldberg & Kaashoek [43], as well as the references therein); however, those generalities do not add value to the analysis of this chapter.

The result of Exercise 11 is due to Gohberg [42] in the meromorphic case. Exercises 12 and 13 are based on results by Magnus & Mora-Corral [94]. They use the asymptotic developments of $\mathfrak{L}^{-1}$ to obtain multiplicity formulae involving certain kernels and ranges of the underlying operator coefficients. Exercise 15 completes the equivalence theory already studied in Section 7.10 by means of the new devices introduced in this chapter.

Establishing interrelations between the Laurent expansion of $\mathfrak{L}^{-1}$ and the left and right Jordan chains of $\mathfrak{L}$ at λ_0 has been shown to be extremely fruitful in spectral theory (see Gohberg, Lancaster & Rodman [50]). These ideas go back to Keldyš [66, 67]. Exercise 11 provides a generalization of some classic results in the context of matrix polynomials. In Exercises 11-13 a series of Toepliz matrices appear. This fact is utterly attributable to the intrinsic nature of the underlying theory, as Toepliz matrices naturally arise in developing formal products. The interested reader is sent to Magnus & Mora-Corral [94] for a general perspective of Toepliz and Hankel matrices in the context of the algebraic multiplicity, and Putnam [109] or Peller [106], for general accessible approaches to these matrices.

Exercise 4 proposes the elaboration of an alternative proof of Theorem 9.3.2, in this occasion through the local Smith form studied Chapter 7.

The real non-analytic theory developed in this chapter does actually make a huge number of results rigorous, obtained independently in mathematical physics by means of *formal* asymptotic developments (see Helffer [58], Hislop & Sigal [60], Ramm [113], and the list of references therein).

Chapter 10

The Spectral Theorem for Matrix Polynomials

This chapter studies the polynomial families of the form

$$\mathfrak{L}(\lambda) = \sum_{j=0}^{\ell} A_j \lambda^j, \qquad \lambda \in \mathbb{C}, \tag{10.1}$$

where $\ell \in \mathbb{N}$ and

$$A_0, \dots, A_\ell \in \mathcal{M}_N(\mathbb{C}), \qquad A_\ell \neq 0,$$

for some $N \in \mathbb{N}^*$. These families are called *matrix polynomials* in most of the available literature. More precisely, the family $\mathfrak{L}$ defined in (10.1) is said to be a matrix polynomial of order N and degree ℓ. The main goal of this chapter is to obtain a spectral theorem for matrix polynomials, respecting the spirit of the Jordan Theorem 1.2.1.

Generically, the spectrum of the family $\mathfrak{L}$ consists of $N\ell$ (different) eigenvalues and, hence, we cannot expect a decomposition of $\mathbb{C}^N$ into the direct sum of all underlying generalized eigenvectors. To state the main result of this chapter, suppose

$$\Sigma(\mathfrak{L}) = \{\lambda_1, \dots, \lambda_p\}, \qquad \lambda_i \neq \lambda_j, \quad 1 \leq i < j \leq p,$$

and, for each $1 \leq j \leq p$, let $\nu_j := \nu(\lambda_j)$ denote the order of λ_j as an algebraic eigenvalue of $\mathfrak{L}$. Then, according to Theorem 7.6.2,

$$\chi[\mathfrak{L}; \lambda_j] = \dim N\left[\mathfrak{L}_{[\nu_j - 1, j]}\right], \qquad 1 \leq j \leq p,$$

where

$$\mathfrak{L}_{[\nu_j - 1, j]} := \operatorname{trng}\{\mathfrak{L}_{0,j}, \dots, \mathfrak{L}_{\nu_j - 1, j}\}, \qquad 1 \leq j \leq p,$$

and

$$\mathfrak{L}_{i,j} = \frac{1}{i!}\mathfrak{L}^{i)}(\lambda_j), \qquad i \geq 0, \quad 1 \leq j \leq p.$$

Set

$$C := \begin{pmatrix} 0 & I_{\mathbb{C}^N} & & & \\ & 0 & I_{\mathbb{C}^N} & & \\ & & \ddots & \ddots & \\ & & & 0 & I_{\mathbb{C}^N} \\ -A_0 & -A_1 & \cdots & & -A_{\ell-1} \end{pmatrix} \in \mathcal{M}_{N\ell}(\mathbb{C});$$

the main theorem of this chapter establishes the existence of p isomorphisms

$$J_j \in \operatorname{Iso}\left(N\left[\mathfrak{L}_{[\nu_j-1,j]}\right], N\left[(\lambda_j I_{\mathbb{C}^{N\ell}} - C)^{\nu_j}\right]\right), \qquad 1 \leq j \leq p,$$

such that

$$\mathbb{C}^{N\ell} = \bigoplus_{j=1}^{p} J_j\left(N\left[\mathfrak{L}_{[\nu_j-1,j]}\right]\right) = \bigoplus_{j=1}^{p} N\left[(\lambda_j I_{\mathbb{C}^{N\ell}} - C)^{\nu_j}\right],$$

very much in the spirit of Chapter 1. This result, referred to as the *spectral theorem* for matrix polynomials, provides us with a generalization of the Jordan Theorem 1.2.1 and of the fundamental theorem of algebra. Its proof follows straightforwardly from Theorem 7.6.2, Proposition 8.3.1 and Theorem 1.2.1. Once proved, we will use the results of Chapter 7 in order to construct a particular basis of $\mathbb{C}^{N\ell}$ through a canonical set of Jordan chains at every eigenvalue of $\mathfrak{L}$, as in Chapter 1.

Essentially, this chapter uses the local theory developed in Part II to obtain a global spectral theorem, in the spirit of Part I, but in this occasion for general matrix polynomials, not just for classical operator families; thus opening a door towards the development of new techniques and results in spectral theory.

The chapter is organized as follows. Section 10.1 shows the existence of a global *linearization* for matrix polynomials. Basically, a global linearization consists of a classical family capturing all spectral invariants of the matrix polynomial itself. Section 10.2 proves the spectral theorem announced above. Section 10.3 derives the fundamental theorem of algebra from the spectral theorem, and establishes some rather obvious connections between the underlying theory and the theory of linear differential equations and systems. Finally, Section 10.4 uses the spectral theorem to construct a basis of $\mathbb{C}^{N\ell}$ consisting of generalized eigenvectors; thus respecting the genuine classical tradition of constructing Jordan basis for classical families.

10.1 Linearization of a matrix polynomial

Let $N \in \mathbb{N}^*$ and $\ell \in \mathbb{N}$; given

$$A_0, \ldots, A_\ell \in \mathcal{M}_N(\mathbb{C}), \qquad A_\ell \neq 0,$$

the map $\mathfrak{L} \in \mathcal{H}(\mathbb{C}, \mathcal{M}_N(\mathbb{C}))$ defined by (10.1) is said to be a *matrix polynomial of order* N *and degree* ℓ. Obviously, matrix polynomials belong to $\mathbb{P}[\mathcal{M}_N(\mathbb{C})]$. The matrix polynomial $\mathfrak{L}$ defined by (10.1) is said to be *monic* when

$$A_\ell = I_{\mathbb{C}^N}.$$

According to the notation and concepts introduced in the presentation of Part II, the *spectrum* of $\mathfrak{L}$, denoted by $\Sigma(\mathfrak{L})$, is given by

$$\Sigma(\mathfrak{L}) = \{\lambda \in \mathbb{C} \;:\; \det \mathfrak{L}(\lambda) = 0\}.$$

Note that $\det \mathfrak{L}$ is a complex-valued polynomial. If $\mathfrak{L}$ is monic, then $\det \mathfrak{L}$ is a monic complex-valued polynomial of degree $N\ell$ and, therefore, according to Theorem 8.2.1,

$$\Sigma(\mathfrak{L}) \subset \operatorname{Alg}(\mathfrak{L}).$$

From now on, all vectors are regarded as column vectors, and we will use the notation trng introduced in (2.24) and (4.34). In addition, square matrices of order $N\ell$ will often be decomposed in square blocks of order N, and, for simplifying notation, for every integer $m \geq 1$, the identity operator of $\mathbb{C}^m$ will be denoted by I_m.

The next result provides us with a *global linearization* of a matrix polynomial, in the sense that there exist a Banach space U, two entire operator-valued functions $\mathfrak{E}, \mathfrak{F} \in \mathcal{H}(\mathbb{C}, \mathcal{L}(V))$, where $V := \mathbb{C}^N \oplus U$, and an operator $C \in \mathcal{L}(V)$ such that $\mathfrak{E}(\mathbb{C}), \mathfrak{F}(\mathbb{C}) \subset \operatorname{Iso}(V)$ and

$$\mathfrak{L}(\lambda) \oplus I_U = \mathfrak{E}(\lambda)(\lambda I_V - C)\mathfrak{F}(\lambda), \qquad \lambda \in \mathbb{C}.$$

In such a case, the space V is called the *phase space* of $\mathfrak{L}$.

Lemma 10.1.1. *Let* $N, \ell \geq 1$ *be integers, and* $\mathfrak{L} \in \mathbb{P}[\mathcal{M}_N(\mathbb{C})]$ *a monic matrix polynomial of the form* (10.1), *for some* $A_0, \dots, A_\ell \in \mathcal{M}_N(\mathbb{C})$ *with* $A_\ell = I_N$. *Consider the matrix*

$$C := \begin{pmatrix} 0 & I_N & & & \\ & 0 & I_N & & \\ & & \ddots & \ddots & \\ & & & 0 & I_N \\ -A_0 & -A_1 & \cdots & & -A_{\ell-1} \end{pmatrix} \in \mathcal{M}_{N\ell}(\mathbb{C}),$$

and, for each $1 \leq j \leq \ell - 1$, *the matrix polynomials* $\mathfrak{e}^j \in \mathbb{P}[\mathcal{M}_N(\mathbb{C})]$ *and* $\mathfrak{E}, \mathfrak{F} \in \mathbb{P}[\mathcal{M}_{N\ell}(\mathbb{C})]$, *defined, for every* $\lambda \in \mathbb{C}$, *by*

$$\mathfrak{e}^j(\lambda) := \sum_{i=0}^{\ell-j} A_{j+i}\lambda^i, \qquad 1 \leq j \leq \ell - 1,$$

$$\mathfrak{F}(\lambda) = \operatorname{trng}\{I_N, \lambda I_N, \dots, \lambda^{\ell-1} I_N\},$$

and

$$\mathfrak{E}(\lambda) = \begin{pmatrix} \mathfrak{e}^1(\lambda) & \mathfrak{e}^2(\lambda) & \cdots & \mathfrak{e}^{\ell-1}(\lambda) & I_N \\ -I_N & 0 & & & \\ & -I_N & \ddots & & \\ & & \ddots & 0 & \\ & & & -I_N & 0 \end{pmatrix}.$$

Then, $\det \mathfrak{E}$ *and* $\det \mathfrak{F}$ *are non-zero constant polynomials, and the following global factorization is satisfied:*

$$\operatorname{diag}\{\mathfrak{L}(\lambda), I_{N(\ell-1)}\} = \mathfrak{E}(\lambda)\,(\lambda I_{N\ell} - C)\,\mathfrak{F}(\lambda), \qquad \lambda \in \mathbb{C}. \tag{10.2}$$

Proof. It is immediate to see that $\det \mathfrak{F}$ is the constant polynomial 1, and $\det \mathfrak{E}$ is the constant polynomial 1 or -1. Identity (10.2) follows easily by a direct calculation. □

10.2 Generalized Jordan theorem

The following result collects some basic consequences of the factorization (10.2).

Proposition 10.2.1. *Let* $\mathfrak{L} \in \mathbb{P}[\mathcal{M}_N(\mathbb{C})]$ *be a monic matrix polynomial of degree* $\ell \geq 1$*, and set*

$$\mathfrak{M} := \operatorname{diag}\{\mathfrak{L}, I_{N(\ell-1)}\} \in \mathbb{P}[\mathcal{M}_{N\ell}(\mathbb{C})].$$

Let $\mathfrak{E}, \mathfrak{F} \in \mathbb{P}[\mathcal{M}_{N\ell}(\mathbb{C})]$ *and* $C \in \mathcal{M}_{N\ell}(\mathbb{C})$ *be the maps given by Lemma* 10.1.1. *Then, for every* $\lambda_0 \in \Sigma(\mathfrak{L})$ *and* $s \in \mathbb{N}$*, the following three maps are linear isomorphisms:*

$$\begin{aligned} \varphi_1 : N[\mathfrak{L}_{[s]}] &\to N[\mathfrak{M}_{[s]}] \\ \operatorname{col}[x_0, x_1, \dots, x_s] &\mapsto \operatorname{col}[x_0, \mathbf{0}, x_1, \mathbf{0}, \dots, x_s, \mathbf{0}], \end{aligned}$$

$$\begin{aligned} \varphi_2 : N[\mathfrak{M}_{[s]}] &\to N[\mathfrak{L}^C_{[s]}] \\ y &\mapsto \mathfrak{F}_{[s]} y, \end{aligned}$$

$$\begin{aligned} \varphi_3 : N[\mathfrak{L}^C_{[s]}] &\to N[(\mathfrak{L}^C_0)^{s+1}] = N[(\lambda_0 I_{N\ell} - C)^{s+1}] \\ \operatorname{col}[y_0, y_1, \dots, y_s] &\mapsto y_s, \end{aligned}$$

where we have denoted

$$\mathbf{0} = (0, \dots, 0) \in \mathbb{C}^{N(\ell-1)}, \qquad \mathfrak{L}^C(\lambda) := \lambda I_{N\ell} - C, \quad \lambda \in \mathbb{C}.$$

Consequently,

$$\mathfrak{J}_{s+1,\lambda_0} := \varphi_3 \circ \varphi_2 \circ \varphi_1 \in \operatorname{Iso}\left(N[\mathfrak{L}_{[s]}], N[(\mathfrak{L}^C_0)^{s+1}]\right). \tag{10.3}$$

Proof. According to Lemma 10.1.1,

$$\mathfrak{M} = \mathfrak{E}\mathfrak{L}^C\mathfrak{F}. \tag{10.4}$$

Taking into account

$$\mathfrak{M}_0 = \mathrm{diag}\{\mathfrak{L}_0, I_{N(\ell-1)}\}, \qquad \mathfrak{M}_j = \mathrm{diag}\{\mathfrak{L}_j, 0\}, \quad j \geq 1,$$

one can see easily that φ_1 is a linear isomorphism.

By the Leibniz rule, we find from (10.4) that, for every $s \geq 0$ and $\lambda_0 \in \Sigma(\mathfrak{L})$,

$$\mathfrak{M}_{[s]} = \mathfrak{E}_{[s]}\mathfrak{L}^C_{[s]}\mathfrak{F}_{[s]}.$$

Moreover, $\mathfrak{E}_{[s]}$ and $\mathfrak{F}_{[s]}$ are isomorphisms, because $\mathfrak{E}_0$ and $\mathfrak{F}_0$ are invertible matrices. Consequently, φ_2 must be a linear isomorphism.

The fact that φ_3 is an isomorphism was proved by Proposition 8.3.1. This concludes the proof. □

Now we fix a monic matrix polynomial

$$\mathfrak{L} \in \mathbb{P}[\mathcal{M}_N(\mathbb{C})]$$

of degree $\ell \geq 1$. Then, $\det \mathfrak{L}$ is a monic complex-valued polynomial of degree $N\ell$. Thus,

$$\det \mathfrak{L}(\lambda) = \prod_{j=1}^{p} (\lambda - \lambda_j)^{m_j}, \qquad \lambda \in \mathbb{C},$$

for some integers $m_1, \dots, m_p \geq 1$ and complex numbers $\lambda_1, \dots \lambda_p$ such that $\lambda_i \neq \lambda_j$ if $1 \leq i < j \leq p$. Necessarily,

$$N\ell = \sum_{j=1}^{p} m_j$$

and, according to Lemma 10.1.1, we have that

$$\{\lambda_1, \dots, \lambda_p\} = \Sigma(\mathfrak{L}) = \Sigma(\mathrm{diag}\{\mathfrak{L}, I_{N(\ell-1)}\}) = \Sigma(\mathfrak{L}^C) = \sigma(C).$$

In the sequel, much like in Theorem 1.2.1, we will work simultaneously with all points of $\Sigma(\mathfrak{L})$, and, so, we ought to refine the notation used previously throughout this book by setting

$$\mathfrak{N}_{i,j} = \frac{1}{i!}\mathfrak{N}^{i)}(\lambda_j), \qquad i \geq 0, \quad 1 \leq j \leq p,$$

and

$$\mathfrak{N}_{[s,j]} := \mathrm{trng}\{\mathfrak{N}_{0,j}, \dots, \mathfrak{N}_{s,j}\}, \qquad s \geq 0, \quad 1 \leq j \leq p,$$

for every complex Banach space U and $\mathfrak{N} \in \mathcal{H}(\mathbb{C}, \mathcal{L}(U))$. Furthermore, the notation introduced by (10.3) will be simplified by simply setting

$$\mathfrak{J}_{s,j} := \mathfrak{J}_{s,\lambda_j}, \qquad 1 \leq j \leq p, \quad s \geq 0.$$

There is no need to emphasize the dependence on $\mathfrak{L}$, because $\mathfrak{L}$ is fixed throughout the remaining of this chapter.

The following generalized version of Theorem 1.2.1 is satisfied.

Theorem 10.2.2 (Generalized Jordan theorem). *Let $\mathfrak{L} \in \mathbb{P}[\mathcal{M}_N(\mathbb{C})]$ be a monic matrix polynomial of degree $\ell \geq 1$ such that*

$$\Sigma(\mathfrak{L}) = \{\lambda_1, \ldots, \lambda_p\}, \qquad \lambda_i \neq \lambda_j, \quad 1 \leq i < j \leq p, \tag{10.5}$$

and, for each $1 \leq j \leq p$, let ν_j denote the order of λ_j as an algebraic eigenvalue of $\mathfrak{L}$. Then,

$$\mathbb{C}^{N\ell} = \bigoplus_{j=1}^{p} \mathfrak{J}_{\nu_j,j}\left(N[\mathfrak{L}_{[\nu_j-1,j]}]\right). \tag{10.6}$$

In particular, by Theorem 7.6.2 and, according to Definition 8.4.1, we have that

$$N\ell = \sum_{j=1}^{p} \dim N[\mathfrak{L}_{[\nu_j-1,j]}] = \sum_{j=1}^{p} \kappa[\mathfrak{L}; \lambda_j] = \mathfrak{m}[\mathfrak{L}; \mathbb{C}].$$

Proof. Let C be the matrix given by Lemma 10.1.1. According to (10.2), we have that $\Sigma(\mathfrak{L}) = \sigma(C)$ and, actually, thanks to Theorems 8.3.2 and 7.3.1, for each $1 \leq j \leq p$, the integer ν_j equals the algebraic ascent of λ_j as an eigenvalue of C. Thus, by Theorem 1.2.1 applied to the matrix C, we find that

$$\mathbb{C}^{N\ell} = \bigoplus_{j=1}^{p} N[(\lambda_j I_{N\ell} - C)^{\nu_j}].$$

Finally, owing to Proposition 10.2.1,

$$\mathfrak{J}_{\nu_j,j}\left(N[\mathfrak{L}_{[\nu_j-1,j]}]\right) = N[(\lambda_j I_{N\ell} - C)^{\nu_j}], \qquad 1 \leq j \leq p,$$

which ends the proof. □

Remark 10.2.3. Suppose $\ell = 1$. Then,

$$\mathfrak{L}(\lambda) = \lambda I_N + A_0, \quad \mathfrak{E}(\lambda) = \mathfrak{F}(\lambda) = I_N, \quad C = -A_0, \qquad \lambda \in \mathbb{C},$$

and, hence, (10.6) becomes

$$\mathbb{C}^{N} = \bigoplus_{j=1}^{p} N[(\lambda_j I_N + A_0)^{\nu_j}].$$

Consequently, Theorem 10.2.2 indeed provides us with Theorem 1.2.1.

10.3 Remarks on scalar polynomials

Suppose $N = 1$ and $\ell \geq 1$. Then $A_j \in \mathbb{C}$ for all $0 \leq j \leq \ell$, and $\mathfrak{L}$ is a monic scalar-valued polynomial. In such a case, (10.6) becomes

$$\mathbb{C}^\ell = \bigoplus_{j=1}^{p} N[(\lambda_j I_\ell - C)^{\nu_j}], \tag{10.7}$$

where

$$C = \begin{pmatrix} 0 & 1 & & & \\ & 0 & 1 & & \\ & & \ddots & \ddots & \\ & & & 0 & 1 \\ -A_0 & -A_1 & \cdots & & -A_{\ell-1} \end{pmatrix} \in \mathcal{M}_\ell(\mathbb{C}). \tag{10.8}$$

By Lemma 10.1.1, the set of roots of $\mathfrak{L}$ equals the set of roots of the polynomial

$$\lambda \mapsto \det(\lambda I_\ell - C).$$

Thanks to (10.7), there exists $j \in \{1, \ldots, p\}$ such that

$$N[(\lambda_j I_\ell - C)^{\nu_j}] \neq \{0\}.$$

Consequently,

$$\det(\lambda_j I_\ell - C) = 0$$

and we conclude that

$$\mathfrak{L}(\lambda_j) = 0.$$

Therefore, Theorem 10.2.2 indeed provides us with the fundamental theorem of algebra.

Now, we consider the linear homogeneous differential equation of order ℓ whose symbol is given by $\mathfrak{L}$, that is to say,

$$\sum_{j=0}^{\ell} A_j D^j u = 0, \qquad D^j u = u^{j)}, \quad 0 \leq j \leq \ell,$$

which can be briefly expressed in the form

$$\mathfrak{L}(D)u = 0. \tag{10.9}$$

By well-known results in the theory of ordinary differential equations, the set of entire solutions $u \in \mathcal{H}(\mathbb{C})$ of equation (10.9), denoted by $\mathcal{H}_e$, is an ℓ-dimensional linear subspace of $\mathcal{H}(\mathbb{C})$. Actually, a basis of $\mathcal{H}_e$ is

$$\{p_{i,j} : \ 0 \leq i \leq \nu_j - 1, \ 1 \leq j \leq p\},$$

where, by definition,

$$p_{i,j}(z) := z^i e^{\lambda_j z}, \qquad z \in \mathbb{C}, \quad 0 \leq i \leq \nu_j - 1, \quad 1 \leq j \leq p.$$

Moreover, the map

$$\begin{array}{rcl} \mathcal{H}(\mathbb{C}) & \to & \mathcal{H}(\mathbb{C}, \mathbb{C}^\ell) \\ u & \mapsto & \operatorname{col}[u, Du, \ldots, D^{\ell-1}u] \end{array}$$

is a linear isomorphism between $\mathcal{H}_e$ and the linear subspace $\mathcal{H}_s$ of $\mathcal{H}(\mathbb{C}, \mathbb{C}^\ell)$ formed by all entire solutions of the first-order homogeneous linear system

$$U' = CU, \qquad U = \operatorname{col}[u_0, \ldots, u_{\ell-1}],$$

where C is the matrix defined in (10.8). Further, the column vector-valued functions of the exponential matrix given by

$$z \mapsto e^{zC}, \qquad z \in \mathbb{C},$$

form a basis of $\mathcal{H}_s$. These properties of the matrix C are paradigmatic in the theory of linear differential equations and systems.

10.4 Constructing a basis in the phase space

This section adopts another perspective for expressing Theorem 10.2.2. It will be shown how the canonical sets of Jordan chains of $\mathfrak{L}$ at each of its eigenvalues can be used to construct an appropriate basis in the phase space $\mathbb{C}^{N\ell}$, in the spirit of the construction of a Jordan basis for an operator A that allows us to express it through its Jordan canonical form.

The following preparatory lemma provides an explicit formula for each of the isomorphisms introduced by (10.3).

Lemma 10.4.1. *Let $\mathfrak{L} \in \mathbb{P}[\mathcal{M}_N(\mathbb{C})]$ be a monic matrix polynomial of degree $\ell \geq 1$ satisfying* (10.5). *For $s \in \mathbb{N}$ and $j \in \{1, \ldots, p\}$, consider*

$$\operatorname{col}[x_0, \ldots, x_s] \in N[\mathfrak{L}_{[s,j]}].$$

Set

$$\operatorname{col}[y_0, \ldots, y_{\ell-1}] := \mathfrak{I}_{s+1,j} \operatorname{col}[x_0, \ldots, x_s] \in (\mathbb{C}^N)^\ell.$$

Then,

$$y_k = \sum_{i=\max\{0,s-k\}}^{s} \binom{k}{i-s+k} \lambda_j^{i-s+k} x_i, \qquad 0 \leq k \leq \ell - 1. \tag{10.10}$$

Proof. Adopting the convention

$$\binom{a}{b} = 0 \quad \text{if} \quad b > a \quad \text{or} \quad b < 0,$$

we still have

$$\binom{a}{b} = \binom{a}{a-b} \quad \text{for every} \quad a, b \in \mathbb{Z}, \tag{10.11}$$

and, in addition, (10.10) can be written in the more appropriate form

$$y_k = \sum_{i=0}^{s} \binom{k}{i-s+k} \lambda_j^{i-s+k} x_i, \qquad 0 \leq k \leq \ell - 1. \tag{10.12}$$

Fix $s \geq 0$ and $1 \leq j \leq p$, and let

$$\mathrm{col}[x_0, \dots, x_s] \in N[\mathfrak{L}_{[s,j]}].$$

Then, according to the definition of $\mathfrak{I}_{s+1,j}$,

$$\mathfrak{I}_{s+1,j}\, \mathrm{col}[x_0, \dots, x_s] = \sum_{i=0}^{s} \mathfrak{F}_{i,j}\, \mathrm{col}[x_{s-i}, \mathbf{0}]. \tag{10.13}$$

To prove (10.12) we will proceed by induction on $s \in \mathbb{N}$. Suppose $s = 0$. Then, according to (10.13), we have that

$$\begin{aligned}
\mathfrak{I}_{1,j} x_0 = \mathfrak{F}_{0,j}\, \mathrm{col}[x_0, \mathbf{0}] &= \mathrm{trng}\{I_N, \lambda_j I_N, \dots, \lambda_j^{\ell-1} I_N\}\, \mathrm{col}[x_0, \mathbf{0}] \\
&= \mathrm{col}[\lambda_j^k x_0, \quad 0 \leq k \leq \ell - 1].
\end{aligned}$$

Here and in the rest of the chapter we use the following notation: if $q \in \mathbb{N}$ and for each $0 \leq n \leq q$ there is a vector u_n in a Banach space U_n, then we define

$$\mathrm{col}[u_n, \quad 0 \leq n \leq q] := \mathrm{col}[u_0, \dots, u_q].$$

The previous calculation shows that

$$y_k = \lambda_j^k x_0, \qquad 0 \leq k \leq \ell - 1,$$

which coincides with the realization of (10.12) at $s = 0$. Therefore, the result holds for $s = 0$.

Assume that (10.12) is satisfied for some particular $s \geq 0$ and consider

$$\mathrm{col}[y_0, \dots, y_{\ell-1}] := \mathfrak{I}_{s+2,j}\, \mathrm{col}[x_0, \dots, x_s, x_{s+1}].$$

Then, taking into account (10.13) and rearranging terms, we can infer that

$$\begin{aligned}
\mathrm{col}[y_0, \dots, y_{\ell-1}] &= \sum_{i=0}^{s+1} \mathfrak{F}_{i,j}\, \mathrm{col}[x_{s+1-i}, \mathbf{0}] = \sum_{i=0}^{s} \mathfrak{F}_{i,j}\, \mathrm{col}[x_{s+1-i}, \mathbf{0}] + \mathfrak{F}_{s+1,j}\, \mathrm{col}[x_0, \mathbf{0}] \\
&= \mathfrak{I}_{s+1,j}\, \mathrm{col}[x_1, \dots, x_s, x_{s+1}] + \mathfrak{F}_{s+1,j}\, \mathrm{col}[x_0, \mathbf{0}].
\end{aligned}$$

According to the induction hypothesis, we have that

$$\begin{aligned}
&\mathfrak{I}_{s+1,j}\,\mathrm{col}[x_1,\dots,x_{s+1}]\\
&\quad=\mathrm{col}\left[\sum_{i=0}^{s}\binom{k}{i-s+k}\lambda_j^{i-s+k}x_{i+1},\ 0\le k\le \ell-1\right]\\
&\quad=\mathrm{col}\left[\sum_{h=1}^{s+1}\binom{k}{h-(s+1)+k}\lambda_j^{h-(s+1)+k}x_h,\ 0\le k\le \ell-1\right].
\end{aligned}$$

On the other hand, differentiating $\mathfrak{F}$ gives

$$\begin{aligned}
\mathfrak{F}_{s+1,j}=\mathrm{trng}\Bigg\{&\binom{0}{s+1}\lambda_j^{0-(s+1)}I_N,\ \binom{1}{s+1}\lambda_j^{1-(s+1)}I_N,\\
&\dots,\ \binom{\ell-1}{s+1}\lambda_j^{\ell-1-(s+1)}I_N\Bigg\}
\end{aligned}$$

and, consequently,

$$\begin{aligned}
\mathfrak{F}_{s+1,j}\,\mathrm{col}[x_0,\mathbf{0}]&=\mathrm{col}\left[\binom{k}{s+1}\lambda_j^{k-(s+1)}x_0,\ 0\le k\le \ell-1\right]\\
&=\mathrm{col}\left[\binom{k}{k-(s+1)}\lambda_j^{k-(s+1)}x_0,\ 0\le k\le \ell-1\right],
\end{aligned}$$

where we have used (10.11). Therefore, adding up these two column vectors shows that indeed

$$\begin{aligned}
\mathrm{col}[y_0,\dots,y_{\ell-1}]&=\mathfrak{I}_{s+2,j}\,\mathrm{col}[x_0,\dots,x_s,x_{s+1}]\\
&=\mathrm{col}\left[\sum_{h=0}^{s+1}\binom{k}{h-(s+1)+k}\lambda_j^{h-(s+1)+k}x_h,\ 0\le k\le \ell-1\right],
\end{aligned}$$

which concludes the proof. □

Now, we will invoke Proposition 7.7.1 in order to construct a basis of the subspace $N[\mathfrak{L}_{[\nu_j-1,j]}]$ for every $j\in\{1,\dots,p\}$. Thanks to Proposition 10.2.1, Theorem 10.2.2, and Lemma 10.4.1, these bases will provide us with a basis of the whole phase space $\mathbb{C}^{N\ell}$, thus respecting the spirit of the construction of a Jordan basis to obtain the associated canonical form. Naturally, for each $1\le j\le p$, the basis of $N[\mathfrak{L}_{[\nu_j-1,j]}]$ comes from a canonical set of Jordan chains, which will be labeled according to Section 7.2.

Setting

$$\begin{aligned}
r_j&:=\dim N[\mathfrak{L}_{0,j}], && 1\le j\le p,\\
\kappa_{ji}&:=\kappa[\mathfrak{L};\lambda_j], && 1\le i\le r_j,\quad 1\le j\le p,
\end{aligned}$$

we have

$$\nu_j = \kappa_{j1} \geq \cdots \geq \kappa_{jr_j} \geq 1, \qquad 1 \leq j \leq p,$$

and, hence, owing to Definition 7.2.5 and Theorem 7.6.2,

$$\kappa[\mathfrak{L}; \lambda_j] = \sum_{i=1}^{r_j} \kappa_{ji} = \dim N[\mathfrak{L}_{[\nu_j - 1, j]}], \qquad 1 \leq j \leq p.$$

Now, for each $1 \leq j \leq p$, let

$$\left[x^j_{0,1}, \dots, x^j_{\kappa_{j1}-1,1}; \cdots ; x^j_{0,r_j}, \dots, x^j_{\kappa_{jr_j}-1,r_j}\right], \tag{10.14}$$

denote a canonical set of Jordan chains of $\mathfrak{L}$ at λ_j, as in (7.5). Then, thanks to Proposition 7.7.1, for every $1 \leq j \leq p$, the set

$$\bigcup_{i=1}^{r_j} \left\{ \operatorname{col}\left[0, x^j_{0,i}\right], \operatorname{col}\left[0, x^j_{0,i}, x^j_{1,i}\right], \dots, \operatorname{col}\left[0, x^j_{0,i}, \dots, x^j_{\kappa_{ji}-1,i}\right]\right\}$$

is a basis of $N[\mathfrak{L}_{[\nu_j-1,j]}]$. Thus, due to Proposition 10.2.1 and Lemma 10.4.1, for every $1 \leq j \leq p$, the set of column vectors

$$\begin{aligned} \mathcal{B}^j := \bigcup_{i=1}^{r_j} \Big\{ & \operatorname{col}\left[x^j_{0,i}, \lambda_j x^j_{0,i}, \lambda_j^2 x^j_{0,i}, \dots, \lambda_j^{\ell-1} x^j_{0,i}\right], \\ & \operatorname{col}\left[x^j_{1,i}, \lambda_j x^j_{1,i} + x^j_{0,i}, \lambda_j^2 x^j_{1,i} + 2\lambda_j x^j_{0,i}, \dots, \lambda_j^{\ell-1} x^j_{1,i} + (\ell-1)\lambda_j^{\ell-2} x^j_{0,i}\right], \\ & \dots, \\ & \operatorname{col}\Big[x^j_{\kappa_{ji}-1,i}, \lambda_j x^j_{\kappa_{ji}-1,i} + x^j_{\kappa_{ji}-2,i}, \dots, \sum_{a=1}^{\min\{\ell,\kappa_{ji}\}} \binom{\ell-1}{\ell-a} \lambda_j^{\ell-a} x^j_{\kappa_{ji}-a,i}\Big]\Big\} \end{aligned}$$

provides us with a basis of $N[(\mathfrak{L}^C_{0,j})^{\nu_j}]$. Therefore, by Theorem 1.2.1,

$$\mathcal{B} := \bigcup_{j=1}^{p} \mathcal{B}^j \tag{10.15}$$

is a basis of the phase space $\mathbb{C}^{N\ell}$. The next result characterizes the matrix whose columns consist of the coordinates of the vectors of the basis $\mathcal{B}$. Given two arbitrary integers $m, n \geq 1$, the set $\mathcal{M}_{m \times n}(\mathbb{C})$ stands for the space of rectangular matrices with m rows and n columns with complex entries. Moreover, for each $1 \leq j \leq p$, we denote

$$m_j := \kappa[\mathfrak{L}; \lambda_j] = \dim N[\mathfrak{L}_{[\nu_j-1,j]}],$$

and consider the matrix, denoted by X_j, whose columns consist of the coordinates of all vectors of the canonical set of Jordan chains (10.14)

$$X_j := [x^j_{0,1}, \dots, x^j_{\kappa_{j1}-1,1}, \dots, x^j_{0,r_j}, \dots, x^j_{\kappa_{jr_j}-1,r_j}] \in \mathcal{M}_{N \times m_j}(\mathbb{C}).$$

Then, we can construct the matrix formed by all these canonical sets of Jordan chains

$$X := [X_1, \dots, X_p] \in \mathcal{M}_{N\times N\ell}(\mathbb{C}).$$

We also consider

$$J_j := \operatorname{diag}\{J^T_{j,\kappa_{j1}}, \dots, J^T_{j,\kappa_{jr_j}}\} \in \mathcal{M}_{m_j}(\mathbb{C}), \qquad 1 \le j \le p,$$

where for each $1 \le j \le p$,

$$J^T_{j,\kappa_{ji}} = \begin{pmatrix} \lambda_j & 1 & 0 & \cdots & 0 & 0 \\ 0 & \lambda_j & 1 & \cdots & 0 & 0 \\ 0 & 0 & \lambda_j & \cdots & 0 & 0 \\ \vdots & \vdots & \vdots & \ddots & \vdots & \vdots \\ 0 & 0 & 0 & \cdots & \lambda_j & 1 \\ 0 & 0 & 0 & \cdots & 0 & \lambda_j \end{pmatrix} \in \mathcal{M}_{\kappa_{ji}}(\mathbb{C}), \qquad 1 \le i \le r_j,$$

as well as the matrix

$$J := \operatorname{diag}\{J_1, \dots, J_p\} \in \mathcal{M}_{N\ell}(\mathbb{C}).$$

Then, the following result is satisfied.

Lemma 10.4.2. *The matrix whose columns are the vectors of the basis* (10.15) *is given through*

$$\mathfrak{B} := \operatorname{col}[X, XJ, XJ^2, \dots, XJ^{\ell-1}] \in \mathcal{M}_{N\ell}(\mathbb{C}). \tag{10.16}$$

Proof. First, note that

$$J^n = \operatorname{diag}\{J_1^n, \dots, J_p^n\}, \qquad n \ge 0,$$

and, hence,

$$XJ^n = [X_1J_1^n, X_2J_2^n, \dots, X_pJ_p^n], \qquad 0 \le n \le \ell - 1.$$

Consequently, we have to calculate the columns of the set of matrices

$$\left\{X_jJ_j^n \; : \; 1 \le j \le p, \; 0 \le n \le \ell - 1\right\}.$$

To accomplish this task, we begin by calculating the powers of a matrix $A \in \mathcal{M}_h(\mathbb{C})$ of the form

$$A = \operatorname{trng}\{\lambda, 1, 0, \dots, 0\} = \lambda I_h + N, \quad N := \operatorname{trng}\{0, 1, 0, \dots, 0\},$$

for some integer $h \ge 1$ and $\lambda \in \mathbb{C}$. According to the Newton binomial formula, for every integer $n \ge 0$,

$$A^n = \sum_{i=0}^{n} \binom{n}{i} \lambda^{n-i} N^i.$$

Moreover,

$$N^i = \begin{cases} \operatorname{trng}\{\underbrace{0,\dots,0}_{i},1,\underbrace{0,\dots,0}_{h-i-1}\}, & 0 \le i \le h-1, \\ 0, & i \ge h. \end{cases}$$

Therefore,

$$A^n = \operatorname{trng}\left\{\lambda^n, \binom{n}{1}\lambda^{n-1}, \dots, \binom{n}{n-1}\lambda, 1, \underbrace{0,\cdots,0}_{h-n-1}\right\},$$

whenever $1 \le n \le h-1$, whereas

$$A^n = \operatorname{trng}\left\{\lambda^n, \binom{n}{1}\lambda^{n-1}, \dots, \binom{n}{h-1}\lambda^{n-h+1}\right\}$$

if $n \ge h$. Note also that

$$(A^T)^n = (A^n)^T, \qquad n \ge 0.$$

With all these features in mind, we can fix $j \in \{1,\dots,p\}$ in order to find the expression of the columns of the matrices

$$X_j J_j^n, \qquad 0 \le n \le \ell - 1.$$

Now we set

$$X_j^i := [x_{0,i}^j, \dots, x_{\kappa_{ji}-1,i}^j], \qquad 1 \le i \le r_j.$$

Then,

$$X_j = [X_j^1, \dots, X_j^{r_j}],$$

and, hence,

$$X_j J_j^n = [X_j^1 (J_{j,\kappa_{j1}}^T)^n, \dots, X_j^{r_j}(J_{j,\kappa_{jr_j}}^T)^n].$$

Fix $i \in \{1,\dots,r_j\}$. Then, we have that

$$X_j^i J_{j,\kappa_{ji}}^T = \left[\lambda_j x_{0,i}^j, x_{0,i}^j + \lambda_j x_{1,i}^j, \dots, x_{\kappa_{ji}-2,i}^j + \lambda_j x_{\kappa_{ji}-1,i}^j\right],$$

$$\begin{aligned} X_j^i \left(J_{j,\kappa_{ji}}^T\right)^2 = \Big[& \lambda_j^2 x_{0,i}^j, \binom{2}{1}\lambda_j x_{0,i}^j + \lambda_j^2 x_{1,i}^j, \\ & x_{0,i}^j + \binom{2}{1}\lambda_j x_{1,i}^j + \lambda_j^2 x_{2,i}^j, \dots, \\ & x_{\kappa_{ji}-3,i}^j + \binom{2}{1}\lambda_j x_{\kappa_{ji}-2,i}^j + \lambda_j^2 x_{\kappa_{ji}-1,i}^j\Big], \end{aligned}$$

and, in general, for each integer $n \geq 0$,

$$X_j^i \left(J_{j,\kappa_{ji}}^T\right)^n = \Big[\lambda_j^n x_{0,i}^j, \binom{n}{1}\lambda_j^{n-1}x_{0,i}^j + \lambda_j^n x_{1,i}^j,$$
$$\binom{n}{2}\lambda_j^{n-2}x_{0,i}^j + \binom{n}{1}\lambda_j^{n-1}x_{1,i}^j + \lambda_j^n x_{2,i}^j,$$
$$\dots, \sum_{h=0}^{\kappa_{ji}-1}\binom{n}{\kappa_{ji}-1-h}\lambda_j^{n-\kappa_{ji}+1+h}x_{h,i}^j\Big].$$

Therefore, the columns of the matrix $\mathfrak{B}$ are given by the ordered set

$$\bigcup_{j=1}^{p}\bigcup_{i=1}^{r_j}\Big\{\operatorname{col}\left[x_{0,i}^j, \lambda_j x_{0,i}^j, \lambda_j^2 x_{0,i}^j, \dots, \lambda_j^{\ell-1}x_{0,i}^j\right],$$
$$\operatorname{col}\Big[x_{1,i}^j, x_{0,i}^j + \lambda_j x_{1,i}^j, \binom{2}{1}\lambda_j x_{0,i}^j + \lambda_j^2 x_{1,i}^j, \dots,$$
$$\binom{\ell-1}{1}\lambda_j^{\ell-2}x_{0,i}^j + \lambda_j^{\ell-1}x_{1,i}^j\Big],$$
$$\dots, \operatorname{col}\Big[x_{\kappa_{ji}-1,i}^j, x_{\kappa_{ji}-2,i}^j + \lambda_j x_{\kappa_{ji}-1,i}^j,$$
$$x_{\kappa_{ji}-3,i}^j + \binom{2}{1}\lambda_j x_{\kappa_{ji}-2,i}^j + \lambda_j^2 x_{\kappa_{ji}-1,i}^j, \dots,$$
$$\sum_{h=0}^{\kappa_{ji}-1}\binom{\ell-1}{\kappa_{ji}-1-h}\lambda_j^{\ell-\kappa_{ji}+h}x_{h,i}^j\Big]\Big\}.$$

Finally, note that this set of vectors coincide with the set of vectors forming the basis $\mathcal{B}$. This concludes the proof. □

10.5 Exercises

1. Let $\mathfrak{L} \in \mathbb{P}[\mathcal{M}_N(\mathbb{C})]$. Prove that either $\Sigma(\mathfrak{L})$ is finite or it equals $\mathbb{C}$.

2. Give an example of a polynomial matrix $\mathfrak{L}$ of the form

$$\mathfrak{L}(\lambda) = A_0 + \lambda A_1, \qquad \lambda \in \mathbb{C},$$

for some $A_0, A_1 \in \mathcal{M}_N(\mathbb{C}) \setminus \{0\}$, for which $0 \in \sigma(A_1)$ and $\Sigma(\mathfrak{L})$ is a finite non-empty set.

3. Let $\mathfrak{L} \in \mathbb{P}[\mathcal{M}_N(\mathbb{C})]$ be a monic matrix polynomial satisfying (10.5). For each $1 \leq j \leq p$ let (10.14) be a canonical set of Jordan chains of $\mathfrak{L}$ at λ_j. Prove that

$$\mathbb{C}^N = \operatorname{span}\{x_{k,i}^j \;:\; 1 \leq j \leq p,\; 1 \leq i \leq r_j,\; 0 \leq k \leq \kappa_{ji}-1\}.$$

4. Let $\mathfrak{L} \in \mathbb{P}[\mathcal{M}_N(\mathbb{C})]$ be the matrix polynomial of degree $\ell \geq 1$ defined by (10.1) with $A_\ell = I_N$. Prove that the set of matrix coefficients

$$(A_0, A_1, \dots, A_{\ell-1}) \in (\mathcal{M}_N(\mathbb{C}))^\ell$$

for which $\Sigma(\mathfrak{L})$ consists of $N\ell$ different eigenvalues is dense in $(\mathcal{M}_N(\mathbb{C}))^\ell$. Prove also that its complement has $N^2\ell$-dimensional Lebesgue measure zero.

5. Discuss the implications of Lemma 10.4.2 when $\Sigma(\mathfrak{L})$ possesses $N\ell$ different eigenvalues.

6. Generalize Theorem 10.2.2 and Lemma 10.4.2 to cover the case when A_ℓ is invertible, not necessarily the identity, and give a proof of the corresponding result.

10.6 Comments on Chapter 10

The spectral theorem for matrix polynomials is attributable to Gohberg, Lancaster & Rodman [47]. Extensions of these results to operator polynomials are due to Gohberg, Lancaster & Rodman [49]. Extensions to analytic matrix functions are due to Gohberg & Rodman [52]. Basic texts in the theory of matrix polynomials are Gantmacher [39], Gel'fand [40], Lancaster [73], and Gohberg, Lancaster & Rodman [50].

The use of linearizations in the context of theory of matrices is classic (see the books by Gohberg, Lancaster & Rodman [50] and Rodman [114]). The first infinite-dimensional results on linearization are due to Gohberg, Kaashoek & Lay [44]; see also Boer [14], Mitiagin [97, 98], and Kaashoek, Mee & Rodman [64] for later results about explicit linearizations. Within the context of the theory of multiplicity, they have also been used by Arason & Magnus [5], and Magnus & Mora-Corral [94].

Two excellent books on the theory of matrix and operator polynomials are Gohberg, Lancaster & Rodman [50] and Rodman [114]. Their contents go much further than this chapter.

Apart from the facts pointed out in Section 10.3, there are many other relationships between the theory of matrix polynomials and the theories of (linear) ordinary differential equations, partial differential equations and difference equations. The reader is sent to Gohberg, Lancaster & Rodman [50, 51], Rodman [114], Thijsse [123], and Wloka, Rowley & Lawruk [127] for further details. These results originated in Lancaster [74, 75].

In most of the available literature (as well as in the classic Jordan canonical form), instead of establishing Theorem 10.2.2, the construction of the basis for the phase space $\mathbb{C}^{N\ell}$ is accomplished by means of Lemma 10.4.2. So the spectral theorem is actually established through the invertibility of the matrix (10.16) (see for example, Gohberg, Lancaster & Rodman [50], and Wloka, Rowley & Lawruk [127]). Consequently, the statement of Theorem 10.2.2 in this book might be considered to be new. This change of methodology in the establishment of the spectral theorem of this chapter is deeply inspired by Chapter 1, where Jordan Theorem 1.2.1 was established as the main spectral theorem, and, at a further stage, a Jordan basis was constructed.

Nevertheless, in most of specialized literature, a great part of the effort is focused on the construction of an explicit basis, despite being unnecessary for ascertaining the inherent canonical form. This clearly illustrates that algebraic techniques are still quite imbued with the spirit of the invariant calculus of the nineteenth century, when papers used to contain hundred of pages constructing complete systems of generating invariants. Although Hilbert became world-famous for solving Gordan's problem through an extremely simple, non-constructive, method, which originated one of the most controversial mathematical debates of that time span, yet in the twentieth-first century, the spirit of Hilbert seems to be far from being utterly adopted.

Chapter 11

Further Developments of the Algebraic Multiplicity

This chapter exposes briefly some further developments of the theory of multiplicity whose treatment lies outside the general scope of this book. Consequently, it possesses an expository and bibliographic character.

Section 11.1 deals with general Fredholm families, not necessarily of index zero. Section 11.2 analyzes meromorphic families. Section 11.3 discusses families of linear, not necessarily bounded, operators. Finally, Section 11.4 considers general families of linear, not necessarily Fredholm, operators.

11.1 General Fredholm operator families

Following the pioneering work of Gohberg [41], Gohberg, Kaashoek & Lay [44] and Gohberg, Lancaster & Rodman [50] studied general Fredholm families of arbitrary index. In such a case, one can still construct Jordan chains and local Smith forms. Precisely, the existence of a local Smith form for analytic families of Fredholm operators can be stated as follows.

Theorem 11.1.1. *Let $\lambda_0 \in \Omega$ and $\mathfrak{L} \in \mathcal{C}^\omega(\Omega, \mathcal{L}(U, V))$ be such that $\lambda_0 \in \Sigma(\mathfrak{L})$ and $\mathfrak{L}_0 \subset \mathrm{Fred}(U, V)$. Then, there exist an open neighborhood $\Omega' \subset \Omega$ of λ_0, two families $\mathfrak{E} \in \mathcal{C}^\omega(\Omega', \mathcal{L}(V))$ and $\mathfrak{F} \in \mathcal{C}^\omega(\Omega', \mathcal{L}(U))$ with $\mathfrak{E}_0 \in \mathrm{Iso}(V)$ and $\mathfrak{F}_0 \in \mathrm{Iso}(U)$, two topological decompositions*

$$U = U_0 \oplus U_1, \qquad V = V_0 \oplus V_1,$$

with $\dim U_0 < \infty$ and $\dim V_0 < \infty$, an isomorphism $\mathfrak{J} \in \mathrm{Iso}(U_1, V_1)$, an integer $n \geq 1$, and n unique $\kappa_1, \dots, \kappa_n \in \mathbb{N}$, with

$$\kappa_1 \geq \cdots \geq \kappa_n \geq 1,$$

such that, for every $\lambda \in \Omega'$,

$$\mathfrak{L}(\lambda) = \mathfrak{E}(\lambda)(\mathfrak{S}(\lambda) \oplus \mathfrak{I})\mathfrak{F}(\lambda)$$

where, according to the sign of the Fredholm index of $\mathfrak{L}_0$, *either*

$$\mathfrak{S}(\lambda) = \left(\begin{array}{c} \operatorname{diag}\left\{(\lambda-\lambda_0)^{\kappa_1}, \dots, (\lambda-\lambda_0)^{\kappa_n}\right\} \\ \hline \mathbf{0} \end{array} \right) \in \mathcal{L}(U_0, V_0), \tag{11.1}$$

or else

$$\mathfrak{S}(\lambda) = \left(\begin{array}{c|c} \operatorname{diag}\left\{(\lambda-\lambda_0)^{\kappa_1}, \dots, (\lambda-\lambda_0)^{\kappa_n}\right\} & \mathbf{0} \end{array} \right) \in \mathcal{L}(U_0, V_0). \tag{11.2}$$

In the context of Theorem 11.1.1, the integers $\kappa_1, \dots, \kappa_n$ are defined as the partial κ-multiplicities of $\mathfrak{L}$ at λ_0. The corresponding equivalence result reads as follows.

Theorem 11.1.2. *Let* U_1, U_2, V_1, V_2 *be four Banach spaces,* $\lambda_0 \in \Omega$, *and*

$$\mathfrak{L} \in \mathcal{C}^\omega(\Omega, \mathcal{L}(U_1, V_1)), \qquad \mathfrak{M} \in \mathcal{C}^\omega(\Omega, \mathcal{L}(U_2, V_2))$$

two analytic operator families such that $\mathfrak{L}_0 \in \operatorname{Fred}(U_1, V_1)$ *and* $\mathfrak{M}_0 \in \operatorname{Fred}(U_2, V_2)$. *Then, the following assertions are equivalent:*

i) *There exist an open neighborhood* $\Omega' \subset \Omega$ *of* λ_0 *and two families*

$$\mathfrak{E} \in \mathcal{C}^\omega(\Omega', \mathcal{L}(V_2, V_1)), \qquad \mathfrak{F} \in \mathcal{C}^\omega(\Omega', \mathcal{L}(U_1, U_2)),$$

with $\mathfrak{E}_0 \in \operatorname{Iso}(V_2, V_1)$ *and* $\mathfrak{F}_0 \in \operatorname{Iso}(U_1, U_2)$, *such that*

$$\mathfrak{L}(\lambda) = \mathfrak{E}(\lambda)\mathfrak{M}(\lambda)\mathfrak{F}(\lambda), \qquad \lambda \in \Omega'.$$

ii) $\mathfrak{L}$ *and* $\mathfrak{M}$ *have the same partial* κ*-multiplicities at* λ_0, *and*

$$\operatorname{ind} \mathfrak{L}_0 = \operatorname{ind} \mathfrak{M}_0.$$

11.2 Meromorphic families

Meromorphic families also have a local Smith form. Precisely, the following result is satisfied (see Gohberg, Kaashoek & Lay [44] and the references therein).

Theorem 11.2.1. *Suppose* $\mathbb{K} = \mathbb{C}$, *and let* $\mathfrak{L} : \Omega \to \mathcal{L}(U, V)$ *be a meromorphic map such that* $\mathfrak{L}(\lambda) \in \operatorname{Fred}(U, V)$ *for all* $\lambda \in \Omega \setminus \{\lambda_0\}$. *Then, there exist an open neighborhood* $\Omega' \subset \Omega$ *of* λ_0, *two families* $\mathfrak{E} \in \mathcal{H}(\Omega', \mathcal{L}(V))$ *and* $\mathfrak{F} \in \mathcal{H}(\Omega', \mathcal{L}(U))$, *with* $\mathfrak{E}_0 \in \operatorname{Iso}(V)$ *and* $\mathfrak{F}_0 \in \operatorname{Iso}(U)$, *two topological decompositions*

$$U = U_0 \oplus U_1, \qquad V = V_0 \oplus V_1,$$

with $\dim U_0 < \infty$ *and* $\dim V_0 < \infty$*, an isomorphism* $\mathfrak{I} \in \mathrm{Iso}(U_1, V_1)$*, an integer* $n \geq 1$*, and* n *unique (up to permutations)* $\kappa_1, \dots, \kappa_n \in \mathbb{Z} \setminus \{0\}$ *such that, for every* $\lambda \in \Omega' \setminus \{\lambda_0\}$*,*

$$\mathfrak{L}(\lambda) = \mathfrak{E}(\lambda)(\mathfrak{S}(\lambda) \oplus \mathfrak{I})\mathfrak{F}(\lambda)$$

where $\mathfrak{S}(\lambda)$ *is given through either* (11.1) *or* (11.2).

As in Section 11.1, the integers $\kappa_1, \dots, \kappa_n$, which may be positive or negative, are defined as the partial κ-multiplicities of $\mathfrak{L}$ at λ_0. The corresponding equivalence theorem reads as follows.

Theorem 11.2.2. *Suppose* $\mathbb{K} = \mathbb{C}$*, and let* U_1, U_2, V_1, V_2 *be four Banach spaces,* $\lambda_0 \in \Omega$*, and*

$$\mathfrak{L} : \Omega \to \mathcal{L}(U_1, V_1), \qquad \mathfrak{M} : \Omega \to \mathcal{L}(U_2, V_2)$$

two meromorphic maps such that $\mathfrak{L}(\lambda) \in \mathrm{Fred}(U_1, V_1)$ *and* $\mathfrak{M}(\lambda) \in \mathrm{Fred}(U_2, V_2)$ *for all* $\lambda \in \Omega \setminus \{\lambda_0\}$*. Then, the following assertions are equivalent:*

i) *There exist an open neighborhood* $\Omega' \subset \Omega$ *of* λ_0 *and two families*

$$\mathfrak{E} \in \mathcal{H}(\Omega', \mathcal{L}(V_2, V_1)), \qquad \mathfrak{F} \in \mathcal{H}(\Omega', \mathcal{L}(U_1, U_2)),$$

with $\mathfrak{E}_0 \in \mathrm{Iso}(V_2, V_1)$ *and* $\mathfrak{F}_0 \in \mathrm{Iso}(U_1, U_2)$*, such that*

$$\mathfrak{L}(\lambda) = \mathfrak{E}(\lambda)\mathfrak{M}(\lambda)\mathfrak{F}(\lambda), \qquad \lambda \in \Omega' \setminus \{\lambda_0\}.$$

ii) $\mathfrak{L}$ *and* $\mathfrak{M}$ *have the same partial* κ*-multiplicities at* λ_0*, and there exists an open neighborhood* $\tilde{\Omega} \subset \Omega$ *of* λ_0 *such that*

$$\operatorname{ind} \mathfrak{L}(\lambda) = \operatorname{ind} \mathfrak{M}(\lambda), \qquad \lambda \in \tilde{\Omega} \setminus \{\lambda_0\}.$$

Naturally, there is a counterpart of the Rouché theorem in the context of general Fredholm meromorphic families of arbitrary index, but we omit its details here, and the reader is sent to Gohberg & Sigal [54].

11.3 Unbounded operators

There are also some available concepts of algebraic multiplicity for operator families whose values are linear, but not necessarily continuous, operators. Among them, the concept introduced by Laloux & Mawhin [71, 72] stands out because of its number of applications to the theory of nonlinear differential equations and systems. Respecting the terminology introduced in this book, given two complex normed spaces U and V, Laloux & Mawhin [71, 72] considered holomorphic families $\mathfrak{L}$ of the form

$$\mathfrak{L}(\lambda) := L - \lambda A, \qquad \lambda \in \mathbb{C},$$

where

$$L : \operatorname{dom} L \subset U \to V$$

is a linear operator, not necessarily continuous, such that $R[L]$ is closed in V and

$$\dim N[L] = \operatorname{codim} R[L] < \infty,$$

while $A : U \to V$ is linear and enjoys certain continuity and compactness assumptions, which will be made precise below. Naturally, in this context, the set of eigenvalues of $\mathfrak{L}$, denoted by $\operatorname{Eig}(\mathfrak{L})$, is defined through

$$\operatorname{Eig}(\mathfrak{L}) := \{\lambda \in \mathbb{C} : N[L - \lambda A] \neq \{0\}\}.$$

Under these general assumptions on L, there exist projections $P \in \mathcal{L}(U)$ and $Q \in \mathcal{L}(V)$ such that

$$R[P] = N[L], \qquad R[L] = N[Q],$$

and the operator L_P defined by restriction of L,

$$L_P := L|_{\operatorname{dom} L \cap N[P]} : \operatorname{dom} L \cap N[P] \to R[L]$$

is invertible. Then, Laloux and Mawhin imposed

$$\Sigma(\mathfrak{L}) \neq \mathbb{C}, \qquad L_P^{-1}(I - Q)A \in \mathcal{K}(U) \quad \text{and} \quad QA \in \mathcal{L}(U, V).$$

Under these assumptions, for each $\mu_0 \in \mathbb{C} \setminus \operatorname{Eig}(\mathfrak{L})$, the operator

$$A_0 := (L - \mu_0 A)^{-1} A$$

is compact and it turns out that $\Sigma(\mathfrak{L})$ equals the set of eigenvalues of

$$A_0 + \mu_0 I_U.$$

Moreover, for any $\lambda_0 \in \operatorname{Eig}(\mathfrak{L})$, the classic algebraic multiplicity of $\lambda_0 - \mu_0$ as an eigenvalue of A_0 is defined and does not depend on μ_0. Therefore, the following concept of multiplicity, attributable to Laloux and Mawhin, is consistent

$$M[\mathfrak{L}; \lambda_0] := m_a(\lambda_0 - \mu_0),$$

where m_a stands for the classic algebraic multiplicity with respect to A_0.

Although this concept is a substantial generalization of the classic concept of algebraic multiplicity of an operator, it turns out that it cannot be generalized to cover the concepts of multiplicity introduced in Chapters 4, 5, 7, 9, and axiomatixed in Chapter 6, because it exclusively works out for a very special type of operator families, under very strong compactness requirements that, in general, are not satisfied.

11.4 Non-Fredholm operators

There also exists a general theory for operator families whose values are arbitrary, not necessarily Fredholm, linear continuous operators. It is the abstract theory developed by Magnus [93], Arason & Magnus [4, 5, 6], and Magnus & Mora-Corral [94]. According to it, given a complex Banach space U, an open set $D \subset \mathbb{C}$, and a holomorphic family

$$\mathfrak{L} \in \mathcal{H}(D, \mathcal{L}(U)), \tag{11.3}$$

the spectrum of $\mathfrak{L}$ is defined as the set $\Sigma(\mathfrak{L})$ of points $\lambda \in D$ such that $\mathfrak{L}(\lambda)$ is not invertible, and, given any bounded open set Ω such that

$$\bar{\Omega} \subset D \quad \text{and} \quad \partial\Omega \cap \Sigma(\mathfrak{L}) = \varnothing, \tag{11.4}$$

the algebraic multiplicity of $\mathfrak{L}$ in Ω, denoted by $m[\mathfrak{L}, \Omega]$, is defined as the isomorphism class of the quotient Banach space

$$\frac{\mathcal{H}(\Omega, U)}{\mathfrak{L}(\mathcal{H}(\Omega, U))}.$$

Naturally, when the quotient is a finite-dimensional space of dimension d, the multiplicity of $\mathfrak{L}$ in Ω is said to be finite, and it actually equals d, of course.

Every pair $(\mathfrak{L}, \Omega)$ satisfying (11.3) and (11.4) is called an *admissible pair.* The following result collects the main properties of this concept of multiplicity, whose significance relies upon its great generality.

Theorem 11.4.1. *Let U, V be complex Banach spaces, $\mathfrak{L} \in \mathcal{H}(D, \mathcal{L}(U))$ and $\mathfrak{M} \in \mathcal{H}(D, \mathcal{L}(V))$. Suppose $\Omega \subset \mathbb{C}$ is a bounded open set such that $(\mathfrak{L}, \Omega)$ and $(\mathfrak{M}, \Omega)$ are admissible pairs. Then, the following properties are satisfied:*

i) $m[\mathfrak{L}, \Omega] = 0$ *if and only if* $\Omega \cap \Sigma(\mathfrak{L}) = \varnothing$.

ii) *If* $U = V$, *then*

$$m[\mathfrak{L}\mathfrak{M}, \Omega] = m[\mathfrak{L}, \Omega] + m[\mathfrak{M}, \Omega],$$

where the sum of two isomorphism classes of Banach spaces is understood as the isomorphism class of the direct sum of any two representatives of these classes.

iii) $m[\mathfrak{L} \oplus \mathfrak{M}, \Omega] = m[\mathfrak{L}, \Omega] + m[\mathfrak{M}, \Omega]$.

iv) *For every pair of open bounded subsets Ω_1 and Ω_2 of $\mathbb{C}$ for which $(\mathfrak{L}, \Omega_1)$ and $(\mathfrak{L}, \Omega_2)$ are admissible pairs and $\Omega_1 \cap \Omega_2 = \varnothing$, necessarily*

$$m[\mathfrak{L}, \Omega_1 \cup \Omega_2] = m[\mathfrak{L}, \Omega_1] + m[\mathfrak{L}, \Omega_2].$$

v) *For any bounded open subset $\Omega \subset \mathbb{C}$ and any continuous map*

$$H \; : \; [0,1] \times D \to \mathcal{L}(U)$$

such that $H(t,\cdot) \in \mathcal{H}(D, \mathcal{L}(U))$ *and* $(H(t,\cdot),\Omega)$ *is an admissible pair for all* $t \in [0,1]$, *necessarily*

$$m[H(t,\cdot),\Omega] = m[H(0,\cdot),\Omega] \qquad \text{for all } t \in [0,1].$$

In this abstract setting there are also available uniqueness theorems, in the spirit of those given in Chapter 6. Moreover, according to Theorem 9.5.1 and Exercises 12 and 13 of Chapter 9, the next result makes it apparent that, in the context of holomorphic families, this abstract concept of multiplicity is indeed a generalization of the one studied in the previous chapters (see Magnus & Mora-Corral [99]).

Theorem 11.4.2. *Let* U *be a complex Banach space,* $\mathfrak{L} \in \mathcal{H}(D, \mathcal{L}(U))$ *and* $\lambda_0 \in \Sigma(\mathfrak{L})$ *such that* λ_0 *is a pole of* $\mathfrak{L}^{-1}$ *of order* $\nu \geq 1$. *Let* $\Omega \subset D$ *be an open and bounded set for which* $(\mathfrak{L},\Omega)$ *is an admissible pair and* $\Omega \cap \Sigma(\mathfrak{L}) = \{\lambda_0\}$. *Then, the following properties are satisfied:*

- $m[\mathfrak{L},\Omega]$ *is represented by*

$$N[\operatorname{trng}\{\mathfrak{L}_0,\dots,\mathfrak{L}_{\nu-1}\}] = R[\operatorname{trng}\{(\mathfrak{L}^{-1})_{-\nu},\dots,(\mathfrak{L}^{-1})_{-1}\}]. \tag{11.5}$$

- *The spaces* (11.5) *equal the range of the projection*

$$\begin{pmatrix} (\mathfrak{L}^{-1})_{-\nu} & & \\ \vdots & \ddots & \\ (\mathfrak{L}^{-1})_{-1} & \cdots & (\mathfrak{L}^{-1})_{-\nu} \end{pmatrix} \begin{pmatrix} \mathfrak{L}_{\nu} & \cdots & \mathfrak{L}_1 \\ \vdots & \ddots & \vdots \\ \mathfrak{L}_{2\nu-1} & \cdots & \mathfrak{L}_{\nu} \end{pmatrix}$$

  and, consequently, $m[\mathfrak{L},\Omega]$ *is also represented by such range.*

- $m[\mathfrak{L},\Omega]$ *is represented by the range of the projection*

$$\begin{pmatrix} \mathfrak{L}_{\nu} & \cdots & \mathfrak{L}_1 \\ \vdots & \ddots & \vdots \\ \mathfrak{L}_{2\nu-1} & \cdots & \mathfrak{L}_{\nu} \end{pmatrix} \begin{pmatrix} (\mathfrak{L}^{-1})_{-\nu} & & \\ \vdots & \ddots & \\ (\mathfrak{L}^{-1})_{-1} & \cdots & (\mathfrak{L}^{-1})_{-\nu} \end{pmatrix}.$$

Part III

Nonlinear Spectral Theory

Part III: Summary

Part III of the book is an attempt to describe the importance of the theory developed in Part II to Nonlinear Analysis. It is, therefore, a short introduction to Nonlinear Analysis from the point of view of Linear Algebra and Linear Operator Theory.

In the mathematical modeling of real-world systems, it is imperative to deal with abstract nonlinear operators between real Banach spaces under minimal regularity conditions. In these models, the underlying maps tend to be of class $\mathcal{C}^r$ for some $r \geq 1$, but they are not analytic in general. In fact, in many applications, the parameter λ is real (by the intrinsic nature of these problems) and the map $\mathfrak{L}$ is merely continuous over some interval.

In the context of Nonlinear Analysis, the most important topological magnitude to be measured is the change of the index $\operatorname{Ind}(\mathfrak{L}(\lambda), 0)$ as λ crosses some $\lambda_0 \in \Sigma(\mathfrak{L})$. In finite dimension, such index equals the sign of the determinant of $\mathfrak{L}(\lambda)$. A general principle in Nonlinear Analysis establishes that if $\operatorname{Ind}(\mathfrak{L}(\lambda), 0)$ changes as λ crosses λ_0, then a connected set of solutions (λ, u), with $u \neq 0$, to the equation

$$\mathfrak{L}(\lambda)u + \mathfrak{N}(\lambda, u) = 0$$

emanates from $\Omega \times \{0\}$ at λ_0, for every continuous and compact map $\mathfrak{N}$ such that

$$\mathfrak{N}(\lambda, u) = o(\|u\|), \qquad u \to 0.$$

Such a value of λ_0 is said to be a *nonlinear eigenvalue* of $\mathfrak{L}$.

This part contains only one chapter, in which the most important abstract concepts of Nonlinear Analysis are introduced, and some of their main properties are established. Under the assumption that $\chi[\mathfrak{L}; \lambda_0]$ is well defined and finite, the main theorem of the chapter establishes that λ_0 is a nonlinear eigenvalue of $\mathfrak{L}$ if and only if the algebraic multiplicity $\chi[\mathfrak{L}; \lambda_0]$ is odd. It turns out that the theory of algebraic multiplicity studied throughout this book provides us with an algorithm to determine the changes of the topological index $\operatorname{Ind}(\mathfrak{L}(\lambda), 0)$. This represents a strong link between Linear Operator Theory and Nonlinear Analysis.

This part will use two basic properties of compact operators. Namely, that $\mathcal{K}(U)$ is a closed subset of $\mathcal{L}(U)$, and that if $K \in \mathcal{K}(U)$ then $I_U + K \in \operatorname{Fred}_0(U)$.

As Part III consists of an expository chapter that transfers information, techniques and results from Linear to Nonlinear Analysis, we have not made a list of exercises. Instead, we have decided to close the chapter and, hence, the book, by making a general discussion about some of the central questions in the study of nonlinear problems. Once the reading of this book has been completed, the interested reader might wish to have a further glance at López-Gómez [82].

Chapter 12

Nonlinear Eigenvalues

Throughout this chapter, the field $\mathbb{K}$ will always be the real field $\mathbb{R}$; we consider a real Banach space U, an open interval $\Omega \subset \mathbb{R}$, a neighborhood $\mathcal{U}$ of $0 \in U$, an integer number $r \geq 0$, a family $\mathfrak{L} \in \mathcal{C}^r(\Omega, \mathcal{L}(U))$, and a nonlinear map $\mathfrak{N} \in \mathcal{C}(\Omega \times \mathcal{U}, U)$ satisfying the following conditions:

(AL) $\mathfrak{L}(\lambda) - I_U \in \mathcal{K}(U)$ for every $\lambda \in \Omega$, i.e., $\mathfrak{L}(\lambda)$ is a compact perturbation of the identity map.

(AN) $\mathfrak{N}$ is compact, i.e., the image by $\mathfrak{N}$ of any bounded set of $\Omega \times \mathcal{U}$ is relatively compact in U. Also, for every compact $K \subset \Omega$,

$$\lim_{u \to 0} \sup_{\lambda \in K} \frac{\|\mathfrak{N}(\lambda, u)\|}{\|u\|} = 0.$$

From now on, we consider the operator $\mathfrak{F} \in \mathcal{C}(\Omega \times \mathcal{U}, U)$ defined as

$$\mathfrak{F}(\lambda, u) := \mathfrak{L}(\lambda)u + \mathfrak{N}(\lambda, u), \tag{12.1}$$

and the associated equation

$$\mathfrak{F}(\lambda, u) = 0, \qquad (\lambda, u) \in \Omega \times \mathcal{U}. \tag{12.2}$$

By Assumptions (AL) and (AN), it is apparent that

$$\mathfrak{F}(\lambda, 0) = 0, \quad D_u\mathfrak{F}(\lambda, 0) = \mathfrak{L}(\lambda), \qquad \lambda \in \Omega,$$

and, hence, (12.2) can be thought of as a *nonlinear perturbation* around $(\lambda, 0)$ of the linear equation

$$\mathfrak{L}(\lambda)u = 0, \qquad \lambda \in \Omega, \quad u \in U. \tag{12.3}$$

Equation (12.2) can be expressed as a fixed-point equation for a compact operator. Indeed, $\mathfrak{F}(\lambda, u) = 0$ if and only if

$$u = [I_U - \mathfrak{L}(\lambda)]u - \mathfrak{N}(\lambda, u).$$

Most integral equations of Mathematical Physics, reaction-diffusion systems in Chemistry, and many paradigmatic models of Mathematical Biology and Ecology admit an abstract formulation of this type. In these models, U is an appropriate space of real functions; for example the space $\mathcal{C}_0(\bar{D})$ of all continuous functions $u : \bar{D} \to \mathbb{R}$ such that $u = 0$ on ∂D, for a given domain D of $\mathbb{R}^N$, for some integer $N \geq 1$.

By the intrinsic nature of real-world mathematical models, in most of them the interest of the researcher is focused on functions measuring empirical real magnitudes, e.g., chemical concentrations, population densities, temperatures, pressures, etc.; consequently, one is naturally led to consider mathematical models over the real field. Moreover, in many circumstances, these models involve nonlinear maps of the form

$$\mathfrak{N}(\lambda, u) := \lambda |u|^p, \qquad (\lambda, u) \in \mathbb{R} \times \mathcal{C}_0(\bar{D}),$$

for a given real number $p > 1$, and, hence, they usually fail to be analytic. Similarly, $\mathfrak{L}$ may not be real analytic, as its dependence on the parameter is not fixed by analysts, but determined from empirical studies. For example, if U is finite dimensional, the map $\mathfrak{L}$ might be given through

$$\mathfrak{L}(\lambda) := |\lambda|^{r+1}, \qquad \lambda \in \mathbb{R},$$

for some integer $r \geq 0$, which is of class $\mathcal{C}^r$ and satisfies Assumption (AL), but, if r is even, it is not analytic. Consequently, in most applications the natural setting is the real non-analytic one, rather than the holomorphic one.

The part of Mathematical Analysis devoted to the problem of ascertaining the structure of the solution set of (12.2) is called Nonlinear Analysis. As $(\lambda, 0)$ is a solution of (12.2) for every $\lambda \in \Omega$, it will be referred to as a *trivial solution*. The most ambitious goal of Nonlinear Analysis is to find as much information as possible about $\mathfrak{F}^{-1}(0)$ from the fact that $\Omega \times \{0\} \subset \mathfrak{F}^{-1}(0)$. To accomplish that goal, the first step consists in determining the values $\lambda_0 \in \Omega$ for which a sequence of non-trivial solutions emanates at $(\lambda_0, 0)$. These are the so-called *bifurcation values* from $\Omega \times \{0\}$ (see Section 12.1 for the precise concept). Surprisingly, there are some bifurcation values $\lambda_0 \in \Omega$ for which whether or not bifurcation occurs from $\Omega \times \{0\}$ at $(\lambda_0, 0)$ is based exclusively on the properties of the family $\mathfrak{L}$ about λ_0, and not on the specific choice of the nonlinearity $\mathfrak{N}$. Such values of λ_0 are called *nonlinear eigenvalues* of $\mathfrak{L}$ (see Section 12.1). Note that $\lambda_0 \in \Sigma(\mathfrak{L})$ if and only if $(\lambda_0, 0)$ is a bifurcation point from $\mathbb{R} \times \{0\}$ for the linear equation (12.3). In such a case, the set of solutions emanating from $\mathbb{R} \times \{0\}$ at $(\lambda_0, 0)$ is a linear manifold.

The main goal of this chapter is to prove that λ_0 is a nonlinear eigenvalue of $\mathfrak{L}$ if and only if the algebraic multiplicity of $\mathfrak{L}$ at λ_0, as discussed in Part II, is an odd natural number. Our proof of this fundamental characterization relies on the topological degree (introduced in Exercise 6 of Chapter 8, in the linear case) and on the existence of the local Smith form established in Chapter 7.

This chapter is organized as follows. Section 12.1 introduces the concept of nonlinear eigenvalue and states the theorem of characterization of nonlinear eigenvalues through the oddity of $\chi[\mathfrak{L};\lambda_0]$. Section 12.2 gives a short self-contained introduction to the topological degree (without proofs). Section 12.3 establishes that $\chi[\mathfrak{L};\lambda_0]$ is odd if and only if the local topological degree of $\mathfrak{L}(\lambda)$ at 0 changes as λ crosses λ_0. Finally, Section 12.4 provides the proof of the characterization theorem presented in Section 12.1.

At a later stage, Nonlinear Analysis studies the local and global structures of the components of the set of non-trivial solutions emanating from $\Omega \times \{0\}$ at $(\lambda_0, 0)$. Singularity Theory and Real Algebraic Geometry provide us with rather satisfactory answers to the local problem, while Global Bifurcation Theory analyzes the global structure of these components. These topics are beyond the specific scope of this chapter and will not be covered in the book.

Throughout this chapter, given a real Banach space U, we denote by $GL_c(U)$ the set of $T \in \mathrm{Iso}(U)$ such that $T - I_U \in \mathcal{K}(U)$.

12.1 Bifurcation values. Nonlinear eigenvalues

The following concept plays a crucial role in Nonlinear Analysis.

Definition 12.1.1. Let $\lambda_0 \in \Omega$. If

$$(\lambda_0, 0) \in \overline{\mathfrak{F}^{-1}(0) \cap [\Omega \times (\mathcal{U} \setminus \{0\})]}$$

then $(\lambda_0, 0)$ is called a bifurcation point of (12.2) from $\Omega \times \{0\}$. In addition, λ_0 is said to be a bifurcation value of (12.2) from $\Omega \times \{0\}$.

Of course, $(\lambda_0, 0)$ is a bifurcation point of (12.2) from the curve of trivial solutions $\Omega \times \{0\}$ if for each natural n there exists

$$(\lambda_n, u_n) \in \mathfrak{F}^{-1}(0) \cap [\Omega \times (\mathcal{U} \setminus \{0\})]$$

such that

$$\lim_{n\to\infty} (\lambda_n, u_n) = (\lambda_0, 0).$$

The next result establishes that the set of bifurcation values from $\Omega \times \{0\}$ must be contained in $\Sigma(\mathfrak{L})$.

Lemma 12.1.2. *Suppose λ_0 is a bifurcation value of (12.2) from $\Omega \times \{0\}$. Then,*

$$\lambda_0 \in \Sigma(\mathfrak{L}).$$

Proof. For each integer $n \geq 1$, there exist $\lambda_n \in \Omega$ and $u_n \in \mathcal{U} \setminus \{0\}$ such that $\mathfrak{F}(\lambda_n, u_n) = 0$ and

$$\lim_{n\to\infty} (\lambda_n, u_n) = (\lambda_0, 0).$$

Then, for every $n \geq 1$,

$$-\mathfrak{L}(\lambda_0)\frac{u_n}{\|u_n\|} = [\mathfrak{L}(\lambda_n) - \mathfrak{L}(\lambda_0)]\,\frac{u_n}{\|u_n\|} + \frac{\mathfrak{N}(\lambda_n, u_n)}{\|u_n\|},$$

and, hence, by Assumption (AN), it is apparent that

$$\lim_{n\to\infty} \mathfrak{L}(\lambda_0)\frac{u_n}{\|u_n\|} = 0.$$

Consequently, since

$$\frac{u_n}{\|u_n\|} = [I_U - \mathfrak{L}(\lambda_0)]\frac{u_n}{\|u_n\|} + \mathfrak{L}(\lambda_0)\frac{u_n}{\|u_n\|}, \qquad n \geq 1,$$

and $I_U - \mathfrak{L}(\lambda_0) \in \mathcal{K}(U)$, there exists a subsequence of $\{u_n\}_{n\geq 1}$, say $\{u_{n_m}\}_{m\geq 1}$, for which

$$\varphi := \lim_{m\to\infty} \frac{u_{n_m}}{\|u_{n_m}\|}$$

exists. Necessarily, $\|\varphi\| = 1$; in particular, $\varphi \neq 0$. Also,

$$\mathfrak{L}(\lambda_0)\varphi = 0,$$

which concludes the proof. □

Simple examples show that the converse of Lemma 12.1.2 is false in general. Indeed, when we make the choice

$$U = \mathcal{U} = \Omega = \mathbb{R}, \qquad \mathfrak{L}(\lambda) = \lambda^2, \qquad \mathfrak{N}(\lambda, u) = u^3,$$

the equation (12.2) becomes

$$\lambda^2 u + u^3 = 0, \tag{12.4}$$

whose unique solutions are those of the form $(\lambda, 0)$; consequently, in this case $\mathfrak{F}$ does not admit a bifurcation value from $\Omega \times \{0\}$, despite $0 \in \Sigma(\mathfrak{L})$. Note that if, instead of (12.4), we consider

$$\lambda^2 u - u^3 = 0,$$

then 0 becomes a bifurcation value from $\Omega \times \{0\}$. Therefore, in case $\mathfrak{L}(\lambda) = \lambda^2$, whether or not 0 is a bifurcation value from $\Omega \times \{0\}$ is based on the special nature of the nonlinearity $\mathfrak{N}$.

Note that if $\lambda_0 \in \Sigma(\mathfrak{L})$, then there always exists a map $\mathfrak{N}$ satisfying Assumption (AN) for which $(\lambda_0, 0)$ is a bifurcation point of (12.2) from $\Omega \times \{0\}$, for example $\mathfrak{N} = 0$. When this occurs for any admissible nonlinearity $\mathfrak{N}$, it is said that λ_0 is a *nonlinear eigenvalue* of $\mathfrak{L}$. More precisely, this concept is introduced in the following definition.

Definition 12.1.3. Let $\mathfrak{L}$ satisfy Assumption (AL), and consider $\lambda_0 \in \Sigma(\mathfrak{L})$. Then, λ_0 is said to be a nonlinear eigenvalue of $\mathfrak{L}$ if λ_0 is a bifurcation value of (12.2) from $\Omega \times \{0\}$ for all $\mathfrak{N}$ satisfying Assumption (AN).

The concept of algebraic multiplicity studied in Part II provides us with the following fundamental characterization. Recall that a *continuum* is a compact and connected metric space.

Theorem 12.1.4 (Characterization of nonlinear eigenvalues). *Suppose $\mathfrak{L}$ satisfies Assumption* (AL) *and $\lambda_0 \in \mathrm{Alg}_\nu(\mathfrak{L})$ for some integer number $1 \leq \nu \leq r$. Then, the following assertions are equivalent:*

i) *λ_0 is a nonlinear eigenvalue of $\mathfrak{L}$.*
ii) $\chi[\mathfrak{L}; \lambda_0] \in 2\mathbb{N}+1$.

Moreover, in such a case, there is a continuum $\mathfrak{C} \subset \mathfrak{F}^{-1}(0)$ satisfying

$$\mathfrak{C} \neq \{(\lambda_0, 0)\} \quad \textit{and} \quad \mathfrak{C} \cap (\Omega \times \{0\}) = \{(\lambda_0, 0)\}.$$

In the general context of this chapter, the set of *non-trivial solutions* of equation (12.2) is defined by

$$\mathcal{S} := \left[\mathfrak{F}^{-1}(0) \cap [\Omega \times (\mathcal{U} \setminus \{0\})]\right] \cup (\Sigma(\mathfrak{L}) \times \{0\}), \tag{12.5}$$

so that it consists of all solutions $(\lambda, u) \in \Omega \times \mathcal{U}$ of (12.2) with $u \neq 0$, plus all possible bifurcation points from $\Omega \times \{0\}$. The last assertion of Theorem 12.1.4 establishes that $\mathcal{S}$ has a component $\mathfrak{C}$ strictly containing $(\lambda_0, 0)$, if λ_0 is a nonlinear eigenvalue of $\mathfrak{L}$. By a *component* it is meant a closed and connected subset which is maximal for the inclusion.

As anticipated in the presentation of this chapter, the proof of Theorem 12.1.4 is postponed until Section 12.4. The following two sections provide some key technical devices that will be used in the proof of Theorem 12.1.4. Now we present a result that states that $\Sigma(\mathfrak{L})$ and $\mathcal{S}$ are closed.

Proposition 12.1.5. *$\Sigma(\mathfrak{L})$ is a closed subset of Ω, and $\mathcal{S}$ is a closed subset of $\Omega \times \mathcal{U}$.*

Proof. First we prove that $\Sigma(\mathfrak{L})$ is closed in Ω. Let $\{\lambda_n\}_{n\in\mathbb{N}}$ be a sequence in $\Sigma(\mathfrak{L})$ such that

$$\lim_{n\to\infty} \lambda_n = \lambda_\omega \in \Omega.$$

Define

$$\mathfrak{K}(\lambda) := I_U - \mathfrak{L}(\lambda), \qquad \lambda \in \Omega.$$

We find from Assumption (AL) that $\mathfrak{K}(\lambda) \in \mathcal{K}(U)$ for every $\lambda \in \Omega$. Moreover, for each $n \geq 1$, there exists $u_n \in U$ such that

$$\mathfrak{K}(\lambda_n)u_n = u_n, \qquad \|u_n\| = 1.$$

Clearly,

$$\mathfrak{K}(\lambda_\omega)u_n = u_n + [\mathfrak{K}(\lambda_\omega) - \mathfrak{K}(\lambda_n)]\, u_n, \qquad n \geq 1. \tag{12.6}$$

As $\mathfrak{K}$ is continuous,

$$\lim_{n\to\infty} [\mathfrak{K}(\lambda_\omega) - \mathfrak{K}(\lambda_n)]\, u_n = 0,$$

and, by compactness of $\mathfrak{K}(\lambda_\omega)$, it follows from (12.6) that there exists a subsequence of $\{u_n\}_{n\geq 1}$ (not re-labeled) such that the limit

$$u_\omega := \lim_{n\to\infty} u_n \in U$$

is well defined. Consequently, passing to the limit as $n \to \infty$ in (12.6) shows

$$\mathfrak{K}(\lambda_\omega)u_\omega = u_\omega.$$

Equivalently, $\mathfrak{L}(\lambda_\omega)u_\omega = 0$. Moreover, $\|u_\omega\| = 1$, since $\|u_n\| = 1$ for every $n \geq 1$; therefore, $\lambda_\omega \in \Sigma(\mathfrak{L})$. This shows that $\Sigma(\mathfrak{L})$ is closed in Ω.

Now we prove that $\mathcal{S}$ is closed in $\Omega \times \mathcal{U}$. Define

$$Z := \mathfrak{F}^{-1}(0) \cap [\Omega \times (\mathcal{U} \setminus \{0\})] .$$

As $\mathfrak{F}$ is continuous, then Z is closed in $\Omega \times (\mathcal{U} \setminus \{0\})$, so

$$Z = \bar{Z} \cap [\Omega \times (\mathcal{U} \setminus \{0\})] .$$

By Lemma 12.1.2,

$$\bar{Z} \cap (\Omega \times \{0\}) \subset \Sigma(\mathfrak{L}) \times \{0\}.$$

Therefore,

$$\left[\bar{Z} \cap (\Omega \times \mathcal{U})\right] \cup (\Sigma(\mathfrak{L}) \times \{0\}) \subset \mathcal{S}$$

and

$$\mathcal{S} = \left[\bar{Z} \cup (\Sigma(\mathfrak{L}) \times \{0\})\right] \cap (\Omega \times \mathcal{U}) ,$$

which shows that $\mathcal{S}$ is closed in $\Omega \times \mathcal{U}$. □

12.2 A short introduction to the topological degree

Essentially, the topological degree is a generalized counter of the number of zeros that a continuous map f has in a bounded open set D such that

$$f^{-1}(0) \cap \partial D = \varnothing.$$

The topological degree was introduced by Brouwer [16] in the special case $U = \mathbb{R}^N$ as a technical device to prove his celebrated fixed point theorem; it was later extended by Leray & Schauder [80] to include general compact perturbations of the identity map in arbitrary Banach spaces. This extension provides us, rather naturally, with the celebrated fixed point theorem of Schauder [118] and Tychonoff [125], as will become apparent in Subsection 12.2.3.

When D is the bounded component enclosed by a plane Jordan curve of class $\mathcal{C}^1$ and f is holomorphic in a neighborhood of $\bar{D}$, with

$$f^{-1}(0) \cap \partial D = \varnothing,$$

then the degree of f in D, subsequently denoted by $\mathrm{Deg}(f, D)$, can be defined through

$$\mathrm{Deg}(f, D) := \frac{1}{2\pi i} \int_\gamma \frac{f'}{f}$$

and, therefore, by the Argument Principle, $\mathrm{Deg}(f, D)$ equals the total number of zeros of f in D counted according to their respective orders.

Although for an arbitrary continuous function f, the degree $\mathrm{Deg}(f, D)$ does not provide us with the exact number of zeros of f in D, it will certainly satisfy some fundamental properties that any suitable generalized counter of zeros should exhibit; in particular:

- The identity map has degree one in D if $0 \in D$.
- The topological degree of f in any disjoint union of open sets equals the sum of its degrees in each of them separately.
- The degree is invariant by continuous deformations provided the maps of the deformation do not have zeros along the boundary ∂D.
- $f^{-1}(0) \cap D \neq \varnothing$ if $\mathrm{Deg}(f, D) \neq 0$.

Strikingly, these four properties determine uniquely the topological degree.

12.2.1 Existence and uniqueness

The most versatile and quickest way to introduce the degree possibly proceeds through the following existence and uniqueness theorem. The existence part goes back to Brouwer [16] in $\mathbb{R}^N$, and to Leray & Schauder [80] in arbitrary Banach spaces. The uniqueness part is attributable to Führer [38] in the context of the Brouwer degree, and to Amann & Weiss [3] in the general setting of Leray and Schauder.

Theorem 12.2.1 (Existence and uniqueness of the topological degree). *Let U be a real Banach space. Let $\mathcal{A}$ denote the set of pairs (f, D) such that $D \subset U$ is open and bounded, $f \in \mathcal{C}(D(f), U)$ with $\bar{D} \subset D(f) \subset U$, the map $f - I_U$ is compact, and $0 \notin f(\partial D)$. Then, there exists a unique mapping*

$$\mathrm{Deg} : \mathcal{A} \to \mathbb{Z}$$

satisfying the following properties:

(D1) *For every open and bounded subset $D \subset U$ with $0 \in D$,*

$$\mathrm{Deg}(I_U, D) = 1.$$

(D2) *For every $(f, D) \in \mathcal{A}$ and every pair of open disjoint subsets D_1, D_2 of D such that $0 \notin f(\bar{D} \setminus (D_1 \cup D_2))$,*

$$\mathrm{Deg}(f, D) = \mathrm{Deg}(f, D_1) + \mathrm{Deg}(f, D_2).$$

(D3) *For every $D \subset U$ and $H \in \mathcal{C}([0,1] \times \bar{D}, U)$ such that $(H(t, \cdot), D) \in \mathcal{A}$ for all $t \in [0,1]$,*

$$\mathrm{Deg}(H(0, \cdot), D) = \mathrm{Deg}(H(1, \cdot), D).$$

The value $\mathrm{Deg}(f, D)$ is called the Leray–Schauder degree (or the topological degree, or simply the degree) of f in D.

Property D1 is usually referred to as the *normalization property*. Note that for every $n \in \mathbb{Z}$, the new map $n\,\mathrm{Deg}$ satisfies Properties D2, D3, but not D1. Thus, Property D1 indeed normalizes the degree of the identity map to make its value equal to its number of zeros in any D containing 0.

Property D2 entails three basic properties that any counter of zeros must satisfy. Precisely, by choosing $D = D_1 = D_2 = \varnothing$, it shows that

$$\mathrm{Deg}(f, \varnothing) = 0, \tag{12.7}$$

thus showing that no continuous map can have a zero in the empty set. Moreover, in the special case when $D = D_1 \cup D_2$, it shows that

$$\mathrm{Deg}(f, D) = \mathrm{Deg}(f, D_1) + \mathrm{Deg}(f, D_2), \tag{12.8}$$

so establishing the *additivity property* of the counter. Finally, in the special case when $D_2 = \varnothing$, (12.7) and (12.8) imply

$$\mathrm{Deg}(f, D) = \mathrm{Deg}(f, D_1),$$

which is usually referred to as the *excision property* of the degree. If, in addition, $D_1 = \varnothing$, the excision property establishes that

$$\mathrm{Deg}(f, D) = 0 \qquad \text{if} \quad f^{-1}(0) \cap \bar{D} = \varnothing.$$

Equivalently,

$$\text{if} \quad (f, D) \in \mathcal{A} \quad \text{and} \quad \mathrm{Deg}(f, D) \neq 0 \quad \text{then} \quad f^{-1}(0) \cap D \neq \varnothing, \tag{12.9}$$

which guarantees the existence of a zero of f in D when the degree of f in D is non-zero. Property (12.9) is usually referred to as the *solution property* of the degree.

Property D3 establishes the *invariance by homotopy* of the counter. This pivotal property shows that the generalized number of zeros of f in D, measured by $\mathrm{Deg}(f, D)$, remains unchanged if the map is deformed continuously in such a way that it does not lose or gain zeros through ∂D. This property endows the degree with the possibility of being computed.

12.2.2 Computing the degree. Analytic construction

This subsection describes a standard procedure for computing the topological degree. Let $\mathcal{R}$ denote the set of $(f,D) \in \mathcal{A}$ such that f is of class $\mathcal{C}^1$ in D, and for every $u \in f^{-1}(0) \cap D$, we have that $Df(u) \in \mathrm{Iso}(U)$. These pairs are usually referred to as *regular* pairs. Using the Inverse Function Theorem and the facts that f is a compact perturbation of the identity, and that $f^{-1}(0) \cap \partial D = \emptyset$, it is easy to see that $f^{-1}(0) \cap D$ must be finite, possibly empty. The following result is a direct consequence of the infinite dimensional version by Smale [120] of the celebrated theorem of Sard [115], and of Property D3 of Theorem 12.2.1.

Theorem 12.2.2 (of regularization). *Let U be a real Banach space. Let $\mathcal{A}$ denote the class of admissible pairs introduced by Theorem 12.2.1 and let $\mathcal{R} \subset \mathcal{A}$ be the associated class of regular pairs. Then, for every $(f,D) \in \mathcal{A}$ there exists $h \in \mathcal{C}([0,1] \times \bar{D}, U)$ such that*

$$h(0,\cdot) = f, \qquad (h(1,\cdot), D) \in \mathcal{R},$$

and $(h(t,\cdot), D) \in \mathcal{A}$ for every $t \in [0,1]$; moreover, for every such h,

$$\mathrm{Deg}(f,D) = \mathrm{Deg}(h(1,\cdot), D).$$

Consequently, the computation of $\mathrm{Deg}(f,D)$ can be reduced to calculate the degree of any homotopic regular pair. To calculate the degree of a regular pair, the following procedure should be implemented. Let $(f,D) \in \mathcal{R}$. If $f^{-1}(0) \cap \bar{D} = \emptyset$ then $\mathrm{Deg}(f,D) = 0$, due to (12.9). If not, let $n \geq 1$ be the number of elements $f^{-1}(0) \cap \bar{D}$, and let

$$f^{-1}(0) \cap \bar{D} = \{u_1, \dots, u_n\} \subset D.$$

Now, let $\varepsilon > 0$ satisfy

$$B_\varepsilon(u_j) \subset D, \qquad 1 \leq j \leq n,$$

and

$$\bar{B}_\varepsilon(u_i) \cap \bar{B}_\varepsilon(u_j) = \emptyset, \qquad 1 \leq i < j \leq n.$$

Then, according to Property D2,

$$\mathrm{Deg}(f,D) = \sum_{j=1}^{n} \mathrm{Deg}\left(f, B_\varepsilon(u_j)\right). \tag{12.10}$$

Moreover, by the excision property, $\mathrm{Deg}(f, B_\varepsilon(u_j))$ is independent of ε as long as

$$B_\varepsilon(u_j) \subset D \quad \text{and} \quad f^{-1}(0) \cap \bar{B}_\varepsilon(u_j) = \{u_j\}.$$

More generally, if $(f,D) \in \mathcal{A}$ and $u_0 \in D$ satisfies that f does not have zeros in a perforated neighborhood of u_0, then $\mathrm{Deg}(f, B_\varepsilon(u_0))$ is well defined and independent of $\varepsilon > 0$ as long as

$$f^{-1}(0) \cap \bar{B}_\varepsilon(u_0) \subset \{u_0\}.$$

The value $\operatorname{Deg}(f, B_\varepsilon(u_0))$ will be referred to as the *index* of f at u_0, and denoted by $\operatorname{Ind}(f, u_0)$. With this concept, (12.10) can be expressed in the form

$$\operatorname{Deg}(f, D) = \sum_{j=1}^{n} \operatorname{Ind}(f, u_j)\,. \tag{12.11}$$

Now, for every $1 \le j \le n$ and $u \in \bar{B}_\varepsilon(u_j)$, we set

$$R_j(u) := f(u) - Df(u_j)(u - u_j).$$

Then, since $(f, D) \in \mathcal{R}$ and $f(u_j) = 0$, it is easy to see that if $\varepsilon > 0$ is small enough, then the homotopy H_j defined as

$$H_j(t, u) := Df(u_j)(u - u_j) + tR_j(u), \qquad (t, u) \in [0, 1] \times \bar{B}_\varepsilon(u_j),$$

satisfies $(H_j(t, \cdot), B_\varepsilon(u_j)) \in \mathcal{A}$ for every $1 \le j \le n$ and $t \in [0, 1]$. Therefore, by the homotopy invariance of the degree (Property D3), we find from (12.11) that

$$\operatorname{Deg}(f, D) = \sum_{j=1}^{n} \operatorname{Ind}(Df(u_j)(\cdot - u_j), u_j)\,. \tag{12.12}$$

Finally, since $f - I_U$ is compact, we have that

$$Df(u_j) - I_U \in \mathcal{K}(U), \qquad 1 \le j \le n,$$

and, owing to the celebrated formula of Leray & Schauder [80], we find that

$$\operatorname{Ind}(Df(u_j)(\cdot - u_j), u_j) = (-1)^{m_j}, \qquad 1 \le j \le n,$$

where m_j stands for the sum of the algebraic multiplicities of all negative eigenvalues of $Df(u_j)$. In particular,

$$\operatorname{Ind}(T, 0) = \operatorname{sign}\det T \tag{12.13}$$

if $\dim U < \infty$ and $T \in \mathcal{L}(U)$. Therefore, we conclude from (12.12) that

$$\operatorname{Deg}(f, D) = \sum_{j=1}^{n} (-1)^{m_j}. \tag{12.14}$$

Conversely, the most popular analytical construction of the topological degree consists in defining the topological degree of a pair $(f, D) \in \mathcal{R}$ by (12.14) and extending it to the class $\mathcal{A}$ through Theorem 12.2.2. In practical applications, to compute the topological degree of a pair $(f, D) \in \mathcal{A}$, one should accomplish the following tasks:

1. Construct a regularizing homotopy to transform f into some map F with $(F, D) \in \mathcal{R}$. Calculate $F^{-1}(0)$.
2. Calculate $DF(u_0)$ for every $u_0 \in D \cap F^{-1}(0)$.
3. Compute the sum of the algebraic multiplicities of all negative eigenvalues of $DF(u_0)$ for every $u_0 \in D \cap F^{-1}(0)$.

12.2.3 The fixed point theorem of Brouwer, Schauder and Tychonoff

To illustrate the scheme of Subsection 12.2.2, we will prove the following special, but significant, version of the fixed point theorem of Brouwer [16], Schauder [118], and Tychonoff [125].

Theorem 12.2.3 (of Brouwer–Schauder–Tychonoff). *Let U be a real Banach space. Let $B \subset U$ be an open ball. Suppose $f : \bar{B} \to \bar{B}$ is a continuous and compact map. Then,*

$$f(u^*) = u^*$$

for some $u^* \in \bar{B}$.

Proof. Let $u_0 \in U$ and $R > 0$ satisfy $B = B_R(u_0)$. If $f(u^*) = u^*$ for some $u^* \in \partial B$, then the conclusion of the theorem is true. So, suppose

$$f(u) \neq u \quad \text{for all} \quad u \in \partial B, \tag{12.15}$$

and consider the map H defined as

$$H(t, u) := u - u_0 + t[u_0 - f(u)], \qquad (t, u) \in [0, 1] \times \bar{B}.$$

Clearly, H is continuous and, for every $t \in [0, 1]$, the map $H(t, \cdot) - I_U$ is compact. Moreover,

$$H(t, u) \neq 0 \qquad \text{for all} \quad (t, u) \in [0, 1] \times \partial B. \tag{12.16}$$

Indeed, by (12.15), property (12.16) holds for $t = 1$. Thus, if (12.16) fails, there must exist $t^* \in [0, 1)$ and $u^* \in \partial B$ such that

$$H(t^*, u^*) = 0.$$

Consequently,

$$u^* - u_0 = t^*[f(u^*) - u_0]$$

and, hence,

$$R = \|u^* - u_0\| = t^* \|f(u^*) - u_0\| \leq Rt^* < R,$$

which is impossible. Therefore, (12.16) is satisfied, and, consequently, according to Property D3,

$$\operatorname{Deg}(I_U - f, B) = \operatorname{Deg}(H(1, \cdot), B) = \operatorname{Deg}(H(0, \cdot), B) = 1,$$

since

$$D(H(0, \cdot)) = I_U,$$

and I_U does not have any negative eigenvalue. Consequently, owing to the solution property (12.9), it is apparent that f possesses a fixed point in B. This concludes the proof. □

12.2.4 Further properties

The following result collects some well-known properties of the topological degree that will be used in the next two sections. Properties ii)–iv) follow straight away from the Leray–Schauder formula and basic theory of linear operators, while the general property of invariance by homotopy formulated in Property i) follows from the invariance by homotopy and the excision property. From now on, given $\mathcal{O} \subset \mathbb{R} \times U$ and $t \in \mathbb{R}$, we define

$$\mathcal{O}_t := \{u \in U : (t, u) \in \mathcal{O}\}.$$

Proposition 12.2.4. *The following properties hold:*

i) *Suppose $a < b$, and $\mathcal{O}$ is an open bounded subset of $[a, b] \times U$. Let $H \in \mathcal{C}(\bar{\mathcal{O}}, U)$ be such that the map*

$$H(t, \cdot) - I_U : \mathcal{O}_t \to U$$

is compact for every $t \in [a, b]$, and

$$H(t, u) \neq 0 \qquad \text{for each} \quad t \in [a, b] \quad \text{and} \quad (t, u) \in \partial\mathcal{O}.$$

Then,

$$\operatorname{Deg}\left(H(a, \cdot), \mathcal{O}_a\right) = \operatorname{Deg}\left(H(b, \cdot), \mathcal{O}_b\right).$$

ii) $\operatorname{Ind}(A, 0) \in \{-1, 1\}$ *for every $A \in GL_c(U)$. Moreover, the mapping*

$$\operatorname{Ind}(\cdot, 0) : GL_c(U) \to \{-1, 1\}$$

is continuous.

iii) *For every $A, B \in GL_c(U)$,*

$$\operatorname{Ind}(AB, 0) = \operatorname{Ind}(A, 0) \cdot \operatorname{Ind}(B, 0).$$

iv) *Let U_1, U_2 be two closed subspaces of U such that $U = U_1 \oplus U_2$. Let $T \in GL_c(U)$ be such that $T(U_i) \subset U_i$, for $i = 1, 2$. Then,*

$$\operatorname{Ind}(T, 0) = \operatorname{Ind}(T|_{U_1}, 0) \cdot \operatorname{Ind}(T|_{U_2}, 0).$$

12.3 The algebraic multiplicity as an indicator of the change of index

The following result (anticipated in Excercise 7 of Chapter 8) shows that the algebraic multiplicity $\chi[\mathfrak{L}; \lambda_0]$ can detect all changes of the topological index $\operatorname{Ind}(\mathfrak{L}(\lambda), 0)$ as the real parameter λ varies. In fact, it shows that the theory of the multiplicity developed in Part II provides us with an algorithm to compute any change of index; thus establishing some deep connections between Algebra, Analysis, Topology and the theory of Dynamical Systems.

Proposition 12.3.1. *Suppose* $\mathfrak{L}$ *satisfies* (AL) *and* $\lambda_0 \in \mathrm{Alg}_\nu(\mathfrak{L})$ *for some integer* $1 \leq \nu \leq r$. *Then, the following assertions are equivalent:*

i) $\mathrm{Ind}(\mathfrak{L}(\lambda), 0)$ *changes as* λ *crosses* λ_0.
ii) $\chi[\mathfrak{L}; \lambda_0] \in 2\mathbb{N} + 1$.

Proof. Since $\mathfrak{L}(\lambda)$ is a compact perturbation of the identity map for every $\lambda \in \Omega$, then $\mathfrak{L}(\lambda) \in \mathrm{Fred}_0(U)$. Let $\{\mathfrak{M}_{\{j\}}\}_{j=0}^{\nu}$ be a derived sequence from $\mathfrak{L}$ at λ_0 with corresponding sequence of (finite-rank) projections $\{P_j\}_{j=0}^{\nu}$. Then,

$$\mathfrak{M}_{\{0\}} = \mathfrak{L}, \quad \mathfrak{M}_{\{\nu\}}(\lambda_0) \in \mathrm{Iso}(U), \quad \mathfrak{M}_{\{\nu\}} \in \mathcal{C}(\Omega, \mathcal{L}(U)),$$

and, by (5.35),

$$\mathfrak{L}(\lambda) = \mathfrak{M}_{\{\nu\}}(\lambda)\mathfrak{C}_{P_{\nu-1}}(\lambda - \lambda_0) \cdots \mathfrak{C}_{P_0}(\lambda - \lambda_0), \qquad \lambda \in \Omega,$$

where $\mathfrak{C}_{P_0}, \dots \mathfrak{C}_{P_{\nu-1}}$ are the operator families defined in (5.33). Now we prove by induction on j that $\mathfrak{M}_{\{j\}}$ is a compact perturbation of the identity for any $j \in \{0, \dots, \nu\}$. Clearly, by definition and Assumption (AL),

$$I_U - \mathfrak{M}_{\{0\}}(\lambda) = I_U - \mathfrak{L}(\lambda) \in \mathcal{K}(U), \qquad \lambda \in \Omega.$$

Suppose that

$$I_U - \mathfrak{M}_{\{j\}}(\lambda) \in \mathcal{K}(U)$$

for all $\lambda \in \Omega$ and a particular $j \in \{0, \dots, \nu - 1\}$. Then, for every $\lambda \in \Omega \setminus \{\lambda_0\}$, we find from the definition of derived sequence that

$$\mathfrak{M}_{\{j+1\}}(\lambda) - I_U = \mathfrak{M}_{\{j\}}(\lambda) - I_U + \frac{1 - \lambda + \lambda_0}{\lambda - \lambda_0}\mathfrak{M}_{\{j\}}(\lambda)P_j,$$

which is compact, as a finite-rank perturbation of a compact operator. In addition,

$$\mathfrak{M}_{\{j+1\}}(\lambda_0) - I_U$$

is compact too, as a limit of compact operators. Consequently, $\mathfrak{M}_{\{j\}}$ is a compact perturbation of the identity for every $j \in \{0, \dots, \nu\}$. In particular, there exists $\delta > 0$ such that

$$\mathfrak{M}_{\{\nu\}}(\lambda) \in GL_c(U), \qquad \lambda \in (\lambda_0 - \delta, \lambda_0 + \delta).$$

Then, by Proposition 12.2.4 iii), for all $0 < |\lambda - \lambda_0| < \delta$,

$$\mathrm{Ind}(\mathfrak{L}(\lambda), 0) = \mathrm{Ind}(\mathfrak{M}_{\{\nu\}}(\lambda), 0) \cdot \prod_{j=0}^{\nu-1} \mathrm{Ind}(\mathfrak{C}_{P_j}(\lambda - \lambda_0), 0).$$

In addition, by Proposition 12.2.4 ii), there exists $\eta \in \{-1, 1\}$ such that

$$\mathrm{Ind}(\mathfrak{M}_{\{\nu\}}(\lambda), 0) = \eta, \qquad |\lambda - \lambda_0| < \delta.$$

Moreover, according to Proposition 12.2.4 iv), (5.33), (12.13), and Theorem 12.2.1 (D1), for every $j \in \{0, \dots, \nu - 1\}$ and $0 < |\lambda - \lambda_0| < \delta$, it is apparent that

$$\begin{aligned}\operatorname{Ind}(\mathfrak{C}_{P_j}(\lambda - \lambda_0), 0) &= \operatorname{Ind}((\lambda - \lambda_0) I_{R[P_j]}, 0) \cdot \operatorname{Ind}(I_{N[P_j]}, 0) \\ &= \operatorname{sign}\det((\lambda - \lambda_0) I_{R[P_j]}) \\ &= \operatorname{sign}(\lambda - \lambda_0)^{\dim R[P_j]}\end{aligned}$$

and, consequently, we conclude from Corollary 5.3.2 (b) that

$$\begin{aligned}\operatorname{Ind}(\mathfrak{L}(\lambda), 0) &= \eta \prod_{j=0}^{\nu-1} \operatorname{sign}(\lambda - \lambda_0)^{\dim R[P_j]} = \eta \operatorname{sign}(\lambda - \lambda_0)^{\sum_{j=0}^{\nu-1} \operatorname{rank} P_j} \\ &= \eta \operatorname{sign}(\lambda - \lambda_0)^{\chi[\mathfrak{L};\lambda_0]},\end{aligned}$$

which ends the proof. □

12.4 Characterization of nonlinear eigenvalues

This section is devoted to the proof of Theorem 12.1.4, in which we use the local Smith form and Proposition 12.3.1. Thus, throughout this section we assume that $\mathfrak{L}$ satisfies Assumption (AL) and that $\lambda_0 \in \operatorname{Alg}_\nu(\mathfrak{L})$ for some integer $1 \leq \nu \leq r$. Then, thanks to Theorem 5.3.1, $\chi[\mathfrak{L}; \lambda_0]$ is well defined and finite. In particular, λ_0 is isolated in $\Sigma(\mathfrak{L})$. Moreover, by Proposition 12.3.1, $\operatorname{Ind}(\mathfrak{L}(\lambda), 0)$ changes as λ crosses λ_0 if and only if $\chi[\mathfrak{L}; \lambda_0]$ is odd. Suppose this is the case, let $\mathfrak{N} \in \mathcal{C}(\Omega \times \mathcal{U}, U)$ be any map satisfying Assumption (AN), and consider the map $\mathfrak{F}$ defined by (12.1), and the equation (12.2).

As λ_0 is isolated in $\Sigma(\mathfrak{L})$, and Ω is open, there exists $\delta > 0$ such that

$$[\lambda_0 - \delta, \lambda_0 + \delta] \subset \Omega \quad \text{and} \quad [\lambda_0 - \delta, \lambda_0 + \delta] \cap \Sigma(\mathfrak{L}) = \{\lambda_0\}.$$

Moreover, according to Lemma 12.1.2, there exists $\eta_0 > 0$ such that, for every $\eta \in (0, \eta_0]$, the cylinder

$$Q_\eta := \{(\lambda, u) \in \mathbb{R} \times U : \; |\lambda - \lambda_0| \leq \delta, \; \|u\| \leq \eta\},$$

satisfies $Q_\eta \subset \Omega \times \mathcal{U}$ and

$$|\lambda - \lambda_0| < \delta \quad \text{if} \quad (\lambda, u) \in Q_\eta \cap \mathcal{S}. \tag{12.17}$$

Recall from (12.5) the definition of $\mathcal{S}$. These cylinders satisfy the following important property.

Lemma 12.4.1. *For every $\eta \in (0, \eta_0]$, we have $\partial Q_\eta \cap \mathcal{S} \neq \varnothing$.*

Proof. Assume that

$$\partial Q_\eta \cap \mathcal{S} = \emptyset \tag{12.18}$$

for some $\eta \in (0, \eta_0]$; we shall derive a contradiction. According to (12.17), by the excision property and the invariance by homotopy of the degree, arguing as in Subsection 12.2.2, it is easily seen that

$$\begin{aligned} \operatorname{Deg}(\mathfrak{F}(\lambda_0 - \delta, \cdot), B_\eta) &= \operatorname{Ind}(\mathfrak{L}(\lambda_0 - \delta), 0), \\ \operatorname{Deg}(\mathfrak{F}(\lambda_0 + \delta, \cdot), B_\eta) &= \operatorname{Ind}(\mathfrak{L}(\lambda_0 + \delta), 0). \end{aligned} \tag{12.19}$$

Moreover, by (12.18), it follows from the invariance by homotopy of the degree that $\operatorname{Deg}(\mathfrak{F}(\lambda, \cdot), B_\eta)$ is constant for every $\lambda \in [\lambda_0 - \delta, \lambda_0 + \delta]$. Therefore, we conclude from (12.19) that

$$\operatorname{Ind}(\mathfrak{L}(\lambda_0 - \delta), 0) = \operatorname{Ind}(\mathfrak{L}(\lambda_0 + \delta), 0), \tag{12.20}$$

which contradicts the fact that $\operatorname{Ind}(\mathfrak{L}(\lambda), 0)$ changes as λ crosses λ_0, and concludes the proof. □

According to (12.17) and Lemma 12.4.1, for each $\eta \in (0, \eta_0]$ there exists

$$(\lambda_\eta, u_\eta) \in \partial Q_\eta \cap \mathcal{S}$$

with $\|u_\eta\| = \eta$. Since $|\lambda_\eta - \lambda_0| \leq \delta$ for every $\eta \in (0, \eta_0]$, there exist $\tilde{\lambda} \in [\lambda_0 - \delta, \lambda_0 + \delta]$ and a sequence $\{\eta_n\}_{n \geq 1}$ of positive numbers such that

$$\lim_{n\to\infty} \eta_n = 0, \qquad \lim_{n\to\infty} (\lambda_{\eta_n}, u_{\eta_n}) = (\tilde{\lambda}, 0).$$

Therefore, thanks to Lemma 12.1.2, $\tilde{\lambda} = \lambda_0$. Consequently, $(\lambda_0, 0)$ is a bifurcation point from $\Omega \times \{0\}$ of equation (12.2) and, hence, λ_0 is a nonlinear eigenvalue of $\mathfrak{L}$. This shows that ii) implies i) in Theorem 12.1.4. To show the existence of the continuum $\mathfrak{C}$ we will use the following result, borrowed from Whyburn [126], whose proof is omitted here.

Lemma 12.4.2. *Let M be a compact metric space and A and B two disjoint nonempty compact subsets of M. Then, some of the following (mutually exclusive) alternatives occur:*

i) *There exists a continuum $\mathfrak{C} \subset M$ such that*

$$\mathfrak{C} \cap A \neq \emptyset \quad \text{and} \quad \mathfrak{C} \cap B \neq \emptyset.$$

ii) *There exist two disjoint compact subsets M_A and M_B of M such that*

$$A \subset M_A, \qquad B \subset M_B, \qquad M = M_A \cup M_B.$$

Now we take $\eta \in (0, \eta_0]$ and consider the set

$$M := Q_\eta \cap \mathcal{S}.$$

According to Proposition 12.1.5, M is closed and bounded. Moreover, since $(\lambda, u) \in M$ if and only if

$$|\lambda - \lambda_0| \leq \delta, \qquad \|u\| \leq \eta, \qquad u = [I_U - \mathfrak{L}(\lambda)]u - \mathfrak{N}(\lambda, u),$$

it follows from Assumptions (AL) and (AN) that M is compact. Now, consider the sets A and B defined as

$$A := \{(\lambda, u) \in M : \|u\| = \eta\}, \qquad B := \{(\lambda_0, 0)\}.$$

By Lemma 12.4.1 and (12.17), A is a non-empty compact subset of M. Clearly, so is B. Moreover, A and B are disjoint. Thus, some of the alternatives of Lemma 12.4.2 are satisfied. We shall prove that ii) does not hold. Suppose, looking for a contradiction, that ii) is satisfied. Then, since M_A and M_A are disjoint compact sets, there exists $\varepsilon > 0$ such that the ε-neighborhood of M_B,

$$M_B^\varepsilon := M_B + \{(\lambda, u) \in \mathbb{R} \times U : \|(\lambda, u)\| < \varepsilon\},$$

satisfies $M_A \cap M_B^\varepsilon = \varnothing$ and

$$|\lambda - \lambda_0| < \delta \quad \text{if} \quad (\lambda, u) \in M_B^\varepsilon; \tag{12.21}$$

this is possible thanks to (12.17). Then, since $M = M_A \cup M_B$, we have that $\partial M_B^\varepsilon \cap M = \varnothing$. Now we show that if $\beta > 0$ is small enough, then

$$Q_\beta \cap M_A = \varnothing. \tag{12.22}$$

Indeed, if not, there would exist a sequence $\{(\lambda_n, u_n)\}_{n\in\mathbb{N}}$ in M_A such that

$$\lim_{n\to\infty} (\lambda_n, u_n) = (\tilde{\lambda}, 0)$$

for some $\tilde{\lambda} \in [\lambda_0 - \delta, \lambda_0 + \delta]$. Since $(\tilde{\lambda}, 0) \in M_A \subset M$ we have that $\tilde{\lambda} = \lambda_0$. So $(\lambda_0, 0) \in M_A \cap B$, which is a contradiction. This proves that (12.22) is true for $\beta > 0$ small enough; fix such a β. Then the open set

$$\mathcal{O} := \mathring{Q}_\beta \cup M_B^\varepsilon$$

satisfies $M_B \subset \mathcal{O}$ and $\partial\mathcal{O} \cap M = \varnothing$. Thus, by Proposition 12.2.4 i), we find that

$$\operatorname{Deg}(\mathfrak{F}(\lambda_0 - \delta, \cdot), \mathcal{O}_{\lambda_0 - \delta}) = \operatorname{Deg}(\mathfrak{F}(\lambda_0 + \delta, \cdot), \mathcal{O}_{\lambda_0 + \delta}).$$

Also, according to (12.21), it is obvious that

$$\mathcal{O}_{\lambda_0 - \delta} = \mathcal{O}_{\lambda_0 + \delta} = B_\beta \subset \mathcal{U},$$

and, by (12.19), equality (12.20) holds true, which is impossible. This contradiction shows that Alternative i) of Lemma 12.4.2 holds, and, hence, the existence of the continuum $\mathfrak{C}$ with the properties described in that lemma concludes the proof of the last assertion of Theorem 12.1.4.

Next, we suppose that

$$\chi[\mathfrak{L}; \lambda_0] \in 2\mathbb{N}.$$

To complete the proof of Theorem 12.1.4 it suffices to construct an operator $\mathfrak{N}$ satisfying Assumption (AN) and such that the unique zeros of $\mathfrak{F}$ in a neighborhood of $(\lambda_0, 0)$ are in $\Omega \times \{0\}$. This would show that λ_0 cannot be a nonlinear eigenvalue of $\mathfrak{L}$, and would conclude the proof.

According to Theorem 7.8.3, $\mathfrak{L}$ possesses a local Smith form at λ_0. More precisely, there exist an open neighborhood $\Omega' \subset \Omega$ of λ_0, two families $\mathfrak{E} \in \mathcal{C}(\Omega', \mathcal{L}(U))$ and $\mathfrak{G} \in \mathcal{C}(\Omega', \mathcal{L}(U))$, with $\mathfrak{E}_0, \mathfrak{G}_0 \in \text{Iso}(U)$, a direct sum topological decomposition $U = U_0 \oplus U_1$, and $n := \dim U_0$ integer positive numbers $k_1 \geq \cdots \geq k_n$, with $k_1 \leq r$, such that

$$\mathfrak{L}(\lambda) = \mathfrak{E}(\lambda)\ [\mathfrak{D}(\lambda) \oplus I_{U_1}]\ \mathfrak{G}(\lambda), \qquad \lambda \in \Omega',$$

where

$$\mathfrak{D}(\lambda) = \text{diag}\left\{(\lambda - \lambda_0)^{k_1}, \ldots, (\lambda - \lambda_0)^{k_n}\right\} \in \mathcal{L}(U_0), \qquad \lambda \in \Omega'.$$

Through the change of variables

$$v = \mathfrak{G}(\lambda)u, \qquad \lambda \in \Omega',$$

it is easy to see that equation (12.2) is equivalent, in a neighborhood of λ_0, to

$$[\mathfrak{D}(\lambda) \oplus I_{U_1}]v + \mathfrak{M}(\lambda, v) = 0, \tag{12.23}$$

where

$$\mathfrak{M}(\lambda, v) := \mathfrak{E}(\lambda)^{-1}\mathfrak{N}(\lambda, \mathfrak{G}(\lambda)^{-1}v),$$

in the sense that $\mathfrak{D}(\lambda) \oplus I_{U_1}$ satisfies Assumption (AL), $\mathfrak{M}$ satisfies Assumption (AN), and $\mathfrak{F}(\lambda, u) = 0$ if and only if $(\lambda, v) := (\lambda, \mathfrak{G}(\lambda)u)$ solves (12.23). In particular, $(\lambda_0, 0)$ is a bifurcation point of (12.2) if and only if it is a bifurcation point of (12.23). Consequently, to complete the proof of Theorem 12.1.4 it suffices to construct an $\mathfrak{M}$ satisfying Assumption (AN) for which all solutions (λ, v) of (12.23) in a neighborhood of $(\lambda_0, 0)$ satisfy $v = 0$. Set

$$v = x + y, \qquad x \in U_0, \quad y \in U_1,$$

and choose any $\mathfrak{M}$ satisfying

$$\mathfrak{M}(\lambda, x + y) = \mathfrak{M}(\lambda, x) \in U_0$$

for all $(\lambda, x, y) \in \mathbb{R} \times U_0 \times U_1$ in a neighborhood of $(\lambda_0, 0, 0)$.

Then, equation (12.23) can be expressed as

$$\begin{cases} \mathfrak{D}(\lambda)x + \mathfrak{M}(\lambda, x) = 0, \\ \hfill y = 0, \end{cases}$$

and, consequently, it suffices to find some $\varepsilon > 0$ and n continuous functions

$$\mathfrak{M}_i : [\lambda_0 - \varepsilon, \lambda_0 + \varepsilon] \times \bar{B}_\varepsilon \to \mathbb{R}, \qquad 1 \leq i \leq n,$$

such that

$$\lim_{x \to 0} \max_{|\lambda - \lambda_0| \leq \varepsilon} \frac{|\mathfrak{M}_i(\lambda, x)|}{\|x\|} = 0, \qquad 1 \leq i \leq n,$$

for which $(\lambda, 0)$ are the solely solutions (λ, x) of the n-dimensional system

$$\begin{pmatrix} (\lambda - \lambda_0)^{k_1} & & \\ & \ddots & \\ & & (\lambda - \lambda_0)^{k_n} \end{pmatrix} \begin{pmatrix} x_1 \\ \vdots \\ x_n \end{pmatrix} + \begin{pmatrix} \mathfrak{M}_1(\lambda, x) \\ \vdots \\ \mathfrak{M}_n(\lambda, x) \end{pmatrix} = 0, \tag{12.24}$$

for $|\lambda - \lambda_0| \leq \varepsilon$, and $\|x\| \leq \varepsilon$; here $x_1, \dots, x_n$ stand for the components of x with respect to the canonical basis of $\mathbb{R}^n$, and the ball B_ε is in $\mathbb{R}^n$. Note that Assumption (AN) is automatically satisfied, since $\mathfrak{M}$ has been chosen with finite rank. Now we make the appropriate choice of $\mathfrak{M}_i$, for each $i \in \{1, \dots, n\}$.

Let $i \in \{1, \dots, n\}$ such that k_i is even. Then, by making the choice

$$\mathfrak{M}_i(\lambda, x) := x_i^3, \qquad |\lambda - \lambda_0| \leq \varepsilon, \quad \|x\| \leq \varepsilon,$$

the ith equation of system (12.24) becomes

$$(\lambda - \lambda_0)^{k_i} x_i + x_i^3 = 0,$$

whose unique solution (λ, x_i) is $(\lambda, 0)$.

According to Theorem 7.6.3, we have that

$$\chi[\mathfrak{L}; \lambda_0] = \sum_{i=1}^{n} k_i.$$

As we are assuming that $\chi[\mathfrak{L}; \lambda_0]$ is even, there is an even number of indexes $i \in \{1, \dots, n\}$ such that k_i is odd. If there are none at all, the proof is done. Suppose $i, j \in \{1, \dots, n\}$ are such that $i \neq j$, and k_i and k_j are odd. Then, making the choice

$$\mathfrak{M}_i(\lambda, x) := x_j^3, \quad \mathfrak{M}_j(\lambda, x) := -x_i^3, \qquad |\lambda - \lambda_0| \leq \varepsilon, \quad \|x\| \leq \varepsilon,$$

the system consisting of the ith and jth equations of (12.24) becomes

$$\begin{cases} (\lambda - \lambda_0)^{k_i} x_i + x_j^3 = 0, \\ (\lambda - \lambda_0)^{k_j} x_j - x_i^3 = 0. \end{cases}$$

As $k_i - k_j$ is even, it is easy to conclude that any solution of the above system must satisfy $x_i = x_j = 0$, which ends the proof of Theorem 12.1.4.

12.5 Comments on Chapter 12

Some specialized monographs about nonlinear integral equations are Krasnosel'skiĭ [69], López-Gómez [82], and Kielhöfer [68]; each of them enjoys a different taste. Some simple prototypical models of great relevance from the point of view of their applications to Mathematical Biology and Ecology can be found in the paradigmatic monographs of Cantrell & Cosner [18], Murray [102] and Okubo [103], and in the paper of López-Gómez & Molina-Meyer [87], which attempts to be an instance of how serious mathematics should be diffused among empirical scientists. The exposition of this chapter is based on López-Gómez [82, 85], Magnus [92], and Mora-Corral [101].

The concepts of bifurcation point and nonlinear eigenvalue are classic and, essentially, go back to Poincaré [108], at least in the context of *nonlinear oscillations* in Celestial Mechanics. However, the notation and terminologies used in this chapter were essentially fixed by Krasnosel'skiĭ [69] and his School of Nonlinear Analysis. In fact, under the assumptions of this chapter, the first general result concerning nonlinear eigenvalues was a celebrated theorem of Krasnosel'skiĭ, which established that if

$$A \in \mathcal{K}(U), \qquad \mathfrak{L}(\lambda) = \lambda I_U - A, \quad \lambda \in \mathbb{R},$$

and $\lambda_0 \neq 0$ is an eigenvalue of A of odd multiplicity $m_a(\lambda_0)$, then λ_0 is a nonlinear eigenvalue of $\mathfrak{L}$. Note that, according to Theorem 8.3.2,

$$\chi[\mathfrak{L}; \lambda_0] = m_a(\lambda_0).$$

Even in the classical setting of Krasnosel'skiĭ, the characterization described in Theorem 12.1.4 goes back to Esquinas & López-Gómez [29, 30, 31, 82], to whom Theorem 12.1.4 is attributable.

The proof of Theorem 12.1.4 in Section 12.4 reveals that λ_0 is a nonlinear eigenvalue of $\mathfrak{L}$ provided $\operatorname{Ind}(\mathfrak{L}(\lambda), 0)$ changes as λ crosses λ_0. Thus, at least for this implication, the condition that λ_0 is an algebraic eigenvalue of $\mathfrak{L}$ is not necessary. Some further (optimal) improvements by Ize [62] and Fitzpatrick & Pejsachowitz [32] showed that λ_0 is a nonlinear eigenvalue of $\mathfrak{L}$ if and only if $\operatorname{Ind}(\mathfrak{L}(\lambda), 0)$ changes as λ crosses λ_0. To prove this deep characterization, they used Obstruction Theory (a non-constructive technique of Algebraic Topology) in order to show that $\operatorname{Ind}(\mathfrak{L}(\lambda), 0)$ changes as λ crosses λ_0 if λ_0 is a nonlinear eigenvalue. Although in these contexts, Algebraic Topology provides us with optimal results, these techniques do not give an explicit nonlinearity for which $(\lambda, 0)$ is the unique solution of (12.2) nearby $(\lambda_0, 0)$ if $\operatorname{Ind}(\mathfrak{L}(\lambda), 0)$ remains constant as λ crosses λ_0; our proof of Theorem 12.1.4 through the Smith form is, in contrast, constructive. That proof goes back to Mora-Corral [101] and, apparently, is substantially shorter than the original one by Esquinas & López-Gómez [29, 30, 31, 82], where, instead of the Smith form, transversalization theory was used straight away. It should not be forgotten that transversalization theory was the main technical device in characterizing the existence of the local Smith form in Chapter 7 in the general non-analytical case. Consequently, the shortening of the lengthy proofs of López-Gómez [82, Chapter 4] has been actually facilitated by the theory developed in Chapter 7 of this book. Incidentally, it was the development of the necessary technical tools for proving Theorem 12.1.4 that originated the theory of transversality studied in Chapter 4. The interested reader is sent to López-Gómez [82, Chapter 4] for any further details concerning this issue. Later, transversalization theory has been shown to be an important device throughout this book. The idea of using the Smith form in the proof of Theorem 12.1.4 seems to go back to Rabier [110].

Lemma 12.4.2 describes a topological property, traditionally attributed to Whyburn [126], although the result seems to be considerably earlier. See Kuratowski [70] for a compendium of results in this spirit on connection and separation, and Alexander [1] for an exposition from the point of view of Bifurcation Theory.

Within the context of local bifurcation theory, the topological degree was first used by Krasnosel'skiĭ [69] to prove his celebrated bifurcation theorem, stated above. As shown in the proof of Theorem 12.1.4, if $\mathrm{Ind}(\mathfrak{L}(\lambda), 0)$ changes as λ crosses λ_0, then λ_0 is a nonlinear eigenvalue of $\mathfrak{L}$, and a continuum of non-trivial solutions emanates from $\Omega \times \{0\}$ at λ_0. This is one of the most important principles of Nonlinear Analysis, due to its huge number of applications.

As discussed in this chapter, the topological degree was introduced by Brouwer [16] and later refined by Leray & Schauder [80], in connection with fixed point theory. However, some pivotal underlying ideas, mostly those related to the *winding number*, might be attributable to C.F. Gauss, L. Kronecker and H. Poincaré, among other significant predecessors. All proofs of the results of Section 12.2 can be easily constructed from Lloyd [81], which is a paradigmatic textbook on topological degree theory from an analytic point of view. In constructing the topological degree, Brouwer [16] founded Algebraic Topology. Naturally, there are a number of textbooks where degree theory is studied adopting the point of view of Algebraic Topology, but this issue is beyond the general scope of this book.

More general developments of the Leray–Schauder degree for wider classes of fixed-point equations, where the operators enjoy weaker regularity conditions than Assumptions (AL) and (AN), were introduced by Browder, Nussbaum and Petryshyn in the late sixties (cf. Lloyd [81], Deimling [26] and Petryshyn [107]) as well as by Mawhin [96], and, more recently, by Fitzpatrick & Pejsachowitz [32, 33, 34], Pejsachowitz & Rabier [104, 105], and Benevieri & Furi [10, 11, 12], who have extended the range of applicability of the classic topological degree to cover large classes of Fredholm operators of index zero, not necessarily compact perturbations of the identity map.

From now on, we suppose that $\Omega \times \mathcal{U} = \mathbb{R} \times U$ and that λ_0 is a nonlinear eigenvalue of $\mathfrak{L}$. Then, according to Theorem 12.1.4, there is a continuum $\mathfrak{C} \subset \mathcal{S}$ such that $\{(\lambda_0, 0)\} \subsetneq \mathfrak{C}$. The set $\mathcal{S}$ of non-trivial solutions of (12.2) has some component $\mathfrak{C}_m$ such that $(\lambda_0, 0) \in \mathfrak{C}_m$. By a *component*, we mean a closed and connected set which is maximal for the inclusion. Global bifurcation theory is the part of Nonlinear Analysis that studies the global topological behavior of these components $\mathfrak{C}_m$. This branch originated from the pioneering works of Rabinowitz [111, 112], who established that $\mathfrak{C}_m$ must satisfy some of the following (not mutually exclusive) alternatives:

A1. $\mathfrak{C}_m$ is unbounded in $\mathbb{R} \times U$.

A2. There exists a nonlinear eigenvalue $\tilde{\lambda}_0 \in \Sigma(\mathfrak{L}) \setminus \{\lambda_0\}$ of $\mathfrak{L}$ such that $(\tilde{\lambda}_0, 0) \in \mathfrak{C}_m$.

Since the publication of these works, this alternative is referred to as the *global alternative of P. H. Rabinowitz*, and represents a paradigmatic result in global Nonlinear Analysis, due to its huge number of applications to the theory of nonlinear Partial Differential Equations. In linear equations, i.e., when $\mathfrak{N} = 0$, Alternative A1 always occurs, as there is a linear manifold of non-trivial solutions at λ_0. But, in nonlinear problems, according to the nature of $\mathfrak{N}$, any of the two alternatives, or even both simultaneously, can occur (see López-Gómez [82, Chapter 7], and López-Gómez & Molina-Meyer [86]).

Some sharp improvements of these pioneering results were later found by Dancer [24, 25], Ize [61], Magnus [92] and López-Gómez [82] in connection with the problem of ascertaining the fine structure of the compact components of $\mathcal{S}$ meeting $\mathbb{R} \times \{0\}$.

The analysis of these and other related problems in connection, for example, with the analysis of the local structure of the set of solutions bifurcating from $\Omega \times \{0\}$ at a nonlinear eigenvalue, lies within the field of Singularity Theory and real Algebraic Geometry, but they are outside the general scope of the present book. We will, consequently, stop this final discussion here.

From a topological perspective, the algebraic multiplicity $\chi[\mathfrak{L}; \lambda_0]$ merely provides us with a finite algorithm to calculate the local topological index $\mathrm{Ind}(\mathfrak{L}(\lambda), 0)$. Therefore, it seems imperative to be able to compute it for a sufficiently large class of operator families, not necessarily analytic, and this task has been widely accomplished in the book.

Bibliography

[1] J.C. Alexander, *A primer on connectivity.* In *Fixed point theory*, pp. 455–483. Lecture Notes in Mathematics 886, Springer, Berlin-New York 1981.

[2] H. Amann, *Gewöhnliche Differentialgleichungen.* De Gruyter Lehrbuch, Walter de Gruyter, Berlin 1983.

[3] H. Amann and S.A. Weiss, *On the uniqueness of the topological degree.* Math. Z. **130** (1973), 39–54.

[4] J. Arason and R. Magnus, *The universal multiplicity theory for analytic operator-valued functions.* Math. Proc. Cambridge Philos. Soc. **118** (1995), 315–320.

[5] J. Arason and R. Magnus, *An algebraic multiplicity theory for analytic operator-valued functions.* Math. Scand. **82** (1998), 265–286.

[6] J. Arason and R. Magnus, *A multiplicity theory for analytic functions that take values in a class of Banach algebras.* J. Operator Theory **45** (2001), 161–174.

[7] D.K. Arrowsmith and C.M. Place, *Ordinary differential equations. A qualitative approach with applications.* Chapman and Hall Mathematics Series, Chapman & Hall, London 1982.

[8] J.A. Ball, I. Gohberg and L. Rodman, *Interpolation of Rational Matrix Functions.* Operator Theory: Advances and Applications **45**, Birkhäuser, Basel 1990.

[9] H. Bart, M.A. Kaashoek and D.C. Lay, *Relative inverses of meromorphic operator functions and associated holomorphic projection functions.* Math. Ann. **218** (1975), 199–210.

[10] P. Benevieri and M. Furi, *A simple notion of orientability for Fredholm maps of index zero between Banach manifolds and degree theory.* Ann. Sci. Math. Québec **22** (1998), 131–148.

[11] P. Benevieri and M. Furi, *On the concept of orientabily for Fredholm maps between real Banach manifolds.* Top. Meth. Nonl. Anal. **16** (2000), 279–306.

[12] P. Benevieri and M. Furi, *Bifurcation results for families of Fredholm maps of index zero between Banach spaces.* In *Nonlinear Analysis and its Applications (St. John's NF,* 1999), pages 35–47. Nonlinear Analysis Forum 6, 2001.

[13] T.S. Blyth and E.F. Robertson, *Linear algebra.* Essential Student Algebra 4. Chapman and Hall, New York 1986.

[14] B. den Boer, *Linearization of operator functions on arbitrary open sets.* Integral Equations and Operator Theory **1** (1978), 19–27.

[15] H. den Boer and G.P.A. Thijsse, *Semistability of sums of partial multiplicities under additive perturbation.* Integral Equations Operator Theory **3** (1980), 23–42.

[16] L.E.J. Brouwer, *Über Abbildung von Mannigfaltigkeiten.* Math. Annal. **71** (1911), 97–115.

[17] D.M. Burton, *Abstract and linear algebra.* Addison-Wesley, Reading-London-Don Mills 1972.

[18] R.S. Cantrell and C. Cosner, *Spatial ecology via reaction-diffusion equations.* Wiley Series in Mathematical and Computational Biology, John Wiley & Sons, Chichester 2003.

[19] S.N. Chow and J.K. Hale, *Methods of bifurcation theory.* Grundlehren der Mathematischen Wissenschaften **251**, Springer, New York-Berlin 1982.

[20] P.G. Ciarlet, *Introduction à l'analyse numérique matricielle et à l'optimisation.* Masson, Paris 1982.

[21] E.A. Coddington and N. Levinson, *Theory of ordinary differential equations.* McGraw-Hill, New York-Toronto-London 1955.

[22] M.G. Crandall and P.H. Rabinowitz, *Bifurcation from simple eigenvalues.* J. Funct. Analysis **8** (1971), 321–340.

[23] M.G. Crandall and P.H. Rabinowitz, *Bifurcation, perturbation of simple eigenvalues, and linearized stability.* Arch. Rat. Mech. Anal. **52** (1973), 161–180.

[24] E.N. Dancer, *On the structure of solutions of non-linear eigenvalue problems.* Ind. Univ. Math. J. **23** (1974), 1069–1076.

[25] E.N. Dancer, *Bifurcation from simple eigenvalues and eigenvalues of geometric multiplicity one.* Bull. London Math. Soc. **34** (2002), 533–538.

[26] K. Deimling, *Nonlinear Functional Analysis.* Springer, Berlin 1985.

[27] N. Dunford and J.T. Schwartz, *Linear operators. Part I.* Pure and Applied Mathematics **7**, Interscience Publishers, New York-London 1958.

[28] V.M. Eni, *Stability of the root-number of an analytic operator-function and perturbations of its characteristic numbers and eigenvectors.* Dokl. Akad. Nauk SSSR **173** (1967), 1251–1254. English translation: Soviet Math. Dokl. **8** (1967), 542–545.

[29] J. Esquinas, *Optimal multiplicity in local bifurcation theory, II: General case.* J. Diff. Equations **75** (1988), 206–215.

[30] J. Esquinas and J. López-Gómez, *Resultados óptimos en teoría de bifurcación y aplicaciones*. In *Actas del IX Congreso de Ecuaciones Diferenciales y Aplicaciones*, pp. 159–162. Universidad de Valladolid, Valladolid 1986.

[31] J. Esquinas and J. López-Gómez, *Optimal multiplicity in local bifurcation theory, I: Generalized generic eigenvalues*. J. Diff. Equations **71** (1988), 72–92.

[32] P.M. Fitzpatrick and J. Pejsachowicz, *A local bifurcation theorem for C^1-Fredholm maps*. Proc. Amer. Math. Soc. **109** (1990), 995–1002.

[33] P.M. Fitzpatrick and J. Pejsachowicz, *Parity and generalized multiplicity*. Trans. Amer. Math. Soc. **326** (1991), 281–305.

[34] P.M. Fitzpatrick and J. Pejsachowicz, *Orientation and the Leray–Schauder theory for fully nonlinear elliptic boundary value problems*. Mem. Amer. Math. Soc. **483**, Providence 1993.

[35] A. Friedman and M. Shinbrot, *Nonlinear eigenvalue problems*. Acta Math. **121** (1968), 77–125.

[36] G. Frobenius, *Theorie der linearen Formen mit ganzen Coefficienten*. J. Reine Angew. Math. **86** (1879), 146–208.

[37] F.G. Frobenius, *Gesammelte Abhandlungen. Band I.* Edited by J.-P. Serre, Springer, Berlin-New York 1968.

[38] L. Führer, *Theorie des Abbildungsgrades in endlichdimensionalen Räumen*. Ph. D. Dissertation, Freie Univ. Berlin, Berlin 1971.

[39] F.R. Gantmacher, *The theory of matrices, vol.* 1, 2. Chelsea, New York 1959. Translated from Russian.

[40] I.M. Gel'fand, *Lectures on linear algebra*. Interscience Tracts in Pure and Applied Mathematics **9**, Interscience, New York – London 1961.

[41] I.C. Gohberg, *On linear operators depending analytically on a parameter* (Russian). Doklady Akad. Nauk. SSSR (N.S.) **78** (1951), 629–632.

[42] I.C. Gohberg, *Certain questions of the spectral theory of finitely meromorphic operator functions* (Russian). Izv. Akad. Nauk. Armjan. SSSR Ser. Mat. **6** (1971), 160–181; corrections in *ibid.* **7** (1972), 152.

[43] I.C. Gohberg, S. Goldberg and M.A. Kaashoek, *Classes of Linear Operators, vol. 1.* Operator Theory: Advances and Applications **49**, Birkhäuser, Basel 1990.

[44] I.C. Gohberg, M.A. Kaashoek and D.C. Lay, *Equivalence, Linearization, and Decomposition of Holomorphic Operator Functions*. J. Funct. Analysis **28** (1978), 102–144.

[45] I.C. Gohberg and M.G. Kreĭn, *Fundamental aspects of defect numbers, root numbers and indexes of linear operators* (Russian). Uspehi Mat. Nauk. (N.S.) **12** (1957), no. 2(74), 43–118. English Translation: *The basic propositions on defect numbers, root numbers and indices of linear operators*. Amer. Math. Soc. Transl. (2) **13** (1960), 185–264.

[46] I.C. Gohberg and M.G. Kreĭn, *Introduction to the Theory of Linear Non-selfadjoint Operators.* Translations of Mathematical Monographs 18, American Mathematical Society, Providence 1969.

[47] I. Gohberg, P. Lancaster and L. Rodman, *Spectral analysis of matrix polynomials, I. Canonical forms and divisors.* Linear Algebra and Applications **20** (1978), 1–44.

[48] I. Gohberg, P. Lancaster and L. Rodman, *Spectral analysis of matrix polynomials, II. The resolvent form and spectral divisors.* Linear Algebra and Applications **21** (1978), 65–88.

[49] I. Gohberg, P. Lancaster and L. Rodman, *Representations and divisibility of operator polynomials.* Canad. J. Math. **30** (1978), 1045–1069.

[50] I. Gohberg, P. Lancaster and L. Rodman, *Matrix polynomials.* Computer Science and Applied Mathematics, Academic Press, New York 1982.

[51] I. Gohberg, P. Lancaster and L. Rodman, *Invariant subspaces of matrices with applications.* Canadian Mathematical Society Series of Monographs and Advanced Texts, Wiley Interscience, John Wiley, New York 1986.

[52] I. Gohberg and L. Rodman, *Analytic matrix functions with prescribed local data.* J. D'Analyse Math. **40** (1981), 90–128.

[53] I. Gohberg and L. Rodman, *Analytic operator-valued functions with prescribed local data.* Acta Math. (Szeged) **45** (1983), 189–199.

[54] I.C. Gohberg and E.I. Sigal, *An Operator Generalization of the Logarithmic Residue Theorem and the Theorem of Rouché.* Math. Sbornik **84(126)** (1971), 607–629. English Trans.: Math. USSR Sbornik **13** (1971), 603–625.

[55] M. de Guzmán, *Ecuaciones diferenciales ordinarias. Teoría de estabilidad y control.* Alhambra, Madrid 1975.

[56] J.K. Hale, *Ordinary differential equations.* 2nd edition, Krieger Publishing, Huntington (N.Y.) 1980.

[57] P. Hartman, *Ordinary differential equations.* John Wiley, New York-London-Sydney 1964.

[58] B. Helffer, *Semi-Classical Analysis for the Schrödinger Operator and Applications.* Lectures Notes in Mathematics 1336, Springer, Berlin 1988.

[59] M.W. Hirsch and S. Smale, *Differential equations, dynamical systems, and linear algebra.* Pure and Applied Mathematics 60, Academic Press, New York 1974.

[60] P.D. Hislop and I.M. Sigal, *Introduction to Spectral Theory With Applications to Schrödinger Operators.* Applied Mathematical Sciences 113, Springer, New York 1996.

[61] J. Ize, *Bifurcation theory for Fredholm operators.* Mem. Amer. Math. Soc. 174, Providence 1976.

[62] J. Ize, *Necessary and sufficient conditions for multiparameter bifurcation.* Rocky Mountain J. Math. **18** (1988), 305–337.

[63] N. Jacobson, *Lectures in abstract algebra, Vol. II. Linear algebra.* Van Nostrand, Toronto-New York-London 1953.

[64] M.A. Kaashoek, C.V.M. van der Mee and L. Rodman, *Analytic operator functions with compact spectrum, I. Spectral nodes, linearization and equivalence.* Integral Equations Operator Theory **4** (1981), 504–547.

[65] T. Kato, *Perturbation Theory for Linear Operators.* 2nd Edition, Springer, Berlin 1976.

[66] M.V. Keldyš, *On the characteristic values and characteristic functions of certain classes of non-self-adjoint equations* (Russian). Doklady Akad. Nauk SSSR (N.S.) **77** (1951), 11–14.

[67] M.V. Keldyš, *The completeness of eigenfunctions of certain classes of nonselfadjoint linear operators* (Russian). Uspehi Mat. Nauk **26** (1971), 15–41. English Translation: Russian Math. Surveys **26** (1971), 15–44.

[68] H. Kielhöfer, *Bifurcation Theory. An Introduction with Applications to PDEs.* Applied Mathematical Sciences 156, Springer, New York 2004.

[69] M.A. Krasnosel'skiĭ, *Topological Methods in the Theory of Nonlinear Integral Equations.* Gos. Izdat. Tehn.-Teor. Lit., Moscow 1956. English trans.: Pergamon Press, New York 1964.

[70] K. Kuratowski, *Topology. Vol. II.* Academic Press, New York-London, PWN Polish Scientific Publishers, Warsaw 1968.

[71] B. Laloux and J. Mawhin, *Coincidence index and multiplicity.* Trans. Amer. Math. Soc. **217** (1976), 143–162.

[72] B. Laloux and J. Mawhin, *Multiplicity, Leray–Schauder formula, and bifurcation.* J. Diff. Equations **24** (1977), 309–322.

[73] P. Lancaster, *Theory of matrices.* Academic Press, New York-London 1969.

[74] P. Lancaster, *A fundamental theorem on lambda-matrices with applications, I. Ordinary differential equations with constant coefficients.* Linear Algebra and Applications **18** (1977), 189–211.

[75] P. Lancaster, *A fundamental theorem on lambda-matrices with applications, II. Difference equations with constant coefficients.* Linear Algebra and Applications **18** (1977), 213–222.

[76] S. Lang, *Linear algebra.* 3rd edition, Undergraduate Texts in Mathematics, Springer, New York, 1987.

[77] P.D. Lax, *Linear algebra.* Pure and Applied Mathematics, Wiley Interscience, John Wiley, New York 1997.

[78] J. Leiterer, *Local and global equivalence of meromorphic operator functions, I.* Math. Nachr. **83** (1978), 7–29.

[79] J. Leiterer, *Local and global equivalence of meromorphic operator functions, II.* Math. Nachr. **84** (1978), 145–170.

[80] J. Leray and J. Schauder, *Topologie et équations fonctionelles.* Ann. Sci. École Norm. Sup. Sér. 3 **51** (1934), 45–78.

[81] N.G. Lloyd, *Degree Theory.* Cambridge Tracts in Mathematics, Cambridge University Press, Cambridge 1978.

[82] J. López-Gómez, *Spectral Theory and Nonlinear Functional Analysis.* CRC Press, Chapman and Hall RNM 426, Boca Raton 2001.

[83] J. López-Gómez, *Ecuaciones diferenciales y variable compleja. Con teoría espectral y una introducción al grado topológico de Brouwer.* Prentice Hall, Madrid 2001.

[84] J. López-Gómez, *Ecuaciones diferenciales y variable compleja. Problemas propuestos y resueltos.* Prentice Hall, Madrid 2002.

[85] J. López-Gómez, *Spectral Theory and Nonlinear Analysis.* In *Ten Mathematical Essays on Approximation in Analysis and Topology* (J. Ferrera, J. López-Gómez and F. Ruiz del Portal, editors), pp. 151–176, Elsevier, Amsterdam 2005.

[86] J. López-Gómez and M. Molina-Meyer, *Bounded components of positive solutions of abstract fixed point equations: mushrooms, loops and isolas.* J. Diff. Equations **209** (2005), 416–441.

[87] J. López-Gómez and M. Molina-Meyer, *The competitive exclusion principle versus biodiversity through competitive segregation and further adaptation to spatial heterogeneities.* Theoretical Population Biology **69** (2006), 94-109.

[88] J. López-Gómez and C. Mora-Corral, *Characterizing the existence of local Smith forms for $\mathcal{C}^\infty$ families of matrix operators.* Contemporary Mathematics 321, pp. 139–151, Amer. Math. Soc., Providence 2003.

[89] J. López-Gómez and C. Mora-Corral, *Local Smith form and equivalence for one-parameter families of Fredholm operators of index zero.* In *Spectral Theory and Nonlinear Analysis with Applications to Spatial Ecology.* (S. Cano-Casanova, J. López-Gómez and C. Mora-Corral, editors), pp. 127–161, World Scientific, Singapore 2005.

[90] J. López-Gómez and C. Mora-Corral, *Finite Laurent developments and the logarithmic residue theorem in the real non-analytic case.* Integral Equations Operator Theory **51** (2005), 519–552.

[91] J. López-Gómez and C. Mora-Corral, *Counting zeros of $\mathcal{C}^1$ Fredholm maps of index one.* Bull. London Math. Soc. **37** (2005), 778–792.

[92] R.J. Magnus, *A generalization of multiplicity and the problem of bifurcation.* Proc. London Math. Soc. (3) **32** (1976), 251–278.

[93] R. Magnus, *On the multiplicity of an analytic operator-valued function.* Math. Scand. **77** (1995), 108–118.

[94] R. Magnus and C. Mora-Corral, *Natural representations of the multiplicity of an analytic operator-valued function at an isolated point of the spectrum.* Integral Equations and Operator Theory **53** (2005), 87–106.

[95] A.S. Markus and E.I. Sigal, *The multiplicity of the characteristic number of an analytic operator function* (Russian). Mat. Issled. **5** (1970), 129–147.

[96] J. Mawhin, *Equivalence theorems for nonlinear operators equations and coincidence degree theory for some mappings in locally convex topological vector spaces.* J. Diff. Equations **12** (1972), 610–636.

[97] B. Mitiagin, *Linearization of holomorphic operator functions, I.* Integral Equations and Operator Theory **1** (1978), 114–131.

[98] B. Mitiagin, *Linearization of holomorphic operator functions, II.* Integral Equations and Operator Theory **1** (1978), 226–249.

[99] C. Mora-Corral, *On the Uniqueness of the Algebraic Multiplicity.* J. London Math. Soc. **69** (2004), 231–242.

[100] C. Mora-Corral, *Axiomatizing the algebraic multiplicity.* In *The first 60 years of Nonlinear Analysis of Jean Mawhin,* (M. Delgado, A. Suárez, J. López-Gómez and R. Ortega, editors), pp. 175–187, World Scientific, Singapore 2004.

[101] C. Mora-Corral, *Algebraic Multiplicities and Bifurcation Theory.* Ph.D. Thesis, Universidad Complutense de Madrid, Madrid 2004.

[102] J.D. Murray, *Mathematical Biology.* Biomathematics Texts, Springer, Berlin 1989.

[103] A. Okubo, *Diffusion and Ecological Problems: Mathematical Models.* Springer, New York 1980.

[104] J. Pejsachowicz and P.J. Rabier, *A substitute for the Sard–Smale theorem in the C^1 case.* J. d'Anal. Math. **76** (1998), 265–288.

[105] J. Pejsachowicz and P.J. Rabier, *Degree theory for C^1 Fredholm mappings of index* 0. J. d'Anal. Math. **76** (1998), 289–319.

[106] V.V. Peller, *Hankel operators and their applications.* Springer Monographs in Mathematics, Springer, New York 2003.

[107] W.V. Petryshyn, *Generalized topological degree and semilinear equations.* Cambridge Tracts in Mathematics 117, Cambridge University Press, Cambridge 1995.

[108] H. Poincaré, *Les méthodes nouvelles de la mécanique céleste.* Tomes I–III. Dover, New York 1957.

[109] C.R. Putnam, *Commutation properties of Hilbert space operators and related topics.* Ergebnisse der Mathematik und ihrer Grenzgebiete 36, Springer, Berlin 1967.

[110] P.J. Rabier, *Generalized Jordan chains and two bifurcation theorems of Krasnoselskii.* Nonl. Anal. TMA **13** (1989), 903–934.

[111] P.H. Rabinowitz, *Some global results for nonlinear eigenvalue problems.* J. Funct. Anal. **7** (1971), 487–513.

[112] P.H. Rabinowitz, *Some aspects of nonlinear eigenvalue problems,* Rocky Mountain J. of Maths. **3** (1973), 161–202.

[113] A.G. Ramm, *Singularities of the inverses of Fredholm operators.* Proc. Roy. Soc. Edinburgh **102A** (1986), 117–121.

[114] L. Rodman, *An introduction to operator polynomials.* Operator Theory: Advances and Applications 38, Birkhäuser, Basel 1989.

[115] A. Sard, *The measure of the critical values of differentiable maps.* Bull. Amer. Math. Soc. **48** (1942), 883–890.

[116] P. Sarreither, *Transformationseigenschaften endlicher Ketten und allgemeine Verzweigungsaussagen.* Math. Scand. **35** (1974), 115–128.

[117] P. Sarreither, *Zur algebraischen Vielfachheit eines Produktes von Operatorscharen.* Math. Scand. **41** (1977), 185–192.

[118] J. Schauder, *Der Fixpunktsatz in Funktionalräumen.* Studia Math. **2** (1930), 171–180.

[119] E.I. Sigal, *The multiplicity of a characteristic value of a product of operator-functions* (Russian). Mat. Issled. **5** (1970), 118–127.

[120] S. Smale, *An infinite dimensional version of Sard's theorem.* Amer. J. Math. **87** (1965), 861–866.

[121] H.J.S. Smith, *On systems of linear indeterminate equations and congruences.* Phil. Trans. Roy. Soc. London **151** (1861), 293–326.

[122] A.E. Taylor, *Introduction to functional analysis.* John Wiley & Sons, Chapman & Hall, London 1958.

[123] G.Ph.A. Thijsse, *On solutions of elliptic differential equations with operator coefficients.* Integral Equations and Operator Theory **1** (1978), 567–579.

[124] V.P. Trofimov, *The root subspaces of operators that depend analytically on a parameter* (Russian). Mat. Issled. **3** (1968), 117–125.

[125] A.N. Tychonoff, *Ein Fixpunktsatz.* Math. Ann. **111** (1935), 767–776.

[126] G.T. Whyburn, *Topological Analysis.* Princeton University Press, Princeton 1958.

[127] J.T. Wloka, B. Rowley and B. Lawruk, *Boundary Value Problems for Elliptic Systems.* Cambridge University Press, Cambridge 1995.

[128] R. Wolf, *Zur Stabilität der algebraischen Vielfachheit von Eigenwerten von holomorphen Fredholm-Operatorfunktionen.* Applicable Anal. **9** (1979), 165–177.

[129] B.F. Wyman, M.K. Sain, G. Conte and A.M. Perdon, *Poles and Zeros of Matrices of Rational Functions.* Linear Algebra Appl. **157** (1991), 113–139.

[130] K. Yosida, *Functional analysis.* 2nd edition, Die Grundlehren der mathematischen Wissenschaften 123, Springer, New York 1968.

Notation

General

$\mathbb{N}$	The sets of natural numbers, including zero
$\mathbb{N}^*$	$\mathbb{N} \setminus \{0\}$
$\mathbb{Z}, \mathbb{R}, \mathbb{C}$	The sets of integer, real and complex numbers, respectively
$\mathbb{K}$	Any of the fields $\mathbb{R}$ or $\mathbb{C}$
δ_{ij}	Kronecker's delta: $\delta_{ij} = 1$ if $i = j$; and $\delta_{ij} = 0$ if $i \neq j$
$\mathcal{C}^r(\Omega, V)$	The space of functions of class $\mathcal{C}^r$ from Ω to V
$\mathcal{C}^r(\Omega)$	$\mathcal{C}^r(\Omega, \mathbb{K})$
$\mathcal{C}(\Omega, V)$	$\mathcal{C}^0(\Omega, V)$
$\mathcal{C}(\Omega)$	$\mathcal{C}^0(\Omega, \mathbb{K})$
$\mathcal{C}^\omega(\Omega, V)$	The space of analytic functions from Ω to V
$\mathcal{H}(\Omega, V)$	The space of the holomorphic functions from Ω to V
$\mathcal{H}(\Omega)$	$\mathcal{H}(\Omega, \mathbb{C})$
$o(\lambda - \lambda_0)^p$	Any function f defined on a perforated neighborhood of λ_0 such that $\lim_{\lambda \to \lambda_0} (\lambda - \lambda_0)^{-p} f(\lambda) = 0$
$\lvert a \rvert$	Absolute value of $a \in \mathbb{R}$, or modulus of $a \in \mathbb{C}$
$\operatorname{Re} e$	Real part of (the complex number or the vector) e
$\operatorname{Im} e$	Imaginary part of (the complex number or the vector) e
$\bar{\lambda}$	Complex conjugate of $\lambda \in \mathbb{C}$
$\bar{e}$	Complex conjugate of the vector e
$\operatorname{Card} A$	Number of elements of the set A, if A is finite, and ∞ if A is infinite
$\binom{a}{b}$	Binomial coefficient
$\operatorname{ord}_{\lambda_0} f$	Order of λ_0 as a zero of f

Operator calculus

$\lVert u \rVert$	Norm of $u \in U$
$\lVert T \rVert$	Operator norm of $T \in \mathcal{L}(U, V)$
$\varrho(A)$	Resolvent set of the operator A; that is, $\mathbb{C} \setminus \sigma(A)$

$\mathcal{R}(z;A)$	Resolvent operator of A at $z \in \varrho(A)$; that is, $(zI-A)^{-1}$
$P(A)$	Action of the polynomial P over the matrix A
$f(A)$	Action of the holomorphic function f over the matrix A
$\operatorname{spr} A$	Spectral radius of the matrix A

Vector spaces

U, V, W	Real or complex Banach spaces
U_0, U_1	Closed vector subspaces of U
$\mathcal{L}(U,V)$	Banach space of the linear and bounded operators from U to V
$\mathcal{L}(U)$	$\mathcal{L}(U,U)$
$\operatorname{Iso}(U,V)$	Set of linear isomorphisms from U to V
$\operatorname{Iso}(U)$	$\operatorname{Iso}(U,U)$
$\operatorname{Fred}(U,V)$	Subset of $\mathcal{L}(U,V)$ formed by the Fredholm operators
$\operatorname{Fred}(U)$	$\operatorname{Fred}(U,U)$
$\operatorname{Fred}_0(U,V)$	Set of operators of $\operatorname{Fred}(U,V)$ with index zero
$\operatorname{Fred}_0(U)$	$\operatorname{Fred}_0(U,U)$
$\mathcal{K}(U,V)$	Set of compact operators from U to V
$\mathcal{K}(U)$	$\mathcal{K}(U,U)$
$GL_c(U)$	Set of $T \in \operatorname{Iso}(U)$ such that $T - I_U \in \mathcal{K}(U)$
$\det A$	Determinant of the square matrix A
$\operatorname{tr} A$	Trace of the finite-rank operator $A \in \mathcal{L}(U)$
$\sigma(A)$	Spectrum of the matrix A or of the linear operator A
$\operatorname{Eig}(A)$	Set of eigenvalues of the matrix A or of the linear operator A
$m_a(\lambda)$	Algebraic multiplicity of λ as an eigenvalue of A
$m_g(\lambda)$	Geometric multiplicity of λ as an eigenvalue of A
u	A vector in U
$\mathcal{M}_{\alpha\times\beta}(\mathbb{K})$	Space of matrices with coefficients in $\mathbb{K}$ with α rows and β columns
$\mathcal{M}_N(\mathbb{K})$	$\mathcal{M}_{N\times N}(\mathbb{K})$
I_U	Identity operator in U
I	Identity operator, when the space is clear from the context
I_N	Identity operator in $\mathcal{L}(\mathbb{K}^N)$, or identity matrix in $\mathcal{M}_N(\mathbb{K})$
$N[T]$	Kernel (null-space) of $T \in \mathcal{L}(U,V)$
$R[T]$	Range (image) of $T \in \mathcal{L}(U,V)$
$\operatorname{rank} T$	$\dim R[T]$
$\operatorname{span}[u_1,\dots u_n]$	Set of linear combinations of $u_1,\dots,u_n \in U$
$\dim U_0$	Dimension of the vector space U_0
$\operatorname{codim} U_1$	Codimension in U of the subspace $U_1 \subset U$, that is, $\dim(U/U_1)$
$\operatorname{ind} T$	Fredholm index of the operator $T \in \mathcal{L}(U,V)$, that is, $\dim N[T] - \operatorname{codim} R[T]$

$\operatorname{diag}\{a_1, \dots, a_n\}$	Diagonal matrix with diagonal elements $a_1, \dots, a_n$ in this order, or block-diagonal matrix or operator
trng	Block-triangular matrix or operator defined in (4.34)
$\mathfrak{L}_{[s]}$	$\operatorname{trng}\{\mathfrak{L}_0, \dots, \mathfrak{L}_s\}$ (see (7.1))
$[a_1, \dots, a_k]$	Block-decomposition of a matrix or of an operator
$\operatorname{col}[a_1, \dots, a_n]$	Matrix or operator defined in (4.33)
$\oplus$	It indicates direct sum of vector subspaces
$\times$	It indicates direct product (or Cartesian product) of vector spaces
u^T	Transposed vector of u
A^T	Transposed matrix of A
P_A	Characteristic polynomial of A
$\langle u, v \rangle$	Skew-linear product of the vectors $u, v \in \mathbb{C}^N$
$B_r(u)$	Open ball of center $u \in U$ and radius $r > 0$
B_r	$B_r(0)$
$D_R(z_0)$	Open disk in $\mathbb{C}$ of radius $R > 0$ and center $z_0 \in \mathbb{C}$
D_R	$D_R(0)$
$C_R(z_0)$	Positively oriented boundary of $D_R(z_0)$
C_R	$C_R(0)$
D_γ	The inner domain of the Cauchy contour γ

One-parameter families of linear operators

Ω	An open subset of $\mathbb{K}$
λ_0	A point of Ω
$\bar{\Omega}$	Closure of the set Ω
$\partial\Omega$	Boundary of the set Ω
$\mathbb{P}[\mathcal{L}(U,V)]$	Set of polynomial operator families from U to V
$\mathbb{P}[\mathcal{L}(U)]$	$\mathbb{P}[\mathcal{L}(U,U)]$
$\operatorname{rank} u$	Rank of $u \in U$ as an eigenvector of a family $\mathfrak{L}$ at λ_0
$\Sigma(\mathfrak{L})$	Spectrum (set of singular values) of a family $\mathfrak{L}$
$\operatorname{Eig}(\mathfrak{L})$	Subset of $\Sigma(\mathfrak{L})$ formed by all eigenvalues
$\operatorname{Alg}_\nu(\mathfrak{L})$	Set of ν-algebraic values
$\operatorname{Alg}(\mathfrak{L})$	Union of all sets $\operatorname{Alg}_\nu(\mathfrak{L})$, for all $\nu \geq 0$
$c(\mathfrak{L}, \lambda_0, j)$	jth coefficient of the Taylor or Laurent development of $\mathfrak{L}$ at λ_0
$\mathfrak{L}_j$	$c(\mathfrak{L}, \lambda_0, j)$
$(\mathfrak{L}^{-1})_j$	$c(\mathfrak{L}^{-1}, \lambda_0, j)$
$\mathfrak{M}_{\{j\}}$	jth derived family, according to Definition 5.1.1
$\chi[\mathfrak{L}; \lambda_0]$	Multiplicity of $\mathfrak{L}$ at λ_0, according to Definition 4.3.3
$\mu[\mathfrak{L}; \lambda_0]$	Multiplicity of $\mathfrak{L}$ at λ_0, according to Definition 5.1.3
$\kappa[\mathfrak{L}; \lambda_0]$	Multiplicity of $\mathfrak{L}$ at λ_0, according to Definition 7.2.5

$\mathfrak{m}[\mathfrak{L};\lambda_0]$	Multiplicity of $\mathfrak{L}$ at λ_0, when no special definition is considered
$\kappa_j[\mathfrak{L};\lambda_0]$	jth partial κ-multiplicity of $\mathfrak{L}$ at λ_0. See Definition 7.2.5
$\mu_j[\mathfrak{L};\lambda_0]$	jth partial μ-multiplicities of $\mathfrak{L}$ at λ_0. See Definition 5.1.3
$\chi_j[\mathfrak{L};\lambda_0]$	$\mu_j[\mathfrak{L};\lambda_0]$
$\alpha[\mathfrak{L};\lambda_0]$	Ascent of $\mathfrak{L}$ at λ_0
$\mathfrak{m}[\mathfrak{L};\Omega]$	Multiplicity of $\mathfrak{L}$ in the open set Ω
$\mathfrak{R}_P$	The family defined by (5.1)
$\mathfrak{C}_P$	The family defined by (5.33)
$\mathfrak{L}^A$	The family defined by (8.1)
$\mathcal{S}^r_{\lambda_0}(U,V)$	The set of families $\mathfrak{L}$ of class C^r from a neighborhood of λ_0 to $\mathcal{L}(U,V)$ such that $\mathfrak{L}(\lambda_0)\in \mathrm{Fred}_0(U,V)$
$\mathcal{S}_{\lambda_0}(U)$	$\mathcal{S}_{\lambda_0}(U,U)$
$\mathcal{S}^r_{\lambda_0}$	The class of families $\mathfrak{L}$ for which there exist Banach spaces U and V with $\mathfrak{L}\in\mathcal{S}^r_{\lambda_0}(U,V)$
$\mathcal{A}^\infty_{\lambda_0}(U,V)$	The set of families $\mathfrak{L}\in\mathcal{S}^\infty_{\lambda_0}(U,V)$ such that $\lambda_0\in\mathrm{Alg}(\mathfrak{L})$
$\mathcal{A}^\infty_{\lambda_0}(U)$	$\mathcal{A}^\infty_{\lambda_0}(U,U)$
$\mathcal{M}_{\lambda_0}(U,V)$	The set of families $\mathfrak{L}$ from a neighborhood of λ_0 to $\mathcal{L}(U,V)$ such that the multiplicity of $\mathfrak{L}$ at λ_0 is defined
$\mathcal{M}_{\lambda_0}(U)$	$\mathcal{M}_{\lambda_0}(U,U)$

Nonlinear operators

$\mathcal{S}$	Set of non-trivial solutions of the equation $\mathfrak{F}(\lambda,u)=0$
$\mathrm{Deg}(f,D)$	Topological degree of $\mathfrak{F}$ in D
$\mathrm{Ind}(f,u_0)$	Leray–Schauder index of f at the isolated zero u_0

Index

GPSR Compliance

The European Union's (EU) General Product Safety Regulation (GPSR) is a set of rules that requires consumer products to be safe and our obligations to ensure this.

If you have any concerns about our products, you can contact us on ProductSafety@springernature.com

In case Publisher is established outside the EU, the EU authorized representative is:

Springer Nature Customer Service Center GmbH
Europaplatz 3
69115 Heidelberg, Germany

Batch number: 10732470

Printed by Printforce, the Netherlands